CONTENTS

SESSION II

ACADEMIC PRESS RAPID MANUSCRIPT REPRODUCTION

This Is the Second Volume of a Series Entitled:
Ecotoxicology and Environmental Quality under the
Editorship of Frederick Coulston, Albany, New York
and Freidhelm Korte, Munich

Other Volumes in Series:
Water Quality: Proceedings of an International Forum
Edited by Frederick Coulston and Emil Mrak

Sponsored by
the International Academy of Environmental Safety
the International Society for Ecotoxicology and Environmental
Safety (Secotox)
Das Forum für Wissenschaft, Wirtschaft, und Politik e.V., Bonn

REGULATORY ASPECTS OF CARCINOGENESIS AND FOOD ADDITIVES: THE DELANEY CLAUSE

REGULATORY ASPECTS OF CARCINOGENESIS AND FOOD ADDITIVES: THE DELANEY CLAUSE

edited by

FREDERICK COULSTON

Institute of Comparative and Human Toxicology
Albany Medical College of Union University
Albany, New York

ACADEMIC PRESS   New York   San Francisco   London   1979
A Subsidiary of Harcourt Brace Jovanovich, Publishers

ACADEMIC PRESS, INC.
111 Fifth Avenue, New York, New York 10003

United Kingdom Edition published by
ACADEMIC PRESS, INC. (LONDON) LTD.
24/28 Oval Road, London NW1 7DX

ISBN 0–12–192750–4

PRINTED IN THE UNITED STATES OF AMERICA

79 80 81 82   9 8 7 6 5 4 3 2 1

SESSION III

ORGANIZING COMMITTEE

Frederick Coulston

Institute of Comparative and Human Toxicology, Albany Medical College of Union University

Emil Mrak

Chancellor Emeritus, University of California, Davis

David Clayson

Eppley Institute for Research in Cancer, University of Nebraska

Morris Cranmer

National Center for Toxicological Research, Jefferson, Arkansas

Albert C. Kolbye, Jr.

Bureau of Foods, United States Food and Drug Administration, Washington D. C.

PARTICIPANTS

Rajender Abraham   Institute of Comparative and Human Toxicology, Albany Medical College, 47 New Scotland Avenue, Albany, New York 12208

Richard H. Adamson   Laboratory of Chemical Pharmacology (ET, DCT, NCI), National Institutes of Health Bldg. 10–6N 119, 9000 Rockville Pike, Bethesda, Maryland 20014

James G. Aftosmis   E.I. Du Pont de Nemours & Co., Haskell Lab, Elkton Road, Newark, Delaware 19711

Elizabeth Anderson   RD 689, US EPA, 401 M Street, SW, Washington, D.C. 20460

Roy Albert   Institute of Environmental Medicine, New York University Medical Center, 550 First Street, New York, New York 10016

Stephanie P. April   Exxon Corporation, P.O. Box 45, Linden, New Jersey 07036

Bernard D. Astill   Eastman Kodak Company, Health, Safety, and Human Factors Laboratory, B-320, Kodak Park, Rochester, New York 14650

Emilio Astolfi   University of Buenos Aires, Ayacucho 1337, Buenos Aires (1111), Argentina

Emilio Astolfi, Jr.   St. Paul's College, Ayacucho 1337, Buenos Aires (1111), Argentina

M. C. Bachman   IMC Chemical Group, Inc., P.O. Box 207, Terre Haute, Indiana 47808

Robert C. Baldwin   Rohm and Haas Company, Toxicology Department, Norristown & McKean Roads, Spring House, Pennsylvania 19477

Donald G. Barnes   Office of Assistant Administrator for Toxic Substances, EPA, 401 M Street, SW, Washington, D.C. 20460

J. D. Behun   Mobil Chemical Company, P.O. Box 240, Edison, New Jersey 08817

Etcyl H. Blair   Dow Chemical Company, 2020 Dow Center, Midland, Michigan 48640

Marvin Bleiberg   Bureau of Foods Division of Toxicology, HFF-185, 200 C Street, SW, Washington, D.C. 20204

Herbert Blumenthal   FDA, Washington, D.C. 20460

Gale Boxill   WYETH Laboratories, P.O. Box 861, Paoli, Pennsylvania 19301

Jackson B. Browning   Union Carbide Corporation, 270 Park Avenue, New York, New York 10017

Jerry Brunton   Animal Health Institute, 1717 K Street, NW, Suite 1009, Washington, D.C. 20006

W. H. Butler   Imperial Chemical Industries Ltd., Pharmaceuticals Division, Experimental Toxicology Section, Safety of Medicines Department, Mereside Alderley Park, Macclesfield, Cheshire SK10 4TG, England

Jim Elfstrum   Stauffer Chemical Company, Environmental Service Department, Westport, Connecticut 06880

B. P. Cardon   Arizona Feeds, P.O. Box 5526, Tucson, Arizona 85703

Thomas W. Carmody   Union Carbide Corporation, 270 Park Avenue, New York, New York 10017

Kathy Chamberlin   Kirkland & Ellis, 1776 K Street, NW, Washington, D.C. 20460

Roger Chamberlin   Dow Chemical Company, 1800 M Street, NW, Washington, D.C. 20036

C. Chappel   FDC Consultants, Inc., 364 Robin Avenue, Beaconsfield, Quebec, Canada H9W1R8

John J. Clary   Celanese Corporation, 1211 Avenue of the Americas, New York, New York 10036

David Clayson   Eppley Institute for Research in Cancer, University of Nebraska, 42nd & Dewey Avenues, Omaha, Nebraska 68105

David J. Clegg   Health Protection Branch, Ottawa Foods Directorate, Toxicological Evaluation Division, Tunney's Pasture, Ottawa, Ontario KIA OL2 Canada

Nicholas Clesceri   Fresh Water Institute, Rensselaer Polytechnic Institute, Troy, New York 12181

James T. Conner   Mobay, 1140 Connecticut Avenue, NW, Washington, D.C. 20036

Eileen Coulston   17 Woodlawn Avenue, RD #2, Rensselaer, New York 12144

Frederick Coulston   Institute of Comparative and Human Toxicology, Albany Medical College of Union University, 47 New Scotland Avenue, Albany, New York 12208

Eugene F. Cox   Carnegie–Mellon Institute of Research, 4400 Fifth Avenue, Pittsburgh, Pennsylvania 15213

Thomas H. Cox   E.I. Du Pont de Nemours & Co., Biochemicals Department, 14360 Brandywine Bldg., Wilmington, Delaware 19899

P. Craig   G.D. Searle & Co., Box 5110, Chicago, Illinois 60680

Morris F. Cranmer   National Center for Toxicological Research, Jefferson, Arkansas 72079

William Darby   The Nutrition Foundation, Inc., 489 Fifth Avenue, New York, New York 10017

Drew M. Davis   National Soft Drink Association, 1101 16th Street, NW, Washington, D.C. 20036

Philip H. Derse   Warf Institute, Inc., Box 7545, Madison, Wisconsin 53707

Frederick J. DiCarlo   Office of Toxic Substances (WH-557), U.S. EPA, 401 M Street, SW, Washington, D.C. 20460

Richard M. Dowd   Science Advisory Board, EPA, Room 1129 West Tower, 401 M Street, SW, Washington, D.C. 20460

David F. Dunning   Calorie Control Council, 64 Perimeter Center East, Atlanta, Georgia 30346

Harold Egan   Department of Industry, Laboratory of the Government Chemist, Cornwall House, Stamford Street, London, England

R. G. Eggert   American Cyanamid Company, P.O. Box 400, Princeton, New Jersey 08540

Howard E. Everson   Diamond Shamrock Corporation, 1100 Superior Avenue, Cleveland, Ohio 44114

Alfred Farah   Sterling–Winthrop Research Institute, Rensselaer, New York 12144

Kelvin H. Ferber   Allied Chemical Corporation & Buffalo Color Corporation, Box 1069, Buffalo, New York 14240

John Foderaro   E.I. Du Pont de Nemours & Co., Wilmington, Delaware 19899

John P. Frawley   Toxicology Department, Hercules Inc., 910 Market Street, Wilmington, Delaware 19899

William Frunkos   FDA, Bureau of Foods, 200 C Street, SW, HFF-185, Washington, D.C. 20204

Robert H. Furman   Eli Lilly & Company, 307 East McCarty Street, Indianapolis, Indiana 46206

Paul J. Garvin   Standard Oil Company (Indiana), 200 East Randolph Drive, Chicago, Illinois 60601

Hyman R. Gittes   FDA, 200 C Street, SW, Washington, D.C. 20204

Harvey S. Gold   Velsicol Chemical Corporation, 910 17th Street, NW, Suite 1000, Washington, D.C. 20006

Louis S. Grasso   Hoechst–Roussel Pharmaceuticals, Inc., Route 202–206, North, Somerville, New Jersey 08876

R. E. Hagen   National Food Processors, 1133 20th Street, NW, Washington, D.C. 20036

Jim Hanson   Dow Chemical Company, 303 Hubbard Street, Midland, Michigan

Anna J. Harrison   Chemistry Department, Mt. Holyoke College, Carr Laboratory, South Hadley, Massachusetts 01075

Yvonne Harrison   Hoffmann–La Roche, Inc., Nutley, New Jersey 07110

Richard Hart   Dow Chemical Company, 2030 Dow Center, Midland, Michigan 48640

Richard H. Hill   Health Effects & Science Policy, Office of Assistant Administrator for Toxic Substances, EPA, 401 M Street, SW, Washington, D.C. 20460

William L. Hollis   National Agricultural Chemical Association, 1155 15th Street, NW, Washington, D.C. 20005

Vilma R. Hunt   Science Advisory Board, EPA, 401 M Street, SW, Washington, D.C. 20046

Peter B. Hutt   Covington & Burlington, 888 16th Street, NW, Washington, D.C. 20006

Jerry T. Hutton   Foremost Foods Company, One Post Street, San Francisco, California 94104

Carol Jennings   Hill & Knowlton, Inc., 633 Third Avenue, New York, New York 10017

Daniel Jones   FDA, Bureau of Foods, 200 C Street, SW, Washington, D.C. 20204

Charles J. Kensler   Arthur D. Little, Inc., 35 Acorn Park, Cambridge, Massachusetts 02140

Georg Kimmerle   Bayer AG, Institute of Toxicology, 56 Wuppertal 1, Friederich-Ebert-Str. 217, West Germany

William H. Kirby   Medical Services, Control Data Corporation, 301 N Charles Street, Baltimore, Maryland 21201

Charles J. Kokoski   FDA, 200 C Street, SW, Washington, D.C. 20204

Albert C. Kolbye, Jr.   Bureau of Foods, HFF-100, Food and Drug Administration, Washington, D.C. 20204

F. Korte   Technische Universität München, Lehrstuhl für Ökologische Chemie, Institute für Chemie, Weihenstephan Am Löwentor, D-805 Weihenstephan, West Germany

Herman F. Kraybill   Carcinogenesis, DCCP, National Cancer Institute, Landow Building, Room C337, Bethesda, Maryland 20014

C. J. Krister   E.I. Du Pont de Nemours & Co. Inc. (Retired), Biochemicals Department, 1007 Market Street, Wilmington, Delaware 19898

William C. Krumrei   The Procter and Gamble Company, 5299 Spring Grove Avenue, Cincinnati, Ohio 45217

W. Lamb   Mobay Chemical Corporation, Chemagro Agricultural Division, Box 4913, Hawthorn Road, Kansas City, Missouri 64120

Roland A. Leimgruber   CIBA-GEIGY Limited, CH-4002 Basle, Switzerland

Bill Brown   Lerch & Co., Inc., 1030 15th Street, NW #840, Washington, D.C. 20005

Ruth R. Levine   Boston University School of Medicine, 80 E. Concord Street, Boston, Massachusetts

William Lijinsky   Chemical Carcinogenesis Program, Frederick Cancer Research Center, P.O. Box B, Frederick, Maryland 21701

Dale R. Lindsay   University of Arizona, Department of Family & Community Medicine, 125 E. Canyon View Drive, Tucson, Arizona 85704

Janice Long   Chemical & Engineering News, 1155 16th Street, NW, Washington, D.C.

Alvin J. Lorman   Covington & Burling, 888 Sixteenth Street, NW, Washington, D.C.

Frank C. Lu   Clinical Professor of Pharmacology, University of Miami, Research and Teaching Center of Toxicology, P.O. Box 248216, Coral Gables, Florida 33124

D. MacDougall   Dow Pharmaceuticals, Dow Chemical of Canada, Ltd., 380 Elgin Mills Road East, Richmond Hill, Ontario, Canada, L4C 5H2

R. C. Mallatt   Standard Oil Company (Indiana), 200 E. Randolph Drive, Chicago, Illinois 60601

James G. Martin   The House of Representatives, Washington, D.C. 20515

William F. Massmann   Dr. Pepper Company, P.O. Box 5086, Dallas, Texas 75222

Alex P. Mathers   1000 Frost Bridge Road, Matherville, Mississippi 39360

David J. Matthews   Australian Government, Embassy of Australia, 1601 Massachusetts Ave., NW, Washington, D.C. 20036

R. M. McClain   Hoffmann—LaRoche, Inc., Nutley, New Jersey 07110

Donald D. McCollister   Dow Chemical, Health and Environmental Research Department, P.O. Box 1706, Midland, Michigan 48640

Bruce L. McDonald   Kirkland & Ellis, 1776 K Street, NW, Washington, D.C. 20006

M. A. Mehlman   Mobil Oil Corporation, 150 E. 42nd Street, New York, New York 10017

Thomas W. Mooney   The Procter and Gamble Company, 5299 Spring Grove Avenue, Cincinnati, Ohio 45217

Victor H. Morgenroth III   Division of Toxicology, FDA, HFF-185, 200 C Street, SW, Washington, D.C. 20204

Emil Mrak   Chancellor Emeritus, University of California, Davis, California 95616

Robert T. Murphy   CIBA—GEIGY Corporation, P.O. Box 11422, Greensboro, North Carolina 27409

Chris Murray   Chemical & Engineering News, 1155 16th Street, NW, Washington, D.C. 20036

A. W. Neff   Agricultural Division, Upjohn Co., 301 Henrietta Street, Kalamazoo, Michigan 49001

Charles A. O'Connor III   Sellers, Conner & Cuneo, 1625 K. Street, NW, Washington, D.C. 20006

A. C. Page   Shell Development Company, P.O. Box 4248, Modesto, California 95352

Dennis V. Parke   Department of Biochemistry, University of Surrey, Guildford, Surrey GU2 5XH, England

Thomas M. Parkinson   Dynapol, 1454 Page Mill Road, Palo Alto, California 94304

Dennis Poller   Exxon Chemical Company, P.O. Box 45, Linden, New Jersey 07036

Vivian Prunier   FDA, Washington, D.C. 20204

W. Quanstrom   Standard Oil Company (Indiana), 200 East Randolph Drive, Chicago, Illinois 60601

Charles E. Ross   Shell Oil Company, P.O. Box 2463, Houston, Texas 77001

Louis Rothschild, Jr.   Food Chemical News, 420 Colorado Building, Washington, D.C. 20005

Randy Roth   Diamond Shamrock, Box 348, Painesville, Ohio 44060

V. K. Rowe   Dow Chemical, 1803 Bldg. - Dow Chemical, Midland, Michigan 48640

Ben C. Rusche   E.I. Du Pont de Nemours & Co., Inc., Administrative Department, 9058 Du Pont Building, Wilmington, Delaware 19898

Georgia Samuel   Lever Brothers Company, 45 River Road, Edgewater, New Jersey 07020

Neil L. Sass   Division of Toxicology, Bureau of Foods, FDA, HFF-190, 200 C Street, SW, Washington, D.C. 20204

F. Schmidt-Bleek   Umweltbundesamt, 1000 Berlin 33, 1 Bismarckplatz, West Germany

George C. Scott   Smith Kline Animal Health Products, Applebrook Research Center, 1600 Paoli Pike, West Chester, Pennsylvania 19380

T. Shellenberger   National Center for Toxicological Research–FDA, Jefferson, Arkansas 72077

George M. Sisson   Mead Johnson Research Center, 2404 Pennsylvania Street, Evansville, Indiana 47721

M. J. Sloan   Shell Oil Company, 1025 Connecticut Avenue, NW, Washington, D.C. 20015

M. B. Slomka   Shell Oil Company, One Shell Plaza, P.O. Box 2463, Houston, Texas 77001

J. Smeets   Service de L'Environnement et de la Protection des Consommateurs, Commission of the European Communities, 200, Rue de la Loi, 1049 Bruxelles, Belgium

J. W. Stanley   Pepsi Co., Inc., Purchase, New York 10577

Richard H. Stanton   Velsicol Chemical, 340 E. Ohio, Chicago, Illinois 60611

Jean M. Taylor   FDA, 200 C Street, SW, Washington, D.C. 20204

Andrew S. Tegeris   Pharmacopathics Research Labs, Inc., 9705 N. Washington Blvd., Laurel, Maryland 20810

J. P. Thorn   Exxon Chemical Company, P.O. Box 3772, Houston, Texas

Cecil Timberlake, Jr.   Phillips Petroleum Company, 1825 K Street, NW, Suite 1107, Washington, D.C. 20006

Edwin Traisman   Food Research Institute, University of Wisconsin, 1925 Willow Drive, Madison, Wisconsin 53706

William M. Upholt   525 East Indian Spring Drive, Silver Springs, Maryland 20901

Leslie M. vanHoose   Schering Corporation—Animal Health Research, P.O. Box 608, Allentown, New Jersey 08501

H. G. S. van Raalte   Shell International Research Maatschappij B.V., Toxicology Division, P.O. Box 162, The Hague, Netherlands

Leonard J. Vinson   Lever Brothers Company, 45 River Road, Edgewater, New Jersey 07020

Ralph C. Wands   Cosmetics Ingredient Review, 1133 Fifteenth Street, NW, Washington, D.C. 20418

Carol S. Weil   Carnegie—Mellon University, 4400 Fifth Avenue, Pittsburgh, Pennsylvania 15213

K. W. Weinstein   Sellers, Conner, & Cuneo, 1625 K Street, NW, Washington, D.C.

Jack Winstead   Cosmetic, Toiletry and Fragrance Association, 1133 15th Street, NW, Washington, D.C. 20005

S. Jackson Wommack   Amoco Chemicals Corporation, 200 E. Randolph Drive, Chicago, Illinois 60601

Ernst L. Wynder   American Health Foundation, 1370 Avenue of the Americas, New York, New York 10019

Ralph E. Yodaiken   National Institute for Occupational Safety & Health, 5600 Fishers Lane, Rockville, Maryland 20852

J. Youngman   Netherlands Embassy, 4200 Linneman Avenue, NW, Washington, D.C. 20008

Ellen Zawel   Zawel Associates Inc., 59 Giles Road, Harrington Park, New Jersey

Robert G. Zimbelman   The Upjohn Company, 301 Henrietta Street, Kalamazoo, Michigan 49001

FOREWORD

The Delaney Clause became a part of the Food and Drug Law in 1958. It was included as an added protection to the consumer from potentially carcinogenic additives. It was also considered by most scientists a statement of scientific principle rather than one of policy established by Congress. Unfortunately, this is still the interpretation of many, if not most, scientists.

The Delaney Clause states in part "that no additive shall be deemed to be safe if it is found to induce cancer when ingested by man or animal, or if it is found after tests which are appropriate for evaluation of safety of food additives to induce cancer in man or animal."

When the Clause was made a part of our food law, chemists were reporting analyses in terms of parts per million. The animal test was then, for all intents and purposes, considered infallible. The United States was about the only country to give serious consideration to the possible relationship of additives to carcinogenesis when the law was passed; others did consider the situation, but not as seriously. Other departments of our government were not intensely interested in safety from the standpoint of carcinogenesis. The Clause, when included in the law, therefore, was for all intents and purposes an instrument of the Food and Drug Administration.

Much has happened since 1958 when the Clause became a part of our food law. Analytical chemists are not only reporting in parts per million but parts per billion and, in some instances, even in parts per trillion. Furthermore, today many consider the animal tests, such as usually used, unreliable and at best indicative or useful only as screening tests.

New procedures, such as the Ames test for mutagenesis, have been developed. The question has arisen as to the possible relationship of mutagenesis to carcinogenesis. Some feel that there is a good relationship; others feel that the Ames test is not at all reliable and should at best be used for screening.

The Commission on Pesticides and Their Relationship to Environmental Health was concerned about protocols used in testing for safety and, accordingly, the National Center for Toxicological Research was established at Jefferson, Arkansas. At the time this facility was established, testing for the safety of additives involved feeding very large doses of the additive, a practice still considered by some to be reliable and the only procedure to follow, whereas others consider it unreliable. The question of the number of animals, and continuous and discontinuous administration of an additive, needed to be studied. Then too, other causes for cancer such as environmental pollution of air, water, food, and smoking have been shown to be important. Furthermore, the use of certain birth control pills has been questioned, as well as alcohol, age, and exposure to sunshine.

The effects of additives in this complex picture have been and still are questioned. It has been said by some workers in the field of cancer that if the carcinogenic effects of smoking, birth control pills, alcohol, and age are taken into consideration and cancer statistics corrected for these, the actual occurrence of cancer is declining or at least not increasing.

Since the Delaney Clause was established, new agencies have become interested in the philosophy of this Clause. Governmental agencies giving serious consideration to carcinogenesis are the Environmental Protection Agency, Occupational Safety and Health Administration, Consumer Products Safety Commission, and the USDA, an old agency that is responsible for regulatory functions pertaining to animal products.

Other countries are also becoming interested in factors influencing carcinogenesis. Hence, the interest of the International Academy of Environmental Safety in holding a conference on the subject has evolved over the years. The Academy concluded, therefore, that a forum on the Delaney Clause should be held with the view of evolving an understanding as to what the Clause means and how countries other than the United States view it, and how departments of the U.S. Government other than the Food and Drug Administration view it.

After twenty years of experience, it is obvious that the Delaney Clause is seldom used by the Food and Drug Administration in making decisions with respect to the safety of foods. As a matter of fact, it is really not needed to protect the consumer, for other provisions of the law can do so without creating undue problems. It is also becoming more obvious that the Clause is a matter of policy and not of science. In this respect, Peter Hutt, formerly General Counsel of the Food and Drug Administration, stated at the forum: "No statutory provision has been the subject of greater discussion, or less understanding, than the Delaney Clause. On one side, scientists have attacked it as a misstatement of scientific fact and indeed as a denial of the most fundamental principles of toxicology and biology. On the other side, consumer advocates have defended it as the last hope for the survival of mankind, without which we would all be poisoned by food laced with potent carcinogens. Neither side to this controversy has bothered to stop and analyze the matter in detail." Hutt went on to say that "The Delaney Clause is a trivial statutory provision that is not worthy of serious public discussion." He may be correct, yet those for and those against the Clause do spend a great deal of time arguing about it, which seems fruitless. In any event, Hutt pointed out that by focusing on exactly the wrong questions, scientists on both sides of this controversy have succeeded in obscuring the real public policy issues that must be resolved in determining food safety for the future. Hutt emphasized that the Delaney Clause never was intended to be and, in fact, is not a statement of scientific principle. It was, and remains, a statement of public policy.

In view of the changes that have taken place during the past twenty years since the Clause was made a part of the Food and Drug Law, it is time to reconsider this policy statement. It appears that a matter of judgment is very important, but who makes the judgment, and what should be considered in making a judgment? In line with these questions, a long list of points that perhaps should be considered in making decisions with respect to safety were mentioned by various forum discussants. A partial list of these points that appear to be so important in making decisions and judgments are given below. These were discussed during regular presentations and during open forums that followed the presentations. They certainly appear to indicate the need for consideration of a new policy with respect to carcinogenesis and safety.

Several discussants stressed the view that there is no such thing as absolute safety and that there is no way of proving a substance is safe, and an attempt to do so would be an attempt to prove the negative.

Safety is a relative concept.

There is a need for special consideration of foods that presumably do not pose a danger, for example, peanuts, corn, and milk, which have been found at times to contain aflatoxins. This apparently raises the question of the safety of substances included in food by man or those present as a result of the activities of nature.

How does one determine the potency of a carcinogen?

How is a carcinogen defined?

Is minimal exposure associated with risk?

Heavy dosages may cause metabolic overload and a breakdown of the system, making it susceptible to cancer.

How much and how often would a particular additive be ingested by an individual?

It is important to balance cost against risk. For example, how much of the product is being consumed, a little or a great amount? There should be a difference in judgment of items eaten every day and those perhaps once or twice a year.

The satisfaction and pleasure of a particular item to the consumer is considered important by some. If the product is relatively safe and yet banned, how much will the consumer miss it?

From the standpoint of economics, how many people may be thrown out of work as a result of banning or restricting the use of an additive?

What is the dollar value of human life, and should this be considered?

What about human needs and human wants?

Can the risk–benefit be defined? Much consideration has been given this point and apparently it needs more attention.

Society apparently needs to evaluate the use of resources. How much can we afford to put into defensive research as compared to constructive research?

If freedom of choice were given serious consideration, how would it be handled? When an additive poses only slight danger, as appears to be the case for saccharin, would it be better to permit the individual to choose, providing there is adequate warning, just as in the case of cigarettes?

Some feel that susceptible people should be warned, but then how does one determine who is susceptible?

The question of dosage–threshold value and manner of feeding are important and should be considered in decision-making.

What about *ad lib* feeding as compared with periodic feeding?

The question of metabolic effects of very high dosages of an additive was raised. Some believe that high dosages may affect the physiologic processes to such an extent that the protective mechanisms of an animal are overcome, thereby enabling the carcinogenic agents to become destructive.

A question was raised about the nature of the additive. Presumably, this means its chemical structure.

Is there a relationship between the nutritional level of the individual—whether it be man or animal—and the effect of the additive?

The real cause of cancer apparently is unknown, but it appears that we are starting to learn something about the mechanisms. What, for example, is the function of the rough endoplasmic reticulum (RER) in protecting against a chemical? When RER is harmed, is a potential protective mechanism impaired?

The relation of age and cancer was mentioned a number of times.

The question was raised as to whether it would be possible to reduce all human risk. During the discussion there appeared to be a feeling that it would not be possible.

The matter of lifetime risk was raised a number of times and extensively discussed.

Is there a clear difference between the pathology of malignant and benign tumors and, at times, even nodules?

In view of recent observations, the question was raised as to what animals and what strains of animals should be used. Should they all be tested or just one, and might that one be more susceptible or more resistant than others?

The use of the mouse was questioned by some, as was the rat.

The significance of animal studies to the human, whether it be mouse, rat, or something else, was questioned. Which is the best animal, and which is most closely related to the human?

How strong and substantial should the scientific data be?

The question as to who makes a judgment arose a number of times. How many scientists and/or nonscientists are involved and, if only a few, is anyone, or a small group, capable of considering the many factors involved in arriving at a sound decision?

If used, where does the administrative law judge fit into the decision-making process?

The international situation with respect to the United States is important. It is important from the standpoint of importation of foods as well as the exportation of foods and the protective measures used in the United States and in other countries.

The importance of chemical–chemical or drug interaction was mentioned a number of times. What might be the effect of one chemical agent or a number of agents on others?

What is the most appropriate procedure for the administration of an additive being tested?

What is the role of the diet nutrition and diet contaminant, such as aflatoxins and PCB, and the type of diet fed animals?

Should the animals be permitted to eat in an *ad lib* manner? *Ad lib* feeding may influence metabolic activities. Is this an important factor?

What about body retention and the role of lyophilic chemicals?

What about metabolic overload and the possibility of stressing the system to the point of making the animal more susceptible to cancer?

The reliability of statistical procedures used in analyzing data was questioned.

Humeral and normal factors and their meaning were discussed at length.

There was much discussion on naturally occurring agents and what to do about them. This takes judgment and consideration.

What about the degrees of safety or danger between safety at one end of the spectrum and danger at the other?

What should the dosage be, and what should the frequency and method of administration be of an additive?

What is a positive or negative result? How many negatives are needed to cancel out a positive and vice versa?

Is cancer pathology well defined?

What is the relationship between mutagenicity and carcinogenicity?

Certain enzymes in the mouse are more active than similar enzymes in the human; this must be taken into account.

The question of environmental chemicals in addition to the one being tested was discussed rather extensively. In some small animal tests, water, air, noise, waste, temperature, and many other environmental factors have been shown to influence results.

The age of the animal appears to be important. A young animal apparently is more susceptible to carcinogens than an older one. The size and sex of the animal are of equal importance. Larger animals seem to be less susceptible than smaller ones.

How long should a test be, and what about continuous and discontinuous feeding?

The question was raised as to which organ to examine in the test. Should it be the liver, bladder, other organs, or all? It appears that we may be overemphasizing the importance of the liver. Other organs are equally, if not more, important.

The rate of metabolism of the compound being tested apparently varies greatly with the animal.

The relation of excretion to dose is important. How much is retained, and how much is excreted, and when?

What about the presence or absence of viruses, bacteria, or parasites in the animal? This apparently is a factor that is becoming more and more important and yet is to a large extent overlooked.

There are immune differences. The question is, how are these taken into consideration? They may vary with the strain of animal and they may vary from one animal to another.

There are differences in repair mechanisms in different animals, in different strains, and different individuals. These should be explored, especially repair of RNA or DNA.

There are xenobiotic differences. The question is, how important are these?

The gastrointestinal flora differ in animal and man and differ from man to man and animal to animal.

Test animals normally used are inbred. Man is not inbred. This makes an enormous difference. How can these differences be taken into consideration?

Geography can be a factor.

Epidemiology is important and requires more study.

Many other factors were mentioned and discussed—perhaps as many as fifty. It is apparent that in considering the safety of an additive, many of these factors should be taken into consideration. Some believe, therefore, that simple tests on one kind of animal are not sufficient; ultimately it is a matter of opinion.

Science is important but, basically, it requires sound scientific, social, and economic judgment to determine whether or not an additive should be used. These are some of the things that became apparent during the conference and are covered intensively and extensively in this book. It became clear that there is a need to review and perhaps revise the present policy used in establishing safety.

Emil M. Mrak

SESSION 1

INTRODUCTORY REMARKS
SESSION I

DR. COULSTON: Welcome to this meeting. This is the seventh meeting of the International Academy of Environmental Safety, established in 1971 in Munich, a group of dedicated individuals limited to 100 people worldwide. Many of the members of the Academy are in the audience, some of whom will speak to you during the meeting. The program beginning today was organized by Drs. Emil Mrak, Al Kolbye, David Clayson, Morris Cranmer, and myself.

We thought that the topic of this meeting was indeed timely. We are not here to discuss the merits of the Delaney Clause. All of us recognize it as a hallmark piece of legislation, which in 1958 said some things in general that I'm sure no one will disagree with. No one wants cancer-producing substances in our food supply. There can be no real argument about that. The argument and discussion rest primarily on the interpretation of the Delaney Clause. On this point experts, politicians, lawyers, scientists disagree very much. It is for this reason that we thought this meeting would be very timely and we hope to hear from all sides on the question, positively, negatively, and somewhere in the middle. Indeed the Organizing Committee tried and did succeed I think in producing what we would call a "neutral meeting," a meeting that does not take sides, except as individual expressions of opinion. This will be the theme of this meeting. Now you may look around and ask where the Naders, etc., are. We did indeed invite Ralph Nader, Sidney Wolfe, and Joe Hyland, but to no avail. We regret that they are not here but I think it only fair to say that we did try everything we could, particularly to get Dr. Sidney Wolfe here, because of the statements made by him, Dr. Lijinsky, Representative Martin, and myself in the *Chemical and Engineering News* symposium, which was published on June 27, 1977.

However, we are very fortunate in having many eminent people here at the meeting who will present their own views relating to the philosophy of the Delaney Clause. We will have, I'm sure, a delightful, stimulating discussion and presentation of some of the facts related to the title of this symposium. At the break during the morning discussion a special statement from Congressman Wampler will be read by his chief assistant.

We have invited the staffs of almost every major Congressman and Senator interested in this subject and many of them will be in the audience, listening to what we have to say.

1

ISBN 0-12-192750-4

I think that we have a chance at this meeting to present
information that will be of great benefit, not only to the
consumer, but to the regulatory officials and the Congress,
who are all vitally interested in the question of what is and
what isn't a cancer, from the point of view of animal testing
and prediction in man, particularly as it relates to chemicals
both in our food and environment.

Now one may ask very rightly what has the Delaney Clause
got to do with EPA or OSHA or NIOSH? These agencies have a
great deal to do with the Clause because it is becoming
obvious that the principles developed for the interpretation
of the Delaney Clause by the FDA have been copied by many
other agencies of government and these are being made into
new regulations. I'm not a lawyer and I may have said that
incorrectly, but I'm sure Peter Hutt will straighten this out
when he talks.

This is the issue that will face this meeting: what is a
chemical carcinogen, what tests are proper, and how do we
interpret these tests in terms of human safety. This is indeed a burning issue of the day, not only here but throughout
the world. In fact, it is probably the most vital issue
today. We as toxicologists are dedicated to the safety evaluation of chemicals and drugs in animals and humans. To
accomplish this we must have facts and then make reasonable
interpretations.

We have many people in the room whom I wish to thank for
making this meeting possible. One of these is Dr. V. K. Rowe,
without whose help we could not have possibly held this meeting. He has given of his time, advice, and counsel, and I
want particularly to honor him on behalf of the Organizing
Committee with this statement. These proceedings will be
taped. You are free to say what you want. All speakers will
receive a transcript of what has been said and they can edit
it as they wish. The program will then be published as is
the custom of the International Academy as a proceedings by
Academic Press. The first of this series, on drinking water
quality was supported by EPA. It presented the proceedings
of what was indeed the first international meeting on drinking
water safety.

I shall now turn the meeting over to the immediate past
president of the Academy, Dr. Emil Mrak.

DR. EMIL MRAK: Thank you. First of all on behalf of the
International Academy of Environmental Safety I say greetings.
We're glad to have you here. Perhaps I should tell you a
little about the International Academy. It is a very unique,
nonprofit organization, incorporated in 1972 under the laws of
West Germany.

Today, literally every major country has a representative in the Academy. It is not a free-wheeling organization. People are elected into it and those who participate, in one way or another in most countries, are close to governmental operations. They have had a number of meetings. Dr. Coulston mentioned one on water. We've had meetings in France, Germany, Finland, Japan, Luxembourg, and of course the United States. The format of the meetings is somewhat as you see here with a limited number of people attending. We try to arrange the meetings so that the participants sit at a table, discuss, argue, etc. The discussions are all recorded, edited by the speakers, omitted if they wish, revised, and then eventually published as the proceedings of the meeting. The Academy is now sponsoring a society--Dr. Smeets of Luxembourg will tell you about it later--the Society of Ecotoxicology and Environmental Safety.

As Dr. Coulston indicated, we tried very hard to get all points of view represented at this meeting. I would like now to introduce some of the people: Dr. Friedhelm Korte from Munich, is the new president of the Academy; president-elect, Dr. Harold Egan, is the Chief Chemist of the Queen of England; Dr. J. Smeets of the Commission of European Communities from Holland is the president-elect of the Society of Ecotoxicology; Dr. Astolfi, from Argentina, is Professor of Toxicology at the University of Buenos Aires; Dr. Schmidt-Bleek, from Berlin, is the Head of Toxic Substances, Federal Environmental Agency for West Germany; Dr. David Clegg, Canada, is Head of the Pesticide Section, Toxicology Evaluation Division, Food Directorate, Canada.

Now I would like to call on Dr. Smeets, president-elect for the Society of Ecotoxicology, to tell you a little about that society.

DR. SMEETS: I shall talk on behalf of the Society and the European Community. I will start with the European Community. When we got the invitation in Europe the first reaction of some of us was that this is a typical American problem, an internal problem, and what have we to do with it in Europe? In fact, when you realize the impact of all of these problems relating to the Delaney Clause, its effect goes far beyond the particular concept (for example, can saccharin or lead be used or not). It is in our opinion a philosophical problem with which we in Europe at the European Commission are confronted daily; we have to translate scientific decisions and data, and political decisions into a final statement of safety evaluation.

In the Commission, on the one hand, political decisions have to be made and the politically responsible people say that there are not enough data available. The next day they say they have enough data, and they will make a decision because they must. This is a topic we are going to discuss. I am sure that the results and the conclusions of this meeting will go far beyond the simple problem of whether saccharin is or is not dangerous. Sooner or later we must find an answer concerning the philosophical approach to the availability of scientific data and political decisions.

As I said, there are two points I want particularly to mention, since we in Europe are also very interested in the conclusions of this meeting. I have already discussed the first part. The second part is that in my capacity as president-elect of the International Society for Ecotoxicology and Environmental Safety, I would like to express my appreciation for the sponsoring of this society by the International Academy of Environmental Safety. In fact this "baby" of the International Academy was guided by the Academy through the difficult period of its formation. The way we have been guided until now has been very successful and we are very happy that we can say that our society already has a good membership, and has proven the need for its existence.

This International Society was established because there was a general need to bring together scientists, research workers, administrators, and others who are studying the scientific basis for implementing regulations to forestall the toxic effects of chemicals and physical agents on living organisms, and especially on populations with defined ecosystems. So we should not confuse this aim of the society with the general toxicological societies. Here, we are dealing with groups of populations, whether they be people, plants, or animals, in living defined ecosystems.

DR. COULSTON: I would like to call on Dr. Korte, President of the International Academy, for a few comments.

DR. KORTE: Ever since its foundation in 1971, the Academy has had the aim to contribute to the solution of urgent scientific problems in the frame of human health and environmental protection. Major emphasis has always been to have experts from academia, scientific agencies, government, and industry enjoy discussions on such specific problems. Consequently, our Academy has organized or sponsored a number of scientific symposia since 1971, collaboratively with governmental authorities and other scientific international organizations such as the International Atomic Energy Agency, the Commission of the

European Communities, the United States Environmental Protection Agency, the Society for Radiation and Environmental
Research in West Germany, the National Center for Toxicological Research, the Institute of Physical and Chemical Research,
Tokyo.

It would take too much time to give a complete inventory
of what the Academy has done so far, but I would like to cite
a few decisions and recommendations of our expert groups.
The key statement of our foundation meeting in 1971, which
was "Second International Symposium on Chemical and Toxicological Aspects of Environmental Quality," was "reliable
priorities for investigating certain products can only be
established when amount, nature, and place of use are known."
Toxicological evaluation of environmental chemicals is of
highest priority but is to be based on well-adopted evaluation
parameters. "As regards the financial effects of the deter-
ioration of environmental quality and the possible maintenance
costs, standards for financial estimates have to be set." In
the context of the scientific criteria for the evaluation of
air quality, the different aspects of setting air quality
standards have been discussed, especially in the case of
benzine, where zero limits had been suggested, specifically
in Germany. "We came to the conclusion that this is not a
practical approach to solve problems if no reliable data are
available." Simultaneously we discussed problems in the
evaluation of environmental chemicals, and aside from other
recommendations, another point of discussion that has been
treated in more detail was the question of the "threshold
value of mutagenic effects of chemicals." At a load of 10µg/
cubic meter lead oxate particles in the air, no chromosomal
changes could be detected, but at levels of about 20 ppm
benzine in the air, chromosomal effects occurred. At lower
concentrations, however, they could not be detected.

These findings demonstrated that chromosomal effects may
be related to the concentration threshold value, for instance,
cytotoxic activity, which only occurs within a certain dose.
Chromosomal changes of peripheral somatic cells do not give
any proof of damage to germ cells. It is known that tissue
specificity plays an important role in chromosomal effects.
The problem of toxic substances of natural origin has been
discussed with the example of aflatoxin. From epidemiologic
data, indications are given that this toxin causes liver cancer. The toxin in peanuts, however, cannot be forbidden, but
an attempt should be made to fix the safe tolerance level
from a benefit/risk evaluation using the necessary technology.

Every food contains natural substances with undesired or
even hazardous effects. When the hazard occurs depends on the
dose, the quantity consumed, so that there never can be an
absolutely hazardless or an absolutely safe substance.
Tolerances must be established for such chemicals as aflatoxin, PCBs, or monosodium glutamate.

In the discussion of the scientific basis for the safety
evaluation of organic chlorines and related compounds, the
Academy made statements on the causes of cancer incidence as
related to the phenomenon of enzyme induction. "The main
problem in this connection should be an understanding of what
happens in the cell nucleus and what can be learned from experiments with animals given high doses compared to those
given low doses." The enzyme activities discussed were not
only liver enzymes but also others. For instance, lung
enzymes should be studied because it was shown that the frequency of lung tumors decreases under the influence of some
foreign compounds. This fact cannot be explained, not only
because there is insufficient knowledge on enzyme induction,
on possible influences of the cell nucleus or the prolifera-
tion of the chromosomes, but also because there is nearly no
knowledge at all on the repair of enzymes. For all these
questions today, we can only discuss the possibilities that
exist for the various mechanisms. The basis for this cannot
be derived from the results of experiments with mice, but
perhaps someday results can be obtained from experiments with
a large number of animal species, especially with nonhuman
primates. What occurs in rodents need not necessarily happen
in other animal species. Furthermore, today it is still usual
to think that the most sensitive animal species is the most
appropriate test organism. Aromatic hydrocarbons, for instance, cause skin cancer in mice and rabbits. This is why
these species are said to be the most appropriate test objects
for such studies. But, extrapolation to humans remains a
very questionable point. Unreasonable doses lead to unrealistic and unreasonable results. For regulatory agencies,
statistical and experimental results with humans would be of
importance. This has been established very clearly in a
meeting in 1974.

Many of us attended another meeting, which was cosponsored
by the International Atomic Energy Agency in Vienna, on the
"Scientific Basis for the Establishment of Threshold Levels
and Dose-Response Relationships of Carcinogens." We concluded
that there are certain conditions in certain animals with
certain chemicals where we can clearly state that a no-effect
level exists, provided the no-effect level is defined. This
means we can say no production of cancer resulted in animals
and we can say this is a no-effect level. We can say there

were no changes in the microsomal enzymes. We tried to
visualize a no-effect level in terms of something that was
reversible. Certainly if we can use certain enzymes such as
processing enzymes or commonly called drug metabolizing enzymes, this may not be a harmful condition, but just a change
in the physiologic state of a cell, which is completely reversible (Coulston's concept). Such a condition would be an
endpoint to a physiological process, meaning a normal process.
The key issue is that it is now clear to most scientists, be
they pathologists, toxicologists, biochemists, or medical
doctors, that there can be a dose-response relationship to a
chemical and that no-effect levels can exist. In some cases,
certain chemicals, even one molecule, might cause change, but
we concluded in this meeting that one should not make a
general statement that any so-called carcinogen will always
cause a cancer at those doses. This has been discussed and
was the main result of the discussion at the Vienna meeting
chaired by Dr. Coulston together with the International Atomic
Energy Agency.

In Munich in 1975, the Academy again discussed "Ecological
and Toxicological Aspects of Organic Chlorines" and I cite
from the recommendations: "An increase incidence of benign
liver tumors in mice exposed to any of the variety of organic
chlorine insecticides would not in the absence of supportive
evidence be regarded as proof of carcinogenic hazard to
humans." Supportive evidence might come from studies of the
mechanisms involved in the use of human liver microsomes,
studies on other animal species, or epidemiological studies
of cancer incidence. In the context of this statement, I
would like to mention that a recent WHO group in Geneva discussing the potential harm of DDT came to the same conclusion
and expressed very clearly that there is no evidence for DDT
being carcinogenic to man.

The highlight of the last Academy meeting in Munich (1977)
was new scientific data on the influence of environmental
chemicals on the endocrine system and on immunological response. Furthermore, we discussed the chemical approach,
namely, the potential of the atmosphere as a final degradation
place for chemicals, specifically for organic chemicals. We
also had a session on the problem of translating scientific
data into regulations. This topic has also been discussed in
an international colloquium cosponsored by the Commission of
European Communities.

This is a very broad field and needs discussion on specific
topics. The last meeting of the Academy was in 1977 in New
Mexico. The subject was "Human Epidemiology and Animal Laboratory Correlations in Chemical Carcinogenesis." We certainly will discuss the result of this meeting in our discussions
here.

The next meeting will be held again in cooperation with the International Atomic Energy Agency in Vienna in August, 1978, and the topic selected was the "Scientific Basis for the Ecotoxicological Assessment of Environmental Chemicals."

I'm very much looking forward to the results of this present conference. I'm convinced that our discussion during the next three days will be a help for science and will serve to benefit our civilization.

Dr. Egan, from England, the Queen's chemist and the President-elect of the International Academy, will say a few words of introduction before I call on the keynote speaker.

DR. EGAN: I did not intend to say very much at this time, rather I intend to listen. I am an analytical chemist and as such I get blamed for many things and so do my colleagues, not least of all in the context of today's meeting, for having methods of analysis, methods of detection, methods of estimation, which are far too sensitive.

An old proverb says that one swallow doesn't make a summer; one paper in the scientific literature which claims the detection of substance X in air and water or cheese at a level of one part in 10^{12} doesn't mean that an analytical chemist can do the same on a regulatory basis in air and water and cheese, or indeed in petroleum exhausts.

PUBLIC POLICY ISSUES IN REGULATING
CARCINOGENS IN FOOD

Peter Barton Hutt

No statutory provision has been the subject of greater discussion, or less understanding, than the Delaney Clause. One one side, scientists have attacked it as a misstatement of scientific fact and indeed as a denial of the most funda- mental principles of toxicology and biology. On the other side, consumer advocates have defended it as the last hope for the survival of mankind, without which we would all be poisoned by food laced with potent carcinogens. Neither side to this controversy has bothered to stop and analyze the matter in detail. At times, it has seemed that the battle was an end in itself, and that the combatants were participating more for the enjoyment of the confrontation than for the purpose of determining rational public policy.

For the past 15 years, I have steadfastly adhered to the view that the Delaney Clause is a trivial statutory provision that is not worthy of serious public discussion. It adds nothing whatever to general safety provisions that have existed in the law for over 70 years. I have declined to participate in the great public debate about the Delaney Clause because I thought, and indeed continue to think, that it is an utter waste of time and effort.

In the past year, however, I have become increasingly concerned that this debate is not only pointless, but indeed harmful. By focusing on exactly the wrong questions, scien- tists on both sides of this controversy have succeeded in obscuring the real public policy issues that must be resolved in determining food safety policy for the future.

The Delaney Clause never was intended, and in fact is not, a statement of scientific principles. It was, and remains, a statement of public policy.

ISBN 0-12-192750-4

The public policy that it embodies, moreover, had been adopted by the Food and Drug Administration before the Delaney Clause was enacted by Congress. That public policy would have been pursued by the Food and Drug Administration, and ultimately by other regulatory agencies, even if the Delaney Clause had never been enacted by Congress. And repeal of the Delaney Clause, without revision of the other food safety provisions in the law, would not change any action that the Food and Drug Administration has taken in the past or may take in the future. Thus, the Delaney Clause is utterly irrelevant to current food safety policy. It is at most a convenient symbol or rallying cry for those who attack and defend it so vehemently.

In my remarks here, I will first trace the history of the development of current food safety policy. I will then point out why, in my opinion, the scientific community has been debating entirely the wrong issues for the past 15 years. Finally, I will describe what I believe to be the public policy issues now facing the country with respect to regulating carcinogens in food, and how they can best be approached in order to develop a consistent rationale for future public policy in this area.

I

Since 1906, Federal food and drug law in this country has provided that a food is adulterated, and therefore illegal, if it contains any poisonous or deleterious substance which may render it injurious to health. The Food and Drug Administration has interpreted this provision, which is contained in Section 402(a)(1) of the present Act, to give it broad authority to prevent the use of potentially harmful ingredients in food. In 1958 Congress amended the Act to include the requirement that all food additives be shown to be safe prior to use. These general food safety provisions are, of course, as applicable to a food ingredient that presents a risk of cancer as they are to a food ingredient that presents any other type of toxicological risk.

In 1974 I conducted, for the House Appropriations Committee, a brief review of some of the important actions of the Food and Drug Administration in removing substances from the food supply since 1950 upon a finding or a suspicion of carcinogenicity. That review demonstrated that, even before the Delaney Clause was being considered by Congress, the Food and Drug Administration had used the general prohibition against poisonous or deleterious substances to prohibit the use in food of ingredients which it concluded to be carcino-

genic in test animals. Two artificial sweeteners (dulcin and
P-4000) were prohibited in 1950, and coumarin was prohibited
in 1954, solely on this ground.

Even after the enactment of the Delaney Clause as part of
the Food Additives Amendment of 1958 and the Color Additive
Amendments of 1960, the Food and Drug Administration has continued to take regulatory action against food components that
it regards as animal carcinogens under the general safety
provisions of the law, rather than under the Delaney Clause
itself. Explicit use of the Delaney Clause has, indeed, been
a rare event.

The legislative history of the Delaney Clause demonstrates
that it was regarded as totally unnecessary by the Food and
Drug Administration. In its report on the House bill that
eventually became the Food Additives Amendment, the Food and
Drug Administration opposed a specific anticancer clause
because:

> We, of course, agree that no chemical should be per-
> mitted to be used in food if, as so used, it may cause
> cancer. We assume that this, and no more, is the aim of
> the sponsor. No specific reference to carcinogens is
> necessary for that purpose, however, since the general
> requirements of this bill give assurance that no chemical
> can be cleared if there is a reasonable doubt about its
> safety in that respect.

In order to assure enactment of the legislation, however, the
Food and Drug Administration later withdrew its objection to
inclusion of a specific anticancer provision. The Agency
concluded to accept the Delaney Clause precisely because it
added nothing whatever to the law. A letter sent to the House
Committee by the Department of HEW stated that the requirement
that a food additive be proved safe would, in itself, preclude
approval of any additive found to be carcinogenic in test
animals, and that the Department would therefore have no
objection to the Delaney Clause, which would simply restate
what would otherwise be true under the general safety provision. The Senate Report made exactly the same point:

> We have no objections to that amendment whatsoever,
> but we would point out that in our opinion it is the in-
> tent and purpose of this bill, even without the amendment,
> to assure our people that nothing shall be added to the
> foods they eat which can reasonably be expected to produce
> any type of cancer in humans or animals... In short, we
> believe the bill reads and means the same with or without
> the inclusion of the clause referred to. This is also the
> view of the Food and Drug Administration.

It is therefore not surprising that most of the controversial action taken by the Food and Drug Administration since 1958 against food ingredients on the ground of animal carcinogenicity studies have been based on the general safety provisions of the law, not on the Delaney Clause. These actions include the bans of cyclamates, DES, and FD&C Red No. 2.

Nor should it be surprising that the Food and Drug Administration has steadfastly stated that, even if the Delaney Clause were repealed, it would nonetheless have banned saccharin under the general safety provision of the law. Those statements simply reiterate the long-standing and consistent Agency position on this matter. Any future change in food safety policy must be accomplished by a revision of all of the food safety provisions in Section 402, 406, and 409 of the Act and not simply by a modification of any one of them.

II

It is equally important to understand that the Delaney Clause is not a statement of scientific fact or scientific principles. That was not the intent of Congress when it was enacted. It is, instead, a statement of public policy. Thus, any debate whether the Delaney Clause is scientific or unscientific is utterly meaningless.

The legislative history of the Food Additives Amendment of 1958 and the Color Additive Amendments of 1960 clearly reveals the basis for this congressional policy. Eminent scientists informed the Committee that there was, at that time, scientific uncertainty about the cause of cancer. In particular, a National Cancer Institute Report presented to Congress in 1960 stated "No one at this time can tell how much or how little of a carcinogen would be required to produce cancer in any human being, or how long it would take the cancer to develop." The Secretary of HEW presented the following testimony to Congress in 1960:

> Whenever a sound scientific basis is developed for the establishment of tolerances for carcinogens, we will request the Congress to give us that authority...
> So long as the outstanding experts in the National Cancer Institute and the Food and Drug Administration tell us that they do not know how to establish with any assurance at all a safe dose in man's food for a cancer-producing substance, the principle in the anticancer clause is sound.

The Secretary made it clear that, even without a specific anticancer clause in the law, the FDA would continue to prohibit any food or color ingredient found to cause cancer in test animals.

The Congressional response was, of course, predictable. The House Report concluded that "in view of the uncertainty surrounding the determination of safe tolerances for carcinogens, the Committee decided that the Delaney anticancer provision in the reported bill should be retained without change."

Thus, it is clear that the Department of HEW did not state and Congress did not determine that there is no threshold for a carcinogen, or that animal studies accurately predict human carcinogenic response, or that high dose feeding studies are appropriate. Congress decided only that, at that time in history, there was scientific controversy and uncertainty surrounding these and many other issues relating to the potential carcinogenicity of chemical substances. Based upon that uncertainty, Congress made a public policy judgment. It concluded that, until the scientific uncertainty is resolved one way or the other, it would require the FDA to regulate on the side of caution by banning all animal carcinogens from the food supply. At that time, of course, the widespread presence of trace amounts of animal carcinogens in food, and the possibility that common food substances might themselves be shown to be animal carcinogens, was not anticipated.

There are therefore only two grounds on which the wisdom of the Delaney Clause can be debated. The first ground involves the question whether the scientific uncertainties that led to the enactment of that clause have now definitively been resolved. The second ground involves the question whether, even if those uncertainties have not been resolved, the public policy judgment made by Congress in the form of the Delaney Clause, and by the FDA and a number of other government agencies in the form of daily regulatory decisions during the past 30 years, represent a wise societal judgment. The first question is solely a scientific determination. The second question is solely a public policy determination.

III

With respect to the first of these two questions, it appears that there is still scientific uncertainty about many important aspects of cancer testing and cancer causation. I am not aware that any scientist has claimed to have experimental proof either that a single molecule of a chemical hitting a single human cell has in fact produced cancer, or that it is impossible for this to happen. If those experiments

have been done, I am not aware that they have been replicated
and that a scientific consensus has now formed around their
results. Similarly, I am not aware that there is now experimental evidence or a scientific consensus definitively resolving the relevance of animal feeding studies in determining
human carcinogenesis, the use of high dose levels, the significance of benign tumors, the variables that will and will not
determine an appropriate animal experiment, or even the correct animal species in which to conduct an experiment. If
that consensus has emerged, it has unfortunately escaped my
attention.

The scientific world abounds, of course, with opinions
about all of these and many other related issues. Those
opinions are fervently, indeed often religiously, held and
expounded. They are advanced on the basis of what each
individual undoubtedly believes, in the best of faith, to be
true and accurate. I doubt that anyone, however, will be so
bold as to suggest that they have direct, irrefutable, experimental evidence that definitively resolves the issues, or
that their position represents a true scientific consensus on
which there is no reasonable dispute among responsible
scientific colleagues.

The Director of the National Cancer Institute addressed
this very matter in congressional testimony just two weeks
ago. He stated that "Unfortunately, today the science of the
matter is not cut and dry. There are honest scientific
differences of opinion about evidence and how one can interpret it... The uncertainties are overwhelming at this stage."
Thus, although I am not competent to judge the scientific
merits of either side of this controversy, it does indeed
appear that uncertainty still exists. If this is true, the
scientific predicate on which Congress made its policy judgment, as embodied in the Delaney Clause, remains unaltered.

IV

Assuming that there still is significant scientific
uncertainty about many critical aspects of cancer testing and
cancer causation, one must confront directly the public policy
issue of how much risk or uncertainty we as a society, and
each of us as individuals, are willing to tolerate. I dealt
with this, in a general way, in my 1973 paper at the Forum
conducted by the National Academy of Sciences on "The Design
of Policy on Drugs and Food Additives." I find that I cannot
improve upon the words used there, and will therefore repeat
it verbatim:

There appears to be no public or scientific consensus today on the risk or uncertainty acceptable to justify the marketing of any substance as a food or drug.

To some, who favor a return to more simple days, no risk or uncertainty whatever is justified for any addition of a chemical to food. They would, indeed, require a showing of some greater benefit to society before any such ingredient is permitted. To others, who see enormous progress in food technology and nutrition from the use of food additives, the usual risk associated with technological innovation are regarded as entirely reasonable. Even in the area of therapeutic drugs, there is intense public dispute about whether, to use one example, the risks of an abortion outweigh the risks that are raised by the use of DES as a postcoital contraceptive.

We must recognize that this type of issue presents fundamental differences in philosophical principles, not simply a narrow dispute on technical details. It raises the most basic questions of personal beliefs and human values--the degree of risk or uncertainty that any individual is willing to accept in his daily life. Attempts to resolve such an issue on the basis of rigorous scientific testing or analytical discourse therefore simply miss the point. A mathematical benefit-risk formula or computer program may eventually be able to quantitate the risk or uncertainty that inheres in a given product, but it is not even relevant to the moral and ethical issues involved in deciding whether this degree of risk or uncertainty is acceptable.

This problem arises whenever new doubts or suspicions are cast upon the safety of an already marketed substance. Those who favor a very low public risk demand that the product immediately be removed from the market. Those who advocate a higher risk demand that it remain on the market until it is shown to be unsafe. If, as I suspect will happen, we eventually prove that many of our basic foods and drugs contain at least trace amounts of highly toxic substances--including carcinogens, teratogens, and mutagens--the public simply will have to face these issues in a more forthright way than it has up to now.

One does not need a degree in science to hold and express deeply felt beliefs on the degree of risk or uncertainty society should accept from food and drugs. Nor, indeed, does a scientific background equip anyone with any greater insight into the intricacies of this type of policy issue or any more impressive credentials or greater authority to act as an arbiter in resolving

these matters. As long as we remain a free society,
these basic philosophical principles will, and properly
should, remain the subject of intense public scrutiny
and debate.

That is the public policy issue we face today. There is no
way to avoid it. The regulatory answers that were once hopefully put forward as solutions have, slowly but surely, been
demolished with the advance of scientific knowledge.

We now have information that we simply did not have when
Congress enacted the general food safety provisions of the
1906 and 1938 Acts and the 1958 and 1960 amendments, and
adopted the Delaney Clause in 1958 and 1960. Virtually every
food product on the market today contains some constituent
that has been shown to be carcinogenic in at least one animal
test. Some of these are unwanted contaminants, such as the
aflatoxin in peanuts, corn, and milk, and the nitrosamines in
bacon. Some are the result of using chemicals for other
important processing purposes, such as the chloroform produced
by chlorination of water. Others are natural constituents of
food, such as the tannin in tea, the caffeine in coffee, and
the safrole in cinnamon. Still others are essential vitamins
and minerals, such as vitamin D, calcium, selenium, and tryptophan. Even egg white, egg yolk, lactose, maltose, and
charcoal-broiled meat have been shown to be animal carcinogens.
It would be difficult to plan a diet for even one day that
would be entirely free of animal carcinogens. It would surely
be impossible to live for any significant time on such a diet,
in light of the utter ubiquity of these substances.

Thus, although the food safety policy embodied in the
current law, and the FDA's implementation of it, may have made
good regulatory sense even as late as a few years ago, it is
obviously no longer sustainable. It is simply not feasible
to remove from the food supply every substance that has been
shown to be carcinogenic in test animals.

Even in the face of enormous scientific uncertainty about
cancer testing and cancer causation then, we must begin to
fashion a new food safety policy for this country. The old
one has fallen victim to our ever-increasing scientific knowledge. We have succeeded in proving that every piece of food
we eat carries with it some risk, even though we are not
certain whether that risk is a theoretical or a real risk, or
how large or serious that risk may be.

In a larger sense, we should not be surprised by that
finding. Everything else that we do in our daily lives
carries with it some risk. Just driving in an automobile or
bus to work, walking from the parking lot or the bus stop to
the building, and riding up the elevator to the office, en-

tails a very definite risk. In retrospect, we can all wonder at our naivete in thinking that eating food would somehow be different from any other aspect of our lives, and would entail no risk whatever.

I believe the FDA deserves great credit for doing its best to handle the current regulatory crises under an obviously outmoded statute. It has attempted, in a number of ways, to smooth over the legal inconsistencies and to find a rationale even when none can any longer be said to exist. Faced with the increasingly untenable safety requirements of the law it has in many instances simply ignored the evidence of animal carcinogenicity of many common food substances and has declined to take action that would be subject to public ridicule. As an interim measure while the basic statutory scheme is reconsidered, this is undoubtedly a laudable holding strategy. As a long-term approach to basic food safety policy, however, it is patently inadequate.

The incident that precipitated the greatest public controversy—the proposed ban on saccharin—simply could not have been ignored by the FDA. It was too many years in the making. Step by step, the FDA had been boxed in by the statutory requirements, the results of animal tests, and its own public statements. The proposed ban was just the last step in a very long process, and was in fact a product of sheer practical necessity. The failure to propose a ban would undoubtedly have resulted in an even greater public outcry and would simply have permitted the underlying policy issues to remain submerged for a little while longer.

Its proposed action in that instance serves to illustrate, quite vividly, how completely our present statutory policy for food safety has crumbled in the face of current scientific knowledge. In order to avoid banning a large portion of our annual production of peanuts and corn, the FDA had earlier determined that the presence of aflatoxin was "unavoidable" and thus not subject to the Delaney Clause or the other general safety provisions of the Act. When faced with the saccharin issue, however, the Agency concluded that saccharin is "avoidable" in soft drinks and thus must be banned. But it is readily apparent that this distinction cannot withstand critical analysis. Obviously, the American diet could be made free of aflatoxin simply by banning all products in which it is found, just as the FDA proposed to make the American diet free of saccharin by banning all products in which it is found. In this respect, aflatoxin is no more unavoidable than saccharin. Moreover, banning saccharin while at the same time permitting aflatoxin is utterly irrational from a public health

standpoint. As evidence mounts of the presence of numerous
other animal carcinogens in the food supply, governmental
regulatory action becomes more and more inconsistent and thus
indefensible.

I have enormous sympathy for the difficulties faced by
the government in this situation. Indeed, I participated in
some of the decisions involved, and wish to emphasis that I
am not criticizing the individuals, their intelligence, or
their integrity. Faced with an outmoded law, they have acted
as responsibly as they could. Nor can one blame Congress for
the status of the current food provisions of the law. When
they were enacted, they represented the best current thinking
about proper public policy. Rather than cast blame on anyone,
we should direct our efforts to determining what should be an
acceptable public policy for the future.

To confront this issue, Congress included in the Saccharin
Study and Labeling Act of 1977 a provision under which the
National Academy of Sciences will be requested to review and
make recommendations with respect to existing Federal food
regulatory policy. On the basis of this report Congress will
consider whether changes in the safety provisions of the law
are now appropriate.

I am confident that we will not emerge from this symposium
with a consensus on a national food safety policy for the
future. That will require participation of a much more
diverse group of people than are represented here, and a much
more detailed consideration than we can give to it during
these three days.

I would, however, offer for your consideration, as a
general conceptual framework for future food safety policy,
the following three general rules that I enunciated last fall
in a paper delivered at MIT.

The first rule is that there is no such thing as absolute
safety. All food components are capable of producing harm in
some people under some circumstances. Some risk must be
tolerated by the public if we are to have any food at all.

One of the most important tasks facing the country is to
review those risks and establish priorities for dealing with
them. The American Industrial Health Council has recently
proposed that the Occupational Safety and Health Administration classify chemicals that may present a carcinogenic risk
into four categories:

        Category I: Proven human carcinogens
        Category II: Proven animal oncogens
        Category III: Require further testing
        Category IV: All others

Within categories I and II, the chemicals would further be subdivided according to whether they are high, medium, or low potency. If this proposal is adopted, it would represent the first serious attempt to set priorities for dealing with carcinogens. It could, and should, be used by the FDA and all other regulatory agencies, as well as OSHA, in establishing appropriate priorities.

The second rule is that safety must be viewed as a relative concept among competing alternatives. If there are several fungible food components that satisfy a particular need, and one poses a significantly greater risk, elemental principles of safety would appear to justify eliminating that component.

Implementation of this second rule will, of course, be very difficult. The determination of acceptable alternatives will in some instances be susceptible to objective analysis, but in a great many instances will depend upon illusive facts and difficult judgments.

The third rule is the need for a special analysis for those food components that do pose significant safety risks but for which there is no fungible alternatives. This special analysis will involve a three-step process. The first step will be to determine whether it is feasible, from both a technological and an economic standpoint, to reduce the risk through further processing. The second step will be to conduct a risk assessment. The third step will be to determine appropriate regulatory action, if any, for that level of risk.

In some instances, the risk can be reduced, but only at a specified cost. It might well be feasible, for example, to reduce or eliminate aflatoxin in peanuts and corn, or chloroform in water, but the technology may not yet be available or the cost of the treatment needed to accomplish this may be prohibitive. Balancing this cost against the reduction in risk will be an extremely difficult regulatory task. Certainly no one would oppose reducing any risk if the cost is trivial. Human exposure to animal carcinogens should be reduced to the lowest feasible level, taking into consideration both technological and economic factors, as long as the potential risk to humans remains uncertain. Where the technology is presently unavailable or the cost is substantial, however, difficult issues of cost/benefit analysis will be raised. It is perhaps here, more than anywhere else, that the scientific uncertainty about the existence of the hazard itself, when the sole evidence consists of animal studies, may be of particular importance. Expenditure of major amounts of money to achieve a relatively small reduction in what can be determined to be only a theoretical human risk may well not be warranted.

Once the risk is reduced as far as is warranted, it is essential to conduct a risk assessment, designed to quantify the risk involved, in order to place it in perspective relative to other risks we face. Without question, a risk analysis is both difficult and' uncertain on the basis of our present scientific knowledge. One of the mathematical models developed to estimate human risk from animal data, or some variation of those models, must be used for this purpose. This approach, while undoubtedly not precise, nonetheless represents a significant improvement over the current use of arbitrary safety factors or levels of toxicological insignificance. Again, it is important to keep in mind that any risk analysis based on animal studies represents a projection only of a potential, not a proven, human risk. Dr. Upton stated this quite forthrightly in his testimony two weeks ago:

> The NCI animal bioassay effort for chemical carcinogens can merely detect a chemical's potential for causing cancer in humans. Its results cannot tell us whether a particular chemical will cause human cancer, but they may alert us to a presumptive risk and thus serve as a basis for further studies of the chemical in question.

Risk assessments are therefore an appropriately conservative method of ensuring public protection.

Once the risk is reduced as far as is warranted and quantified, an appropriate regulatory response must be determined. In the past, this has been done on an entirely ad hoc basis. In the future, a consistent rationale for all government action with respect to carcinogenic risk must be enunciated if the public is to respect these decisions.

Conceptually, there are three levels of risk. For lack of better terminology, I shall call them high, moderate, and low.

For high risks, we impose a ban. Examples of high risks that are banned in our society include putting outright poisons in the food supply, attempted suicide, and going over Niagra Falls in a barrel. In each instance the probability of death is extremely high. The only high risks that we might tolerate would be those with an extraordinary potential benefit, such as the use of a highly toxic anticancer drug for otherwise terminal cancer patients.

For moderate risks, we provide information and appropriate warnings to the public, but do not ban the activity, where individual choice is feasible. We advise mountain climbers, canoers, swimmers, and Indianapolis Speedway drivers to take all appropriate safety precautions, and we warn them about the risks of serious injury and death, but we do not ban that kind of activity. We place appropriate warnings on aerosol-

ized food containers about intentional misuse that can result
in death, but we do not ban them. Many other consumer products bear appropriate directions for use and warnings against
misuse. Even though none of these activities and products is
essential to human life, individuals are permitted to make
their own choice whether or not to accept the risk involved.
Only in situations where individual choice is not feasible--
such as the purity of the air we breathe and the water we
drink--are moderate risks banned as unacceptable to society
as a whole.

For low risks, there is little or no attempt even to dis-
seminate information about the risks or how to avoid them.
Virtually all consumer products can produce low-level allergic
injuries, and undoubtedly all of them could be misused in one
way or another to result in serious injury or death. For
none of these types of risks, however, is there any significant effort to provide warnings or information about safety
precautions.

At least until now, the differentiation between these
three levels of risk has been made on an entirely ad hoc
basis. Regulatory agencies have determined which regulatory
response should be invoked more on the basis of intuition
than by application of any systematic policy rationale.
Perhaps the largest task facing us in the future will be at
least a rough approximation of the dividing lines between
these three risk categories in order to forge a consistent
rationale for regulatory action.

Dr. Richard Wilson of Harvard has suggested that any substance or activity that represents a risk of 10^{-2} or greater
should be banned; 10^{-5} or less should raise no concern; and
between 10^{-2} and 10^{-5} should be the subject of appropriate
information to the individuals involved so that they can make
their own benefit-risk decisions. This concept deserves
serious consideration. Whether or not it or some variation
is ultimately accepted, it unquestionably represents an important attempt to approach the issue on a rational basis,
rather than the more emotional and intuitive approaches that
have been used up to now.

The FDA has experimented with a mathematical approach by
advancing a modification of the Mantel-Bryan model to determine unacceptable residue levels of carcinogenic animal drugs
in human food. That particular approach suffers, however,
from an all-or-nothing philosophy. It does not permit a
middle ground where information is given to consumers and they
are allowed to make their own free choice. Its application
should therefore be restricted to those situations where individual choice is not feasible.

Whatever method is ultimately chosen to differentiate between these three risk categories, it is essential that it
result in consistent handling of equivalent risks. The public
will not understand or tolerate a continuation of our present
ad hoc approach, under which one chemical is banned, another
is labeled with a warning, and a third is allowed to be marketed freely, even though they present essentially equivalent
risks.

I have purposely avoided, in this discussion, any attempt
to define or even describe the benefit side of the benefit-
risk analysis for moderate risks where individual choice is
feasible. It is plainly inadequate to describe benefit solely
in terms of medical or nutritional essentiality. Although
vitamins, minerals, and protein are essential to human life,
probably any single food could be eliminated from the diet
without detrimental nutritional or medical effects. Virtually
all food is consumed as much for pleasure as it is for nutrition. To say that aflatoxin-contaminated peanuts provide
more "benefit" to society than saccharin-sweetened soda pop
is, of course, absurd on its face. The concept of "benefit"
remains a wholly subjective determination that defies logic
or even objective analysis.

For these reasons, it appears to me that the essential
element in future food safety policy consists of a risk analysis, rather than a benefit analysis. For those substances
for which there is a moderate risk and no safer alternative,
and individual choice is feasible, the best approach may well
be to permit the individual consumer to make a fully informed
choice on the basis of his or her own perceptions of benefit,
rather than to impose an arbitrary societal determination of
benefit on all of our citizens. As I said in my 1973 paper
to the National Academy of Sciences, the determination of
what each individual concludes to be an acceptable risk, or
an acceptable level of uncertainty, is intensely personal and
is based upon deeply felt beliefs and human values that perhaps cannot and should not be dictated by any government.

V

It is these public policy issues, rather than the scientific issues, that will determine our approach to food safety
policy, and indeed our approach to regulation of all carcinogenic risks, in the future. And it is these issues, rather
than the technical controversies surrounding animal testing
for carcinogenicity, that I urge you to consider during these
three days.

This does not mean, of course, that the technical scientific issues surrounding testing for carcinogenicity are either unimportant or should be ignored. They are of extreme public importance, and deserve the highest priority. It is particularly urgent to obtain agreement on the scientific criteria for determining animal carcinogenicity. When these controversies are ultimately resolved, it may well be appropriate to alter public policy in order to accommodate the scientific consensus that is reached. But the current need for reconsideration of public policy governing food safety cannot await the experimentation and debate that must precede that resolution. Policy must be forged today in the midst of scientific uncertainty and controversy.

By concentrating on the purely scientific issues in the past, the scientific community has misled Congress and the public into believing that the resolution of the public policy issue depends entirely upon resolution of the scientific debate. However inadvertent and well-intended it may have been, the great scientific debate over the Delaney Clause has unquestionably obscured the real policy issues that lie at the heart of the matter. The real issues are not whether a threshold exists, or how accurate animal models can predict human response, but rather what represents the most rational public policy during the uncertain period of time that it will take to provide definitive answers to those scientific questions. If we cannot have a no-risk food policy--as we now know we cannot--how can we set priorities among the totality of risks that we face, reduce the potential risks to the lowest level feasible, quantify them, and determine whether they are acceptable to each of us as individuals and to our society as a whole?

These are the real issues. Today we are floundering in our attempt to grapple with them. It is time that we began a systematic approach to these matters that will result in the development of a consistent and rational public policy for the future.

REMARKS

DR. MRAK: Thank you very much. I might say that some years
ago I think three of us here were involved with the DHEW
Secretary's Commission on Safety of Pesticides, and out of
that we came to the conclusion that maybe we needed to develop
some generally acceptable protocols for testing for safety.
Out of this came the National Center for Toxicological
Research.

CONGRESSMAN JAMES MARTIN: How important it is that we have
assembled here a conference combining research and technical
specialists with government officials who have responsibilities
in the relevant areas, to discuss essential policy questions.
How important it is, and how timely that this important conference will address the basic question of what is the best
approach for limiting public exposure to carcinogens!
     The Delaney Clause is one answer to that policy question,
an answer that was arrived at in 1958. The Delaney Clause is
one answer for food additives, but as has been explained, it
does not yet apply to natural food carcinogens. However, its
principles do after all guide regulation under the general
safety provision as well as provide a basis of reference for
the discussions that are going on now among several other
agencies of the federal government. So this is a vital,
timely, and important question and I look forward to the conclusions and developments that will come from this conference.
     Before proceeding to my task, I do want to share with you
one somewhat related development of a more general nature.
This topic that we have today and other scientific questions
and the discussion of those in Congress has led me to believe
that there's a great need within the United States Congress
for improved communication or scientific issues and controversies, scientific attitudes, scientific capabilities.
     What I'm suggesting as a problem is best illustrated
every year when we deal with the appropriations bill for the
National Science Foundation, because that becomes a field day
for not only members of Congress but other commentators to
rush into public criticism of the curious titles of basic
research. Accordingly with the American Academy for the Advancement of Science and my colleague Ray Thornton of Arkansas,
who is Chairman of the Subcommittee on Science, we are establishing for members of Congress a science forum. I would say
to you that this is not intended to be an advocacy body; it
will not take positions on questions, but rather will seek to
generate and stimulate discussions and in that way counter

antiintellectual tendencies regarding scientific subjects.
It will seek to improve the quality of scientific discussion
through greater access between members of Congress and their
staff and the professional scientific societies. It's expected at this point that there will be some 30 or 35 members
of the House who will participate, bringing together a range
of partisan and ideological backgrounds, also combining a
variety of committee and geographic representation. It seemed
to me that this new development is something you would want to
know about and to follow as it gets off the ground and into
operation.

THE DELANEY CLAUSE:
AN EXPERIMENTAL POLICY DECISION
THAT HAS BEEN MISINTERPRETED

William M. Upholt

Those few words in the Federal Food, Drug and Cosmetic
Act that are known as the Delaney Clause have probably caused
more debate and disagreement among regulatory toxicologists
than any similar number of words in any other federal law.
In essence, they tell the Food and Drug Administration that
they may not establish a tolerance for any food additive that
reasonably can be considered a carcinogen in man or experimental animals. Why has such a prohibition caused such
controversy?

Scientists have taken exception to many technical aspects.
It has been claimed that it gives legal status to the one-hit/
no-threshold theory of carcinogenesis and that Congress should
not take positions on such controversial scientific theories.
It has been claimed that it assumes that experimental animals
(under reasonable conditions) provide a sound basis for showing carcinogenicity in humans. Again, this a hypothesis upon
which good scientists may reasonably disagree. Some people
have pointed out that there are a number of unavoidable
chemicals in foods that may be carcinogenic. In fact, some
naturally occurring food ingredients are probably carcinogenic
and under the Delaney Clause philosophy all such foods should
be banned to be consistent.

Others say that the principle should be extended to certain
other health effects such as teratogenesis and mutagenesis.
The argument would seem to be that these adverse effects meet
most of the same criteria as, carcinogenesis, i.e., that they
are irreversible, have no known threshold, and may well be
produced by a single target cell.

Another group of individuals agree that the Clause is inappropriate in a law because it is so inflexible that it
deprives the responsible administrator the flexibility to use
reasonable judgment in interpreting the law.

ISBN 0-12-192750-4

It is not my intention to defend or to take issue with any of these arguments. Each of them has a reasonable basis. The only objection that I have to any of them is the frequent implication that the Clause was intended or should be interpreted to apply to substances other than food or feed additives we defined in that Act. It is this common implication that I maintain is a misinterpretation of the facts.

I consider it highly significant that the Clause appears in Section 409 of the Act and not in Section 406 or 208. Of course, there is a similar provision in Section 412, applying to new animal drugs. But that does not weaken my argument. Had Congress intended it to apply to naturally occurring ingredients of food or to unavoidable contaminants--or even to pesticides used in the production of raw agricultural products --it seems reasonable that Congress would have inserted it into some other section of the Act. Since it occurs only in Section 409, the Food Additive section, and the related Section 412, I think it reasonable to assume that Mr. Delaney and the Congress were in fact making an important and rational administrative decision on a specific case--namely, that food additives are not essential to human foods and that though their use may be desirable under many conditions, none of them contribute enough to the quality of our food to justify any avoidable risk of cancer. If this was in fact their intention then it is certainly clear that they were not saying that our food should always be free of carcinogens. They recognized, apparently, that some unavoidable carcinogens in food may well be acceptable if the only alternative is to seize and destroy that food.

Of course it can be argued that even this interpretation deprives the Commissioner of the privilege of making a reasonable interpretation of how much risk a certain food additive might justify. In fact, I think it is fairly obvious that at least some members of Congress now feel that this particular decision may have been too absolute, for example, when it became apparent that it covered saccharin as a food additive. Nevertheless I, for one, would be wary of condemning Congress for making certain policy decisions regarding the risks that the public is willing to take for specific benefits. I may disagree with such decisions but I have no desire to put the wisdom of my judgment as an individual against that of Congress. (I say that without abandoning my right to tell Congress what my judgment might be on certain issues and I may continue to disagree with certain decisions by Congress). But Congress is a direct instrument for the public to express its judgment. I am not.

What then of some of the other arguments about the Delaney Clause? Does it in fact enter the scientific argument regarding the one-hit/no-threshold theory of carcinogenesis? I think not, though I realize that many toxicologists may disagree with me on this. It is certainly true that tolerances for pesticides and food additives have been set in part by adding a safety factor to a "no observed effect" level as found in experimental animals. Without a doubt this procedure assumes a true threshold in the sense of a true "no effect" level of exposure. On the other hand, the law itself says nothing about this procedure. It refers to "conditions under which an additive may be safely used," but it does not define "safe." Under the section on definitions, it simply says "the term 'safe' ... has reference to the health of man or animals." Moreover, the law also specifies that no food additive tolerance shall be fixed at a level higher than the FDA finds to be reasonably required to accomplish the effect for which such additive is intended. A similar provision is included in the pesticide tolerance section. This would seem to be a superfluous provision if a higher tolerance were consistent with absolutely no risk. So even though I believe that the current process involving the concept of a threshold is consistent with the law, it is not specifically required. In fact, the usefulness limitation on a tolerance as provided for by the law implies that Congress recognized that there is no absolute safe level and that any risk should be justified in terms of benefits. Thus, I am persuaded that the Delaney Clause reflects a risk/benefit consideration in which the decision was made by Congress that by policy no benefit associated with food additives justifies any risk of cancer. It specifically excludes pesticides used in the production of agricultural crops from this "policy decision." Therefore, I find no evidence in the Act itself (I have not read the legislative history) that requires a threshold concept for noncarcinogens not a one-hit/no-threshold concept for carcinogens. It simply implies risk/benefit and says no risk for carcinogens.

If my explanation of the intent of Congress is correct, it is consistent with the failure of Congress to apply this decision to other toxicants for which the one-hit/no-threshold theory is also reasonable; namely, teratogens and mutagens. I would not suggest that Congress considered these classes of toxicants as less hazardous to health, for I suspect that they simply saw no reason at the time to consider them at all.

The next misunderstanding is that the Clause establishes appropriate tests with experimental animals as sound evidence of carcinogenicity in humans. From a regulatory standpoint, to delay regulation of a substance until it has been demon-

strated to produce cancer in man is akin to locking the door
after the horse has left. The latency period of cancer in
man is so long that by the time the epidemiological evidence
is statistically significant an additional portion of the
exposed population would already be condemned to cancer.
Properly conducted tests in animals may constitute evidence
of potential risk to man that a prudent regulator cannot
afford to ignore; but this does *not* say that it proves the
chemical to be carcinogenic *in* humans.

For the above reasons, I am convinced that the Delaney
Clause represents a policy decision by Congress that food
additives do not convey a benefit that justifies any risk of
cancer. If I knew more about food additives I might well
disagree with this policy decision. In general, I believe
that it is better to leave such decisions to a responsible
administrator to make on the merits of each particular case.

Nevertheless Congress often provides clear policy guidance
to the Excutive branch and I believe it appropriate for Congress to do so. In the case of Federal Insecticide, Fungicide, and Rodenticide Act, as amended in 1972; and in the
case of the Toxic Substances Control Act, Congress clearly
stated that the regulatory agency must consider benefits as
well as risks in administering these two laws. In both cases
it generally avoids the difficult term "safe" and substitutes
the negative term "unreasonable risk." It then proceeds to
indicate the necessity of considering social, economic, and
technical factors in determining what is an "unreasonable
risk." This is in accord with currently popular terminology
such as that of Lowrance, which defines "safety" as "of acceptable risk." This recognizes that absolute safety is a will-
of-the-wisp and that the job of a regulatory agency is to
make judgmental decisions for a public as to what is an unacceptable risk in the light of the total impact on society
of demanding a further reduction in that risk.

Thus, there may be some specific cases in which the benefits of a substance do not justify any risk of adverse effects
of a nature that can be produced by that substance. In such
cases, a responsible decision-maker will avoid use of that
substance completely. Individuals make such decisions on
their own volition frequently. Where the use of the substance
is such that the exposed individual cannot implement that
decision on his own, then the government must make a decision
on the behalf of the public. To refuse to make a decision in
such a case is a decision in itself for it permits continued
exposure to an old or denies the benefit of a new substance.
Occasionally Congress gives clear guidance to the Executive
branch as to what guidelines they should follow in such cases.
In other cases, such as FIFRA and TSCA, the framework is

described but the only guidelines are to be reasonable and prudent. In the case of the Federal Food, Drug and Cosmetic Act some guidelines are given for pesticides and for most food additives. The guidelines are "do not set the tolerances any higher than are needed to achieve the desired function of the substance." It does not give as clear guidelines for deciding what level of risk would justify refusing a tolerance except that in Section 402 it uses the phrase "quantity of such substance in such food that does not ordinarily render it injurious to health" and it uses the phrase "to the extent necessary to protect the public health" in Section 408 (the pesticide section).

Taken in this context I am persuaded that the Delaney Clause must be considered to be a policy decision on a specific case—that of food additives that show evidence of carcinogenicity in humans or animals. As such, it must be obeyed until Congress or the courts change it. But it should not be interpreted as guidance for any other cases. It recognizes the uncertainty of science both in mode of action and in *interspecific variation*. It recognizes food additives to be different than other adverse effects. It is inflexible but that is Congress' perogative.

REMARKS

CONGRESSMAN MARTIN: Thank you Dr. Upholt. It's always, of course, a source of great comfort and reassurance to you that Congress will use its prerogatives regardless.

I had mentioned just before we took recess that Dr. Roy Albert of the Institute of Environmental Medicine of New York City, who had intended to be here to speak to us on the subject of regulation of carcinogens at the EPA, had through no fault of his own been snowbound in New York City and unable to be here with us.

He has arranged for, and we're delighted to have in his place, Dr. Elizabeth Anderson of the Environmental Protection Agency who will present remarks based upon not only her knowledge of the field, but also the work, and the close association which she has had with Dr. Albert.

She is the Director of the Carcinogen Assessment Program for the Environmental Protection Agency, and I'm delighted to introduce to you Dr. Elizabeth Anderson.

THE REGULATION OF CARCINOGENS AT THE
ENVIRONMENTAL PROTECTION AGENCY

R. Albert and Elizabeth Anderson*

Dr. Albert and I agreed that the experience we have had
at the Environmental Protection Agency (EPA) might serve as
an interesting basis for consideration in the context of this
meeting. We thought we would give you a brief history of
EPA's experience regulating carcinogens.

Of course EPA is a relatively young agency, established
in December 1970 by executive order. It set about the business of regulating carcinogens starting in 1971 through 1975;
it took several major actions against some pesticides such as
aldrin, dieldrin, DDT, chlordane, and heptachlor. In the
course of these actions there was quite a stir created because
the scientific homework on these cases had not been done internally in the agency by agency-sponsored committees and so
forth, but rather the basis for judging the public health
impact had derived from earlier work by people who had worked
outside EPA. In the course of the administrative hearing,
which had been requested by the registrar, the lawyers were
faced in this adversarial arena with presenting the basis for
the actions on the health side of the case. The judge, of
course, was faced with hearing scientists in an adversary
arena giving various kinds of highly technical information to
him.

In the course of these hearings, attorneys summarized
their expert witnesses' testimony in the form of simple statements that they referred to as EPA's cancer principles. These
statements were never intended as scientific fact, but they
were perceived that way and it created quite a howl. I am
sure many of you heard about those principles over the years.

---

* *Dr. Ray Albert, originally scheduled to deliver this
paper, was unable to do so. Dr. Elizabeth Anderson did so in
his absence.*

ISBN 0-12-192750-4

So, it was decided that EPA really had to get its house
in order. We had to set about establishing a very sound basis
internally for laying out the health risks side of the case
before a decision was made, so that it could become a significant part of the risk and benefit balancing in the pesticide
act. The Toxic Substances Act requires the same kind of risk/
benefit balancing as do most of our other significant pieces
or legislation, in one way or another. Even if they are
technologically based, it is well to have a complete picture
of the public health impact situation when you are thinking
of what you are going to require in the way of technologic
controls.

In order to strengthen this assessment of the health risk,
we set about writing guidelines that the agency would follow
in laying out its case. This took about six months of intense
effort and many people here worked on that committe, including
Drs. Hutt, Albert, and myself. We involved a number of people
outside EPA and this effort resulted in publication of the
guidelines in May of 1976 in the Federal Register.

These guidelines did set a procedure by which the agency
would judge evidence of carcinogenicity. We recognized right
away that rarely does evidence of carcinogenicity come in
nice, discreet packages that are clearly labeled. We had a
very complicated situation on our hands and it was in this
context that we proceeded. Our guidelines were underscored I
think by two main ideas: One, that the evidence that an agent
is likely to be a human carcinogen is a kind of evidence that
can be found in many places and comes from many different
types of studies such as human epidemiology, long-term animal
bioassay studies, and short-term *in vitro* tests. The kinds
of information that can contribute to an understanding of this
likelihood are vast, indeed, and they are very complicated.
Also, we found that in many of the studies the quality ranged
from good to bad.

The second thing that underscored the guidelines I think
was the fact that it seemed quite appropriate that considerations of this kind of information be made in the context of
science, insofar as we possibly can remove it from the regulatory policy side of the agency. We have made every effort
to do that. The guidelines set up, in fact, the carcinogen
assessment group for purposes of having a scientific committee
to provide oversight and to do the actual risk assessment that
was necessary. The guidelines attempt to set up a circumstance
where we answer two questions about the likelihood that something is a human carcinogen. The first is exactly that. A
qualitative statement of how likely an agent is to be a human
carcinogen, based on all the evidence available. The second
one is a quantitative statement. What is the likely impact

if the agent goes unregulated? So, we set up the carcinogen assessment group about a year and a half ago. The agency as a whole has been following these guidelines and this group has been deeply involved in helping with this effort. That is to lay out in detail the risk associated with agents suspected of causing human cancer.

However, once we have produced a risk assessment, it has not been entirely clear as to how it can best be plugged into the regulatory decision. It's nice to say that we have this very clear, separate piece of information, hopefully a very clear objective, scientific statement, as best we can make it regarding the likely risk. It has been less clear exactly how we utilize this to arrive at a decision of appropriate and best control, under the various pieces of legislation.

Perhaps it is worth mentioning here that I have already suggested that the Pesticides Act and the Toxic Substances Act clearly provide for risk-benefit balancing. EPA also has a Drinking Water Act, the legislation that enables us to control water effluents and the Clean Air Act. In most cases, as I said, these acts provide for something that can weigh benefit against risk.

One exception is the Clean Air Act, Section 112, which is generating some current discussion and consideration in EPA. This is one section of one act, which does not mention the other side of the coin at all, but rather says that human health should be protected with an adequate margin of safety. Since it has not been possible for scientists to identify thresholds for carcinogens, we get back to the very significant problem of defining what is an adequate margin of safety when dealing with a possible human carcinogen in the context of a section of an act, such as this.

EPA has an enormous regulatory burden in this area. We have multiple numbers of chemicals to control under these acts. I am sure that I need not go into detail on this at all, because we know that we have an enormous burden in our atmosphere of pollutants, many of which are likely carcinogens. Certainly, we have the whole world of pesticides to deal with in the Toxic Substances Act. We are faced with just a huge number of chemicals out there in the environment for which we have the regulatory responsibility.

One thing that's been particularly noticeable from our viewpoint at EPA is the incredible inconsistency that has evolved both within EPA and across agency lines in the regulatory actions that have been taken thus far. There seems to be no sound or consistent basis for making judgments. Much of this is due of course to differences in legislation, the Delaney Clause being I think the most noticeable exception and that is why you are all here to discuss scientific basis.

But to give you an example of what I mean by the inconsistencies we have started picking out as one tool from risk assessment, the average individual solius average lifetime excess
risk of getting cancer from a potential carcinogen that we are
regulating. For example, we are able to construct most con-
sistently in a case like this, the dosage so that we do have
some basis for comparing apples and apples and not apples and
oranges. We have started taking a look at this individual
lifetime risk as an indicator of where we are ending up in our
public health protection endeavors relative to other regulatory actions that have been taken, both within EPA as well as
relative to other agencies' actions.

FDA under its obligation to set a safe level for residues
and animal foodstuffs has arrived at a definition of one
possibility in a million that a person would get cancer from
a particular residue.

The National Council for Radiation Protection set a guidance level of one possibility in a thousand for whole-body
radiation exposure.

EPA when it set its action levels for kepone in fish, fin
fish, crabs, and so forth coming from the Chesapeake Bay,
James River area, set an action level that permits a risk of
four chances in 10,000 of a person getting cancer from this
source, for people who eat these fish every day.

Recent haloform standards for drinking water, have set a
risk that is about 2 chances in 10,000. Clearly, this is a
very difficult circumstance when we are regulating haloforms
in people's drinking water that arrive partially from the fact
that we are protecting people by chlorination. It is necessary
to chlorinate in order not to have large outbreaks of diseases
that are caused by organisms that are killed by the chlorination process.

With regard to vinyl chloride, we have set a standard that
has an associated individual lifetime risk. Again, this is if
you are the person living beside that vinyl chloride plant
breathing the ambient levels of emissions from that plant at
the standard we set. Your individual lifetime risk of getting
cancer from that source is one chance in 100,000. We are being
petitioned in court now to set a lower standard.

You see we are in a range from one chance in a million to
one chance in 100,000. We are being challenged to go lower.
We are setting standards that are a good deal higher in some
cases. In short, I think this means we need to gain some consistency in the way we look at protection of public health
from carcinogens and find some way to stabilize an approach.
EPA is struggling with this whole idea internally and I think
it is something that the whole federal government is certainly
attempting to focus on now.

Several possibilities come to mind when thinking about this and we have been thinking about this now for a year and a half in the carcinogen assessment group. Of course, legal statutes do not always allow flexibility to consider any of these possibilities that you want to consider, but just thinking, as you are getting yourself out of the context of the verbiage in the acts, what would be possible. One possibility, of course, is to always aim for zero exposure, a complete ban across the board for all suspected carcinogens. Immediately, the fact comes to mind that this would be highly impractical. The EPA is dealing with far too many agents to hope to make a dent in the problems that we have. There is no good reason to think we are creating an impact on public health, by attempting to go to zero exposure for everything. Such a stringent goal is likely to end up with the regulation of only a very few things and no regulation of by far the majority of substances. The second possibility is to take a look at the qualitative data. That is, as I said before, the available human epidemiology. We rarely see it, but that is something we certainly like to have in hand: the animal bioassay data, any other supporting information, and then to simply say that we are dealing with a likely human carcinogen based on this information that will go to the lowest feasible level. We have a feeling that this approach is in a sense dealing with an undefined problem. In other words, the word "feasible" certainly has many definitions and, when trying to decide how stringent you want your controls to be, it is well to be aware of the magnitude of your public health problem.

Another possibility is to simply use risk assessments and use the idea of describing the impact, the population at risk, or possibly at risk, and your exposure levels or this individual lifetime risk (which we are finding useful), and simply set an acceptable level of risk and just decide you are always going to shoot for this level.

Of course this raises some major difficulties because (1) it is very hard to decide what that level should be, and (2) it might not work out particularly well. In some cases it may have the effect on a particular part of society of requiring the same thing as zero emission. In other words, you could be putting people out of business as they aim for that level; in other cases you might find that you could go very much lower without enormous stress.

Another possibility that has come up and something that I think we really must mention here is the idea of placing a dollar value on human life. This has come up in the context of the risk/benefit balancing at EPA. It is done by taking the individual lifetime risk; then if you are fortunate enough to know how many people are really in fact exposed to pollu-

tants and at what levels they are exposed, you can actually
take a stab at calculating the number of cancers you would
expect in a year from that particular pollutant. Needless to
say, these numbers are uncertain depending on what you are
dealing with. In many cases they are very uncertain indeed.
Sometimes we have better information, sometimes we don't. We
do these calculations, by the way, just to give a feel for
the urgency of the situation, but you can tell we have a
particular bias. We do not like them to be used if they have
any real certainty with regard to actual numbers and practical
context. With the idea of placing a dollar value on human
life, we then use these very uncertain numbers, which we intend only to give a feel for the situation. On the other side
of the coin, where the economist and the engineer dealing with
technology have arrived at a cost to society in one way or
another, we divide one uncertain number into another uncertain
number and come up with a dollar value on human life. I think
it is distasteful in concept and I also think that it is very
unworthy at this time to approach doing business this way,
because we just do not have enough information to come up with
a meaningful quotient.

Since we are committed to doing these full risk assessments, we are now struggling very much with the idea of exactly how we can best use these to reach some sound regulatory
decision with regard to the regulation of a particular agent
under a particular statute. I think that EPA has come this
far. I think we can say that we have decided insofar as our
statutes permit that we will combine the consideration of the
public health impact and whatever cost to society might mean
in the context of a particular act. But we are doing that and
we have published in our guidelines and also in a supporting
article the notion that we will certainly aim for the lowest
level of exposure possible within the context of what the
costs are to society.

Again I am using that as a catchall, a phrase to cover
economics, technology cost, other kinds of social burdens, and
so forth. We will go as low as possible. If we find that we
can not get down very low with regard to human risk, then
there had better be a good reason why not. I think one example
is the proposed haloform standard, where the risk is not as
low as we would like to see it. There are very convincing
arguments on the other side, which I have not been involved
with, but the people making that case I think have made a very
convincing argument for doing the best they can right now
under the circumstances. I think this is a major step forward.

There are clearly circumstances that just do not permit
the very nice conclusion of clearly being able to reduce all
human risk, or even to reduce all possible human risk even to

very low levels, if you are thinking in the context of one possibility in a million for instance. The other thing I think that EPA certainly is doing now is taking a look at this average lifetime risk as an indicator of where we are ending up. Then we can compare where we are on our scale, and where we are going, when we achieve a certain level of control.

I think it is the overall goal of the federal government to have the federal agencies achieve as much improvement of public health as we can. This is likely to happen, of course, if we can control exposure to as many agents as possible to the lowest possible level. We certainly have many agents that are suspected of causing human cancer, so we have a very large area to cover. We need to do this as quickly as possible. From a purely practical standpoint, it may be better to regulate many agents to a very low level of risk, if your regulatory circumstances is such that this can be achieved, rather than achieving a total ban on only a few.

I think we would all agree that our circumstance now does call for the regulation of many agents. I think this circumstance clearly requires a good deal of thought and work on the part of the federal regulatory agencies, but if we can keep the goal in mind of controlling exposure to as many agents as possible as really the final way of achieving or the most promising way of achieving improvement in public health, we are likely to derive probably our best approach. Clearly, we have a circumstance posed for federal regulatory agencies that is quite a challenge.

GENERAL DISCUSSION

MR. ADAMS: I want to thank you all for the opportunity to present the views of Congressman William Wampler, the ranking minority member of the Committee on Agriculture.

Congressman Wampler's concern is that regulatory policies, parts of current law, such as the Delaney Clause and different principles for correlation of tests for carcinogenicity of chemicals in experimental animals and the extrapolation of these results to man, are at such variance as to cause confusion in the minds of the regulators, those being regulated, and the public who must use the end products, and the scientists advising all three.

The problem is quite complex, as you are all aware. Decisions by the Environmental Protection Agency regarding the banning of certain pesticides have been extended to other chemicals. The basis for these regulatory decisions concerning health risk and economic impact assessments of suspected carcinogens has been published by EPA. These assessment protocols indicate that EPA has taken a position that, briefly summarized, indicates that occurrence of cancer, whether malignant or benign, in test animals, would trigger a high level of suspicion that the substance would be considered dangerous to man and therefore there was a high level of probability that the substances would warrant removal from the market.

At the same time, the Occupational Safety and Health Administration has proposed its own series of assumptions for setting regulations to protect worker exposure to suspected cancer-causing chemicals.

Third,   National Cancer Institute, which is engaged in an extensive bioassay effort to identify carcinogenic materials, has indicated in some of its policy papers that there are a series of precise evaluations that must be followed in order to reach conclusions that a substance actually poses a threat of cancer induction in man. The steps are protocols enumerated by the National Cancer Institute, and their policy statement suggests a slightly different perspective on evaluations to determine cancer potential in man with the distinction between benign and malignant tumors being treated in more detail.

The FDA presents a different view to the public, because it may make a determination permitting the use of potential carcinogens in a drug, while under other legislations on foods such as food additive amendments of the Food, Drug and Cosmetic Act, no flexibility is permitted.

41

ISBN 0-12-192750-4

There is a great deal of scientific controversy about the validity and interpretation of tests of carcinogenicity in animals. Extrapolation of tests demands the significance of certain protocols for a series of confirming tests in which the level of exposure should be considered conclusive evidence for regulatory purposes. Other factors are also associated with the identification of a carcinogen and the implication of these determinations to man's health and well-being. Added to this background of scientific difficulty, the increasing capability to detect the presence of very low concentrations of substances with highly sophisticated instruments, the slight difference in calculating lifetime exposures, and other technical points often can be misleading if taken out of context during the regulatory process.

Because of the complexity of the sciences involved and the very high level of public concern and confusion about these issues, it does appear that a Congressional mandated analysis by an independent and highly prestigious body like the National Academy of Sciences (NAS) would be a timely task. The evolving cancer principles of the several regulatory agencies suggest that these policies might serve as a focus of attention in such an analysis.

There is ample precedent to support Congressman Wampler's proposal to charge an agency like the NAS to conduct this study with the objective being: to conduct an evaluation of the issues associated with the desirability of developing a federal policy for the determination of the potential carcinogenicity in man of chemicals tested primarily in nonhuman test systems. Such an evaluation should give thorough consideration the status of test systems, the validity of different methods of statistical analysis, the techniques of bioassay and their validity, the scientific logic associated with subtle differences among the several policies on criteria for assessing the carcinogenicity of chemicals, and similar scientific issues.

Health and risk assessment policies of particular interest are those evolving from already established regulations by EPA, FDA, the Occupational Safety and Health Administration, and the Consumer Product Safety Commission. The issues should be clearly enumerated, the differences identified, and the scientific basis for these differences in policy examined and evaluated. If it appears possible on the basis of available scientific evidence for a standard federal cancer assessment policy to be constructed for regulatory purposes, then a recommendation or recommendations as to the possible courses of action should be provided the Congress.

If the scientific problems of statistical analysis, correlation of animal tests with potential effects in man (particularly issues dealing with dosage rates and thresholds and
other factors) are still not adequately developed, to assume
that reasonable standard conclusions can be constructed, these
factors should be fully discussed to permit an evaluation of
the need for legislative action. All aspects of the scientific issues associated with the task of assessing the validity
of animal test data on carcinogens as extrapolated to man
should be summarized in sufficient detail to clearly illumi-
nate the problems for public policy consideration.

The NAS, various agricultural groups, and the chemical
industry have informed Congressman Wampler (in a recent report
by the House Agricultural Committee, they also went on record
in support of the current pesticide legislation) that such an
assessment policy was needed, and that a single federal policy
should be established.

The NAS has stated it can perform the study in a year's
time. Congressman Wampler will shortly introduce legislation
in Congress to establish a National Cancer Assessment Policy.
He would be very appreciative of your consideration of his
proposal and any recommendations this body might have to improve it.

CONGRESSMAN MARTIN: Thank you very much for bringing this
review of a very important and timely legislative proposal
that you and Congressman Wampler are presenting.

DR. MRAK: This morning Dr. Smeets pointed out, and I think
others did too, that the Delaney Clause philosophy is not
only a United States situation, it is an international situation. Other nations seem to follow very often things we are
doing. They may do better; they may do worse. I cannot but
wonder, if there shouldn't be consideration on the part of
the Academy to look into an international policy. I haven't
heard much said this morning about Japan or some of the other
countries in conflict with what we are doing. I'm wondering
if there is a good reason for that. Maybe Mr. Adams can
comment on that to start with, or maybe Congressman Martin.

MR. ADAMS: Congressman Wampler's proposal is specifically
for a national assessment policy, and I think in light of the
fact, as he pointed out, that we have not only the EPA and
FDA, but also OSHA, as well as the Consumer Products Safety
Commission, and others, who are anxious--I think that's the
word--to fulfill their responsibility for developing an appropriate regulatory policy within their own area of responsibility. It is, I think, vitally important, as Commissioner

Barbara Franklin of the Consumer Products Safety Commission
mentioned just a few days ago, that we have some coordination
of these policies, so that there's some semblance of similarity in the principles that are used to guide each of the
agencies. Therefore, to begin with, a national assessment
policy is on its own very defensible without denying the importance of going beyond that internationally. Informally,
we have begun discussions of these questions and they will
have to be translated into appropriate legislation by the
legislative bodies in each country.

CONGRESSMAN MARTIN: There are a number of them here. Maybe
they'd like to comment.

DR. SCHMIDT-BLEEK: I think it's very appropriate that Dr.
Mrak brought out this point. Let me just state very briefly
that as far as the Federal Republic of West Germany is concerned, we pay a great deal of attention to what is
being done in your country with respect to filling out the
legislative demands of the Toxic Substance Control Act, with a
hundred billion mark industry in our country in chemicals.
It is quite obvious that we are concerned, and it doesn't
take much basic knowledge in economics to realize, that when
testing methodologies are being established in a regulatory
way they must be harmonized from one major country to another.
In terms of industrial trade and particularly trade in chemicals, we may find ourselves within a rather short time in a
bad situation, where nontariff trade barriers are going to
prevent trade on a rather major scale.

     So, I think now and in the next few years we must pool
our ideas and our experts and come to an agreement on testing
schemes, generally.

     Now in this context I should point out very strongly that
as far as the European Community is concerned our first
priority in legislation is directed toward new chemicals. As
most of the experts in this room know, of course, this puts
a slightly different touch on the approach that we are taking.
Obviously, with a new compound, there is no reference literature, only a more or less large body of information that you
may fall back upon. We must develop from scratch the kinds of
testing step sequence plans that are necessary to develop the
knowledge that is needed to prevent hazards and risks.

     And to us in Europe it is quite important. Two weeks ago
in Brussels, we had a major discussion with EPA on this point,
and were able to harmonize our thinking in this area, too.
That is what we can do on both sides of the Atlantic with new
chemicals. We must not drift apart in the basic ideas that

we develop for testing, if we wish the evaluation of chemicals
with respect to carcinogenicity and other hazardous properties
to be similar.

So let me very quickly summarize by saying, yes, indeed,
we are most anxious to share our ideas with the United States,
and of course with Japan, Canada, and others, for instance
OECD member states, as to what we have arrived at, evaluate
this information, and make judgments, not only because of
trade barriers, but for many other reasons. We are most anxious to see that this work be put in motion and lead to an end
which we all can look back to in five years and say that we
did the right thing.

DR. EGAN: I also wanted to comment, briefly, following Dr.
Mrak's observation. The International Agency for Research on
Cancer (WHO), of course, publishes a series of chemical carcinogenic risk evaluation monographs. Between 130 and 150 chemicals have so far been evaluated in the course of five or six
years in some 15 or 16 monographs, and only a relatively small
number of chemicals, about 20 or so, are clearly identified as
presenting a positive risk to man. There are many others for
which further information is called for. I know full well
that agencies and institutes in the United States have contributed fully to those evaluations, but I would be interested
to know from Dr. Anderson or representatives of other agencies
how far the policy of these regulatory groups is aligned with
the philosophy behind the IARC evaluation of carcinogenic
risk. Because here is an international agency that has already
made a platform for the harmonization evaluation.

DR. ANDERSON: I think that the thing to remember here is,
given the overwhelming obligation to be a caretaker in a very
large sense, a regulatory agency in this country gives us an
operational arm that the IARC monographs can comfortably
ignore. They provide a very useful basis for answering our
first question and that is the qualitative basis for answering
the question: How likely is a substance to be a human carcinogen? Incidentally, we do not say that we regard any agent as
actually being a human carcinogen unless we have convincing
epidemiologic data coupled with our animal studies. But, we
do say that we have suspected carcinogens based on other kinds
of information and we are in the protection business, which
means we have to regulate on the basis of data that give us a
feeling of grayness. I think the IARC monograph series provides a nice description in a scientific way of what that data
base is, but it does not say what you do with it; that is what
we have to get into. It is being on the firing line and, in
short, it is deciding exactly how to plug it in to regulatory
decisions.

DR. COULSTON: I promised myself I'd be very quiet this morning, but Dr. Anderson has provoked me a little bit so I will
ask some questions.

I'm very impressed with what you said, and I hope that this
is now the policy of your agency, but I'm going to test it for
us. I want to discuss the question of the decisions made by
EPA that were made earlier with DDT and, lately, with heptachlor and chlordane, where the case went to a hearing examiner,
a judge, and the judge in both cases said that he found that
DDT, heptachlor, and chlordane were not imminent carcinogenic
hazards to man. After many months and millions of dollars of
effort in the hearing, the EPA through its administrator
overrode this decision of the hearing examiner and declared
that these compounds were indeed imminent carcinogenic hazards
to man.

I don't understand this, but I am not a lawyer. I can't
understand how we go to all the trouble of having hearings
and spend time and money, and then literally by fiat, the
decision is made to override this by the very agency that
called for the hearing. The competent testimony, adversary
and nonadversary, on all sides of the question, went on for
months in each case and yet a decision was made to override
the hearing judge.

Now this causes an international problem and comes back
to what Mr. Adams said and the question raised by Dr. Mrak.
There are international bodies--WHO, FAO, IAEA--that meet
regularly and discuss these very kinds of chemicals. The
joint WHO-FAO group of experts just recently, for example,
said that chlordane is not an imminent carcinogenic hazard to
man.

Even the National Cancer Institute reported that chlordane
is not an imminent carcinogenic hazard to man, as far as rats
are concerned. They did find that hepatic nodules were observed in mice and that gets into the benign/malignant controversy, I am confused. Please help us, Dr. Anderson.

DR. ANDERSON: I think that the cases you were referring to
were cases that I referred to in the historical development
of EPA's guidelines and that was certainly a situation where
we had science being derived on a firing line in an adversary
proceeding. The floor judge was not a technically trained
individual and it certainly does lead to enormous difficulties
and the pulling and tearing of hair and wringing of hands and
so forth.

I think what we recognized is that it is the responsibility
of the EPA, insofar as we have regulatory duties, and I know
our sister agencies feel the same way, to clearly lay out for
public to scrutinize and in fact comment on (we asked for

comments in our guidelines) exactly what kinds of evidence we
would judge and how. This is what we did when we published
the guidelines in May, 1976. We received comments from many
places, other government agencies, private universities, and
industry, and I think generally we are now reaching an understanding with the public as to how we are going to proceed.

I'm not sure exactly what you mean by imminent carcinogenic
hazard. I know that in my agency there is a very strong feeling that we rarely are dealing with something that we know is
an imminent carcinogenic hazard. We are most often dealing
with something we suspect might be. And that's where the
difficulty lies. We only know, I believe, about 22 substances
which actually are cancer-causing substances, and we are
clearly regulating quite beyond those 22, because we don't
feel we can afford to wait.

But with regard to the value of certain kinds of evidence
in judging carcinogenicity I think that we attempted in our
guidelines and are continuing to make every effort to explain
exactly how we do plan to use animal bioassay studies in
making regulatory decisions. I don't know whether that's
directly helpful but I hope so.

DR. COULSTON: This is just for the record. The point is that
the term "imminent carcinogenic hazard to man" is not mine.
This was used by EPA in these cases and I would hope the
policy of the agency has been changed now. I still would like
to hear an expression of opinion as to why we go through a--
I'm going to use a hard word--a farce of a hearing. You can
not say that a judge is not technically trained. My goodness,
all judges are trained to judge.

DR. ANDERSON: I meant scientifically.

DR. COULSTON: Some of the lawyers I know are better scientists
than some of the scientists I know. And I think we're going
to get Mr. Hutt to say a few words very shortly. But I want
to emphasize that, if indeed what you have said to us is the
EPA currently policy, I'm very encouraged.

DR. UPHOLT: I would only say, as I think Dr. Anderson implied,
that there have been no decisions, so far as I know, that have
gone through a public hearing since we established this carcinogenic assessment procedure. So, I think it is fair to say
this does represent a new procedure, a new policy within EPA,
which has not yet gotten to the point, Dr. Coulston, where it
would deal with the specific type of situation you have described. Part of the reason that the agency arrived at this
decision was because of the complexities involved in the earlier procedure.

MR. HUTT: I think the answer to Dr. Coulston is that the
point in question illustrates the difference between science
and public policy that I was talking about earlier.   The
judge was not attempting to set public policy when he made
his determination.   He was attempting merely to preside over
a proceeding at which scientific evidence was admitted into
the public record, on the basis of which he attempted to sort
out the signs.   The administrator of EPA on the basis of that
sign made, as he is charged by law to make, a public policy
determination on which I would have to say as a lawyer he
could have gone either way and been upheld by the courts.
Because that is the nature of the regulatory function in every
governmental agency, not just in EPA.
       Turning to the question of what an imminent hazard is,
the state defines imminent hazard in what I would have to say
is a nonscientific way.   It defines imminent hazard to mean
something that will occur if it takes a long time to make a
regulatory decision.   It does not mean an extremely serious
hazard or a hazard that is definitely going to occur immediately.   So, you have to look, as was mentioned earlier by
Dr. Anderson, at the statutory words and not at your own per-
ception of what you think the statute should mean.
       Now just one final word, as I expressed earlier, I have a
great deal of sympathy for the process that EPA is going
through on this.   What you referred to, Dr. Coulston, were some
earlier decisions at either the beginning or certainly in the
midst of the development in the government of a consistent
policy.   We haven't seen the end of that practice by any
stretch of the imagination and I think that, indeed, Dr.
Anderson would say that where the agency is now is not where
it will be two years or five years from now.   This is an ongoing process and certainly not a static one.

DR. PARKE: I'd like to say something about the scene in the
United Kingdom and the impact of your FDA and the Delaney
Clause on things there.   I thought Dr. Egan was going to
address this, but he has left,me, I think, to deal with it.
       In the United Kingdom, we tend to evaluate each chemical
individually.   We occasionally are alerted, even alarmed, and
sometimes panicked, by the decisions that you make here in the
United States.   We have urgent meetings at which we consider
your decisions and whether we ought to implement a similar
decision, but usually we make up our own minds and we take
quite an individual line.
       On the first occasion, which was I suppose cyclamate, we
were caught unaware, and this was very largely due to the fact
that before the appropriate government departments or advisory
committees meant to consider this, industry itself had made or

sections of industry had made decisions and there was a great
deal of commercial one-up-manship--companies saying, "We are
selling this commodity without cyclamates, therefore it must
be safe"--and so our hand was forced.

Since then over saccharin, chloroform, oral contraceptives,
and a host of other things, we have considered these quite
individually and have taken into consideration the reasons for
which the FDA has perhaps made certain decisions, but we made
up our own minds. In making up our own minds, we pay much
greater attention to dose, the level of human exposure, the
dose-response in the animal species studied, the particular
animal species that have been studied, and last but not least
by any manner of means, the mechanism by which these chemicals
is thought to induce carcinogenesis.

So really the situation in the United Kingdom is that at
the moment, at least, we depend more on science and less on
legislation. I hope we are going to keep it that way!

DR. COULSTON: I would like to ask one more question regarding
this possible national or international policy on carcinogens.
It is quite clear to me at this point exactly what is intended
as a policy. It seems to me the term policy needs to be pretty
well described before it is possible to react very positively
or negatively, either way, to Mr. Adam's proposal. Is the
policy designed as a decision-making procedure? Is it designed as a selection, say, between the four options that Dr.
Anderson mentioned? Is it designed purely to be a method of
testing to determine carcinogenicity, as I think was implied
perhaps by Mr. Egan's decision? What is the intent of your
committee in talking about the need for a national policy?

MR. ADAMS: I think first of all Mr. Wampler's concern was
based on the differences between agencies in all of the aspects
that you talked about. I think he is also concerned about the
level of science. We know, for instance, that the Academy has
recently published a number of volumes. In one especially, on
pesticide decision-making, they recommended to EPA that they
should accomplish certain scientific tests or changes in their
own scientific procedures. Mr. Wampler has requested EPA to
institute those recommendations. Several weeks ago, we felt
we were going to get a reply to the letter to implement those
recommendations of the Academy, which covered all the aspects
that you are talking about now. We have been promised this
reply, but we have not received it back from EPA. We have the
same problem in certain things that FDA does. We do not know
whether we are going to go the Delaney Clause route or we are
going to go the EPA rule route or the National Cancer Institute or whatever it might be.

Our concern is that the federal government should use the best science available and should follow generally the recommendations of the Academy, because we feel it is the most prestigious body in the United States; if it isn't, well, we'll go for a better group. But someone should come up with something to give us guidance. I think Congressman Martin would agree that the staff of the Congress and the members of Congress have very little in the way of scientific knowledge. Therefore, when we get the various government agencies coming forward to Congress to tell us what these problems are, we have no way of verifying or helping in solving this problem.

DR. FURMAN: Further to the matter of an international policy for establishment of carcinogenicity and new chemical entities, I would urge that we include new chemical entities destined for use in human medicine. When the American pharmaceutical industry seeks registration of a new chemical entity for use in medicine abroad, in many countries that company is obligated to repeat long-term, expensive animal testing. This delays the emergence sometimes of important new therapeutic agents into the marketplace and, of course, increases their cost to the consumer. I hope that any formulation of policy would embrace those chemical entities of a therapeutic nature.

DR. SCHMIDT-BLEEK: I can only very quickly support the first point. In fact it seems to be very important that one not limit the thoughts that we are developing here; also I fully recognize that one cannot enlarge the scope endlessly of any particular property of a compound or any particular class of compounds. In fact, I think one should recognize beyond that that the systematic testing of chemicals not only includes carcinogenicity, but must also include other things. I was trying to allude to that before.

But I do have a leading question, Mr. Chairman. We have information on the interagency records or liaison. We have information on the Council on Environmental Quality, Toxic Substance Strategy Group. We hear that Congressman Wampler is apparently about to introduce legislation. We hear that the American Industrial Health Council has certain plans. OSHA and Barbara Franklin have made certain statements. I could go on and on with everything that has happened with the last six weeks. It would be most interesting to us on the European side to try to find out from you, as to how you judge at this time, what development is likely to be during the next six or eight months, and what the timing may be for the legislation to be introduced by Congressman Wampler.

CONGRESSMAN MARTIN: As you have indicated, indeed we appear
to be on the verge of a breathless and incredible array of
different proposals from different agencies as to how to deal
with carcinogens. I take it that the impact of your question
is to say, please, somehow, find a way to find harmony and
some uniformity in that policy as it evolves.

DR. SCHMIDT-BLEEK: I didn't want to imply or give you any
advice at all. It's just that I would like to express my
confusion by trying to follow what is going on and, as I said
before how important it is to us and I'm sure to other
countries that we try to understand what you are trying to do
and how you should go about this.

CONGRESSMAN MARTIN: Yes, I understand that, but I generally
interpret that anytime people acknowledge that they are confused they are usually advising me that they would rather not
be confused, and would hope that we work out some solution for
clarity and harmony, so that they won't be confused in the
future. I take your question to be that.

DR. ANDERSON: If this offers any comfort, the group that's
meeting in West Virginia this week is a group made up of the
four agencies: the Consumer Products Safety Commission, the
FDA, the EPA, and OSHA. The idea is for us to get together
and see if we can come up with a consistent approach for the
four agencies given our current circumstance. This is the
four agency group that you're referring to, the inner agency
regulatory liaison group. I might add that I know the four
agency heads have put a good deal of their own personal effort
and time into this and, if they can make it work, they are
very intent on this as a major project in the new administration. They meet every two weeks and they have the rest of us
scurrying about like mad. I cannot promise you what we will
be producing, but we are trying.

DR. LU: I would like to say something following Dr. Mrak's
comments and also Dr. Coulston's remarks about WHO and the
international aspects. Some of you know that WHO and FAO have
been involved in the evaluation of food additives and pesticides for some 15 or 20 years. I have been fortunate to be
associated with the program officially for 13 years and unofficially for another five years. In the evaluation of food
additives and pesticides, we have encountered a number of
substances that have been shown to have produced tumors or
increased tumor incidences in one or more species of animals,

but not in every case did the committee recommend that the
substance be banned, notwithstanding the basic principle that
no proven carcinogen should be used as a food additive.

I think I want to emphasize the word "proven." In some
cases, such as the subcutaneous injection of chemicals into
the rat producing tumors, earlier meetings considered this
very serious. Later, with more scientific evidence, the
meetings decided this kind of testing for carcinogenesis was
not important. So I just want to say that not all chemicals
which have been found to produce or increase the tumor incidence have been banned.

DR. CLAYSON: I would like to ask three questions briefly and
make two very brief comments. The first question: Are we
right to divorce consideration of carcinogenicity from other
aspects of toxicology? In other words in Mr. Hutt's terms,
is our public policy position correct? Personally, I have
equal concern, for example, for neurotoxicity and teratogenicity which may cause individuals to lead handicapped lives
for quite long periods.

The second question: Is it really possible to quantify
human risks from animal data? I have seen many attempts to do
this, but most of them strike me as being exceedingly naive
and unrealistic.

The third question: What emphasis in carcinogenicity and
in other toxicology do we give to what we make when more than
one chemical interacts with another? Should we take any cognizance of the fact when we give our animals a single agent,
that man is exposed to many agents that may diminish or enhance the effect of the one agent we are trying to regulate.

The first comment: I was extremely interested to hear Dr.
Anderson talk about putting a dollar value on human life. I
think it is a realistic approach but I'm reminded of the
United Kingdom and the Ministry of Transport, who in fact did
this when they were trying to get money for road improvements.
I think the dollar value which they put on human life at that
time worked out somewhere between $20,000 and $25,000.

The second comment is directed toward what Mr. Adams told
us and this is the question of whether we should consider
animal experiments that give us only benign tumors or evidence
of carcinogenicity. Apart from the semantic problems in saying that benign tumors alone are adequate as indicators of
carcinogenicity, I think it's important to notice that two
recent committees, one just been held in Lyon by the International Agency for Research in Cancer, and the Environmental
Subcommittee of the National Cancer Advisory Board, have both
recently come out very clearly to say that *benign tumors
should not be considered as evidence of carcinogenicity*.

The key here is surely that we have to consider in each case when we see a collection of benign tumors, is what is the frequence of progression of these tumors to malignancy, in the individual case? And if the frequency is high, it is unlikely that we'll see only benign tumors. Unless we take that into consideration I think our accusation that certain things are carcinogens in practice is going to become extremely vague.

DR. DARBY: I think that Peter Hutt has very nicely separated two issues that we have confused even in this discussion and these are indeed interrelated. That is the resolution of scientific information and then the setting of public policy. The latter it seems to me is something that is less easy to do, not only between multiple agencies, such as we are discussing here in the United States, but also from country to country because the considerations differ between the needs of a particular society. I think that one can only set forth guidelines on the setting of public policy, whereas I believe we could agree on desirable types of procedures that can be used to generate scientific data for all of us.

DR. SMEETS: I should like to come back to the international aspect of all these problems which we are discussing now. As I mentioned this morning I'm working in the European Community. I have to deal with nine countries. And from one point of view, we are certainly very happy that there is a certain form of independent relation in the nine countries, which means that everyone has its own point of view. On the other hand I must recognize that it is sometimes very hard to establish regulations in the framework of this community. We have no legislative power in the community, and so we must try to harmonize legislations and countries. We must try to establish legislation on community levels. We have some major points in this particular field, since as I said the individualism of the countries forces us sometimes to make less stringent regulations in order to achieve unified decisions. One of the reasons is very often that the available scientific data are not strong enough, when politicians consider them, to establish rules on the community level. From our own experience in this particular field we have learned that when you want to make problems in a particular field, it is very easy to criticize scientific data. We all know that very well in the carcinogenic field, the mutagenic field, the teratogenic field. There is usually a lack of data when you want to establish rules.

Another aspect of this problem is the fact that the federal legislation here in the United States has an impact on the legislation of the individual states. What are the criteria of federal legislation that will be taken also by the states themselves.

For example, we are puzzled by the problem that the legislation with respect to cars in California is different from the legislation elsewhere in the United States. The problem is that now you want to internationalize these problems and I do not say that I disagree with the idea, but how do you harmonize different regulations inside the United States in this particular situation? If we are going to talk at the international level, then, for example in Europe when we have testing data that correspond with the aims imposed by EPA, where is the guarantee that one or another state here in the United States does not impose more or less stringent measurements?

We need to know what is the implementation of legislation in the United States, before we start to talk on the international level. I have some more problems with this view, but these are the main questions for the time being.

MR. HUTT: This is a response both to that and to the prior statement by Dr. Darby.

I would rephrase Dr. Darby's formulation of it just slightly, rather than talking about the scientific and the public policy. What we are talking about here specifically is initially the risk assessment of a particular substance, or category of substances, and then the issue of the regulatory response. On an international level, I do agree with you, we can get to risk assessment. The most interesting thing about the OSHA policy of last October was that it was the first attempt on a systematic basis to proceed with some type of risk assessment. I think all of us ought to applaud that. We might individually disagree with the precise mechanics of our proposal, and a lot of suggestions are being made to improve it, but the *concept* of a systematic approach has got to be applauded by everyone. Ultimately we have to proceed with a systematic risk assessment. That ought to proceed, as I said earlier, not just with OSHA but with the IRLG, and ultimately it can proceed internationally. There is no question about that. Moreover, if we proceed with risk assessment on an international basis, I am less concerned about whether we can ultimately get to the regulatory response on an inter-

national basis.   The reason we have had different regulatory
responses among nations up to now is because we have had no
agreement among countries on the risk assessment.

   If we begin with the scientific risk assessment and once
reach agreement on that, I think the issue of regulatory response and harmonization will flow from the first step.   I
may be too optimistic.

DR. BUTLER: I wonder if we could come back to the EPA?   In
listening to what was said I wonder about the rationale of
their policy.   It seems to be based upon the statement that
there are many agents that are deleterious to public health.
Well, I would like to see some evidence of that.   Of course
some agents are deleterious to public health and individual
health, but I wonder whether many are.   It is a true general-
ization that the more technologically advanced a society is,
the better expectation of life there is.   So what is deleterious to public health?   It is admitted that the risk assessment
is vague, and then following that, it is said that this assessment must be made as quickly as possible.   Well if you can't
recognize compounds as deleterious and you can't recognize risk,
does it have to be done as quickly as possible?   Surely you
should possibly do something about something like aflatoxin,
but where one cannot recognize a hazard does one have to estimate risk very quickly?

   It is possible that the policy, if you're concerned with
policy, is correct and I'm wondering whether the implementation of that policy is right at the moment.

DR. ANDERSON: With regard to the rationale behind the guidelines I think it can be stated very briefly.   The rationale
is to set forth a series of procedures that EPA would follow
in making risk assessments.   We said quite clearly that we
would use human epidemiology as the strongest indication that
something is going to cause human cancer.   We would also rely
on animal studies as substantial evidence that something is
likely to be a human carcinogen, and on down the line to
short-term *in vitro* testing as suggestive evidence.   The idea
here is that, in the test data we see, we are dealing with
homogeneous species and not with many factors that are present
in the environment, such as factors of synergism and certainly
major differences in susceptibility.   That is the reason, we
feel, that we have to proceed with some degree of caution,
because of the major uncertainties that we cannot even hope
to address when we're doing a risk assessment.

   So the rationale is to use the information available to
make a statement about the likelihood that something causes
human cancer.   Once we have that statement then it's hoped

that we can get some handle on how many people are being exposed and what the exposure levels are. Just that information
alone is very helpful, indeed, when we think of trying to reduce exposure. When I say reduce exposure to as many things
as quickly as possible, I certainly do not mean tomorrow, but
I'm looking over the regulatory history of perhaps seven
years or more, when we have really gained very little control
of very many substances. When I say as quickly as possible,
I mean we hope to step this up so that we are capable of reducing exposure to many different agents. I do not mean in
such a painful way that society howls and simply cannot stand
the stress and strain of improving public health in this
fashion, but rather trying to find the balance where we can
make some major inroads in this business. I think there are
substantial data on many agents now on the records for which
we have not gained any degree of control and we would certainly like to set this kind of thing in motion. I hope that re-
sponds in part.

DR. ZAWEL: First of all I am not a scientist. My background
is in consumer affairs and I come out of the public interest
community. Since this is an international conference I'd
like to position the United States a little realistically.
We've got to recognize that public policy in this country is
made very publicly and the differences among the agencies and
the kinds of demands placed on the government to make decisions
with relative due speed have as much to do with sociology and
psychology as sciences, as they do with toxicology and medicine.
      It is within that context that I'm disturbed that there
isn't more representation of that. I agree with Mr. Hutt.
What I hear him saying, if I can rephrase it, is that we must,
to the greatest extent possible, separate and quantify risk,
so that we can identify it across all social conditions and
within a broad psychological and social framework. We are
having difficulty doing that here in the United States. I
believe a lot of the problems we have and a lot of the demands
that are being made are a direct response to the growth of a
technological society, where people feel they should have
something to say about the conditions in which they live.
They feel out of touch with how those decisions are made,
which is the reason why we need an FDA covering food and food
additives in a society where individual choice may be a
factor. But we cannot use the same principles for water
quality control where there is no individual choice. That is
distinctly American and it is distinctly within the context
of this society. I'd like to move back and really reinforce

the need for some agreement on how we are going to quantify scientific risk, but we have to bring in synergism as a major issue among public advocates.

There is no scientific data upon which we can have a dialog. Psychological and behavioral ramifications of carcinogenicity have an impact, but we cannot discuss it. Dr. Feingold's findings on hyperkinesis and additives have not been adequately addressed within the context of our scientific community, but they have had enormous social impact in terms of the way people view the questions that we are asking here. I just wanted to position a great deal of the dialog regarding American policy and the international community in those kinds of terms.

DR. MRAK: Dr. Zawel, you know how much I encouraged you to come to this meeting and I'm delighted you did. What you do not know is how much some of the other advocates were encouraged to come yet did not respond. So I don't think it's fair of you to say that there are not enough of them here. They were invited and I might say, too, that Mr. Hutt insisted we get all points of view and every effort was made by Dr. Coulston to do that. Once again, thank you for coming.

DR. ZAWEL: I am in favor of articulate scientific debate. There are social scientists, there are behavioral scientists, there are medical scientists. There is a broad spectrum of science not presented in this room which should be represented, because those sciences are directly related to the science that you are expert in as well.

So I am glad I am here. I am not sure I can address the questions that need to be addressed, because I do not have that kind of expertise.

DR. BUTLER: I'm very concerned about the idea of things being deleterious to public health. I quite accept that if you look at the history of medicine, its advances have come through improvement in public health. This is absolutely unarguable. My only concern is that when that is taken further and one is making assumptions about deleterious effects of chemicals in public health without evidence, I don't think that one should implement policy on that basis.

MR. HUTT: I would like briefly to respond to that. I do not think that when one acts on the basis of animal data that one is making an assumption of a deleterious effect on human health. As I tried to state in my remarks earlier, we are regulating in total uncertainty. It is not a scientific issue but it is a public policy issue when the issue is uncertain,

whether there will or will not be a deleterious effect.   Do
you regulate on the side of caution or do you, for example,
not regulate at all?   The public policy determination has
been made in the United States that in uncertainty we will
regulate on the side of caution.   Congress made that policy,
I might add, and in that sense every American citizen in this
room made that policy by voting for the people who enacted
the law.

MR. ADAMS: Well I was simply going to say that in Congressman Wampler's request to the National Academy that all fields
of scientific endeavor be included in the Academy it would
be presumed that all of the areas that Dr. Zawel mentioned
would probably be asked for comments.

CONGRESSMAN MARTIN: I know that Dr. Smeets raised the question a little earlier of the impact of federal legislation on
state legislation, and, Mr. Hutt, maybe you will address yourself to that.

MR. HUTT: In the United States, basically, federal legislation has set the rule for the entire country when we are
dealing with safety matters in foods, drugs, cosmetics, and
similar products.   In the area of economic legislation, the
question is almost just the opposite.   The states have regulated extensively in the economic area, contrary on occasion
to federal laws.   I am not aware of many situations in the
United States where, for example, the FDA has decided either
to permit or not to permit a food additive and a state has
ruled in a different way.   Laetrile is not a food additive.

CONGRESSMAN MARTIN: One amendment that might be added to that
is that while federal legislation does preempt state legislation on safety, there are, from time to time, provisions for
states to go beyond the standards that are set by the federal
government.   This is true in the Environmental Protection Act.
There is also the consideration of the Toxic Substances Act
and various legislation, now in debate in the state of New
Jersey, with regard to carcinogens.

MR. HUTT: This is true, but again I stick to my statement as a
practical matter; the states only very infrequently will enact legislation, which in effect overrides federal conclusions.
It has happened very infrequently in this country.

DR. SMEETS: I think that was a very important statement you
made Mr. Hutt, but that is the whole crux of the problem, when
you are distinguishing between public health legislation and
economic legislation. Our problem in the European Community
with nine countries, each having its own legislation, is
trying to harmonize this legislation. I must say it is very
hard for us who have to deal with this problem even though we
are open to discussion where there is public health or an
economic problem. The word has been used already this morning,
we called it "technical trade barriers," and I must say we
cannot always solve the problem. Here in the United States
it is considered an economic problem and in Europe it is considered a public health problem.

DR. ZAWEL: I do think that increasingly we will see state
economic concerns entering into the public health debate as
we learn that more and more different states have different
kinds of economic bases. I believe federal regulation will
run directly in conflict with individual state economic health
factors and that is probably the up and coming public policy
challenge to us. But I think Dr. Smeets covered it.

CONGRESSMAN MARTIN: I think we have enjoyed a spirited discussion of two dominant questions: how to develop a harmonization of policy within the United States, and how to extend
that objective insofar as our international dealings are concerned.

BY APPROPRIATE METHODS: THE DELANEY CLAUSE

H. F. Kraybill

I.   INTRODUCTION

In order to evaluate the hazard of a chemical to man, one must either have direct evidence from epidemiological studies or attempt to extrapolate from animal evidence. This view was embodied in a legislative provision, in the field of food additives, which is referred to as the "anticancer clause" or Delaney Clause as an amendment to the Federal Food, Drug and Cosmetic Act. This clause stipulates "that no additive shall be deemed to be safe if it is found to induce cancer when ingested by man or animals, or if it is found, *after tests which are appropriate* for the evaluation of the safety of food additives, to induce cancer in man or animal." The italicized words are emphasized here in developing the theme of this presentation.

Some may debate that this phrase does not refer to the ingestion concept. Nevertheless, appropriate or valid methods must be used to permit correct scientific interpretations and ultimately meaningful evaluation of risk. Others may argue that the anticancer provision removes the scientist's right to exercise judgment. In some sense this is incorrect because we have not yet developed a mechanism to ensure that all scientific interpretations and appropriate data are submitted before a regulatory/legislative tribunal renders a decision. Full discretion and the opportunity for judgment should be provided to the scientific community to assess whether a substance or chemical has been shown to produce cancer when added to a diet given to test animals. Such decisions should be shielded from political pressure and emotional stampede.

Certainly the legislative provision does not mandate that evidence be considered that is not relevant, i.e., carcinogenicity induced by a series of subcutaneous injections or via the inhalation route is not a priori evidence that the same chemical would be a carcinogen via the oral route. Most of

61

ISBN 0-12-192750-4

the arguments with respect to discretionary factors are
oriented too much around the arguments on whether one can or
cannot prove a threshold or safe dose for an established carcinogen. The denial of a threshold is one of the fundamental
views invariably advanced by some. In the absence of evidence
to the contrary, the existence or denial of a threshold must
at this moment be considered speculative. There is no need to
dwell on this aspect of a problem that has been debated for
many years since as yet we have no absolute scientific procedure to prove, on a broad base, the existence of a threshold,
although recently there are pieces of information to suggest
that one might envisage the existence of one. Rather one's
energy might better be devoted to a challenge of methodological approaches used, which should be critically evaluated as
to their relevance from a biochemical/pharmacological/meta-
bolic standpoint.

   With the above introductory comments, wherein we indicate
the need for innovative mechanisms for evaluation of methods
and data, would it not be reasonable to set up a national
council of senior scientists for the purpose of this important
task? Such a council should be instructed to critically evaluate all the data developed and the methods or procedures
used, both biologically and statistically, in the absence of
administrative and political pressures and in a timeframe that
will provide a well-considered evaluation and decision on
safety and/or hazard.

II. GENERALLY ACCEPTED CRITERIA--SIGNIFICANCE IN
EXTRAPOLATION OF ANIMAL DATA TO MAN REFLECTING A
POTENTIAL HUMAN CARCINOGENIC HAZARD

   There is always a level of uncertainty in making intra-
or interspecies extrapolations, but at best animal experimental data provide some level of concern relevant to carcinogenicity. The literature is replete with background material
on principles recognized as important in achieving uniform
agreement on the significance of a finding on carcinogenicity.
These factors will be briefly enumerated in order to provide
a basis for a later critique on some of the procedures and
methods used in biological experiments and statistical treatment of data derived from these experiments.

A. *Unequivocal Statistical Approaches/Results*

A carcinogenesis bioassay with an ultimate assessment of positive findings on carcinogenicity implies a statistical relationship. That is, the tumor incidence in the test chemical group should, or may, be significantly in excess over that of the control group. The statistical properties of an assay may be unchallenged. If challenged, one usually questions whether hypothesis testing is a proper use of statistics, and thus alternatives may be proposed.

B. *Reproducibility of Findings*

Extrapolation of findings within a species and strains of species with a high index of reproducibility from laboratory to laboratory relevant to derived biological data is most important. Reinforcement of positive findings with other species and strains of species involving different target organ(s) over a wide range of doses provides a high index of credibility. If such reinforcements prevail, the quantum jump to the possibility of such events occurring in many may indeed be a high probability. The fact that such reproducibility does not always prevail introduces a problem for debate and disagreement. Factors accounting for such lack of reproducibility will be discussed later since the question arises as to which experimental findings and which reports can be accepted as a basis for informed decision.

C. *Confirmation of Pathological Diagnosis*

Variance in diagnosis is always a problem. That is, agreement must be reached as to whether the lesions are hyperplastic or benign or malignant tumors. Concurrence on a diagnosis on a broad scale by many pathologists leaves little opportunity for debate.

D. *Prevalence of Dose-Response Relationships*

For wide acceptance of findings on carcinogenicity usually a dose-response relationship prevails. It is unnecessary to go into detail on this point other than to state that there are numerous studies verifying dose-response curves for chemical carcinogens. Testing a chemical where the response increases with increasing dose generally presupposes that the response is not overshadowed by overt toxicity. Such is not

always the case, which demonstrates the complexities inherent
in carcinogenesis mechanisms. If the higher regions of the
dose-response curve plateau then one may question whether
there is a metabolic overload or whether an exposure or stress
is producing an overt toxicity wherein frank neoplasia may
not be observed.

In the development of concepts on that group of diseases
called cancer wherein there is a continuum of biological
events, is it not conceivable that one could erroneously
generate the position that cancer starts with a nonthreshold
event and progresses in a non-dose-response fashion?

Additionally, in the consideration of response data,
negative findings reveal nothing. Positive findings, while
acceptable unequivocally must still be scrutinized in terms
of comparative biology. Thus, the quantitative extrapolation
of laboratory-derived dose-response data to man is of
questioned utility.

E. Time-to-Tumor Incidence Data

Strong carcinogens with increasing doses usually reflect
a decrease in latency period. However, with chemicals that
have a potential for low potency or questionable carcinogenic
activity, the latency period could be well beyond the lifespan of the animal.

F. Experimental Dose Range within Probable Range of Exposure

In order to increase the "power of the test" and minimize
the possibility for obtaining false negative results, the
doses of a chemical administered are often far beyond the
actual range of human exposure. When carcinogenic data are
reported from experimental studies in which the administered
dose falls within the range of human exposure to that chemical, the findings and conclusions developed from such studies
are usually more readily accepted.

G. Species and Strains with Low and Narrow Range in Spontaneous Tumor Incidence

Animal species and strains with a low incidence of
spontaneous tumors for the control population, exhibiting a
low standard deviation from the mean, usually introduce less
of a problem in reproducibility of results and confirmation
of findings from laboratory to laboratory. Conflicting re-

ports from various laboratories on carcinogenic properties of
a chemical may be explained on the basis of wide deviations
in the tumor response for the rodent population and strains
used from time to time in different locations. The aspect of
genetic drift in the control animal population may also play
a great role.

A summarization of these procedural criteria and predictive factors that have been developed in animal models for
assessment of carcinogenicity and that imply some possible
significance of experimental findings to man are presented in
Table I.

Although possibly ignored, in carcinogenesis bioassay
there is one fundamental concept that must be remembered.
This concept, set forth by Shubik (1), is that "carcinogenicity procedures should be require a gross departure from
established principles of toxicity." One might broaden this
admonition to emphasize that carcinogenicity testing, a component and reflection of chronic toxicity, must envision the
role and significance of biochemical, metabolic, and pharmacokinetic mechanisms in the experimental design and ultimate
acceptance of the derived experimental data and findings.

III. PROCEDURES AND FACTORS THAT MAY CONFOUND PROPER ASSESSMENT OF CARCINOGENICITY STUDIES AND DATA

The more recent literature is replete with critiques in
this general subject area. Therefore, only a brief account
will be given of some of the controversial items that are
considered. A fuller description of these problem areas has
been discussed elsewhere (2).

A. Inappropriate Route of Administration

Earlier reference was made to the use of subcutaneous injections, intraperitoneal administrations, and inhalation procedures for chemicals, specifically those in foods or water,
which would gain entrance into the body through oral ingestion. Frequently, bladder implant studies have been conducted,
which may be scientifically interesting but not relevant to
conditions represented, for example, in the normal exposure
to dietary chemicals.

TABLE I

Procedural Criteria and Predictive Factors Applicable to
Animal Models in Assessment of Carcinogenicity and
Significance of Extrapolation to Man

1. Unequivocal statistical approaches and results
2. Reproducibility and confirmation in various species
    and strains
3. Confirmation in pathological diagnoses
4. Tumor incidence increase on dose-response relationship
5. Comparability in experimental dose to human exposure
6. Significance of time-to-tumor data
7. Narrow range in spontaneous tumor incidence
    (control animal population)--reproducibility of data
8. Protocols consistent with established principles of
    toxicity
9. Recognition of biochemical, pharmacokinetic, and
    metabolic mechanisms in interpretation of findings

B. Improper Test Species and Strains

Background information on spontaneous tumor incidence in
various animal models is essential. In some instances,
strains of animals with high spontaneous tumor incidence are
used especially where the test agent may induce tumors in
the same category or for the same target organ. While such
pure bred strains with high tumor incidence have been used
believing that the differential in tumor incidence between
control and test groups is a satisfactory approach, Roe and
Tucker (3) regard such approaches as tantamount to recommending that an analytical chemist use a dirty test tube.

There have been several workshops in recent years about
the mouse as an experimental model in carcinogenesis bioassay. A rather extensive report on this subject has been
presented by Grasso et al. (4). Their critique is oriented
around four types of tumors: lymphomas, lung, liver, and
mammary. They maintain that when tumor response involves
any of the four commonly occurring tumors there are reasons
for questioning the validity of this procedure when statistical procedures are used to delineate significant differences
in strains of mice. An exception seems to be made for skin-
painting experiments to demonstrate skin tumors.

The primary objections to adaptation of this animal model
in broad-scale testing is the problem of interpretation of
results that concurrently are affected by various biological
factors. They claim that certain inherent mechanisms for
tumor production prevail in the mouse (genetic, viral, and

hormonal) that may confound the end result. They conclude
that while the mouse is a unique animal for studying mechanisms, it is not a satisfactory species for routine bioassay
because of its particular susceptibility. Other species
appear to be less susceptible to these variants. Some investigators may disagree with this concept.

C. *Role of Diet, Nutrition, and Diet Contaminants*

    While much has been written previously on this subject
[Tannenbaum and Silverstone (5), Kraybill (6)], it is amazing
that little attention is given to such details. The effects
of diet are shown in Tables II and III.
    Assuming that the role of calories and state of nutrition
are recognized in the planning of experiments, one feature
that is an inherent error or variant in most bioassays is the
contaminants not only in the laboratory chow but in the
drinking water furnished from municipal supplies. Municipal
water supplies may have 300-600 biorefractories of which
about 50 carcinogens and 65 mutagens have been listed by
Kraybill *et al.* (7). One might assume that these contaminants
are in common for the control group of animals, but in the
test group the chemical is added, which could lead to a
synergistic effect influencing the response of the test
animals. It is well recognized that semisynthetic rations
that have lower levels of contaminants really yield different
tumor response data than laboratory chows that are composed
of fishmeal, plant proteins, beef scraps, etc., all ideal
carriers of environmental contaminants. The removal of contaminants in the diet may alter the degree of tumor response
and the end result of a study. Numerous examples have been
reported in the literature.

D. *Contaminants in Test Agents (Chemicals)*

    Many of the organic chemicals bioassayed for carcinogeni-
city would fall in the category where a systematic examination
of the chemical for contaminants should be required. This is
not always the case. It is often desirable to test the
"technical grade" chemical and one that has been purified.
Frequently purified chemicals evoke a different biological
response than the so-called unpurified or technical grade.
Recently, in some studies on chloroform, analytical chemists,
through the use of gas liquid chromatography and mass spectrometry, identified contaminants in the reagent grade solvent
(8). Several of the contaminants have previously been ident-

TABLE II

Effects of Caloric Restrictions during the Two Stages of Carcenogenesis [a]

Group	Diet in period of carcinogen application (10 weeks)	Diet in period of tumor formation (52 weeks)	Tumor incidence
HH	High calorie	High calorie	69
HL	High calorie	Low calorie	34
LH	Low calorie	High calorie	55
LL	Low calorie	Low calorie	24

[a] Benzo(a)pyrene in skin carcinogenesis in mice. From Tannenbaum, A., Cancer Res. 4: 673–677 (1974).

TABLE III

Cancers in Affluent Mice[a]

| Feeding Procedure | Total tumors (18 months) | Tumor site | | | Other neoplasms |
		Liver	Lung	Lymphoreticular	
Diet/day 4 gm 1 mouse/cage	4	1	1	2	0
Diet/day 5 gm 1 mouse/cage	4	2	0	1	1-testis
Diet ad libitum 1 mouse/cage	32	15	2	11	2-testis 1-kidney 1-thyroid
Diet ad libitum 5 mice/cage	23	8	6	9	0

[a]From Roe and Tucker (3).

ified as carcinogens (carbon tetrachloride, trichloroethylene).
Diethyl carbonate occurred at a level of 1 ppm in chloroform.
Chloroform has been identified as a carcinogen in the mouse
and rat in National Cancer Institute studies (9).

E. Body Retention--Lipophilic Chemicals

Chemicals that tend to deposit or are retained in the
adiposity create problems in the design of certain studies.
Of special interest here are the organochlorine compounds
such as DDT, PCB, and the "tetradioxins." While failure to
recognize the problem of this buildup or the "sink phenomena"
may result in an early termination of the bioassay due to
high mortalities from release of the compound into the lean
body mass and toxic effects, other considerations exist.
For instance, chemicals with long residence time may interact
with other chemical moieties in the tissues or conjugate.

F. Metabolic Overload--Overdosing

One of the major criticisms in the design of some experimental studies is that of failure to recognize all the factors
involved in prescribing massive doses for testing. The
reason for selecting a so-called MTD (maximum tolerated dose)
is mainly to increase the "power of the test," that is, to
ensure that no false negatives will appear, especially in the
case of a borderline carcinogen or established weak carcinogen. This goal is fully appreciated. Overlooked, however,
is that the following may be operable:

(a) the large dose administered and results obtained in
terms of response may not be reflective of probable response
or nonresponse at lower doses comparable to human exposure;
(b) different metabolic pathways may be followed when the
dose of a chemical is high compared to a low-dose insult;
(c) injury to target organ or cells from a pathological
or histochemical and biochemical standpoint may not be at all
comparable at the high dose and low dose;
(d) high dose administration increases the probabilities
of achieving overt toxicity or toxicosis without inducing
neoplasia, thus yielding the chance for a false negative;
(e) occasionally, in the subchronic testing, a dose may
be established for the long-term testing of a chemical, which
may be far beyond the dose set for another environmental
chemical.

TABLE IV

Comparison of Starting and Final Levels of Test Chemical Pesticides[a]

Group[b]	Compound	Starting dose (ppm)	Final dose (ppm)	Starting dose (%)
A	Aldrin	120	0	0
	Dieldrin	80	0	0
	Endrin	15	7.5	33
	Photodieldrin	10	10	100
	Chlordecone	60	10	17
	Heptachlor	160	20	13
	Chlordane	800	100	13
	Lindane	640	320	50
	Toxaphene	2,560	640	25
B	Parathion	80	60	75
	Azinphos-methyl	250	125	50
	Phosphamidon	160	160	100
	Dimethoate	500	250	50
	Dichlorvos	1,000	300	33
	Gardona	16,000	8,000	50
	Malathion	16,000	8,000	50
C	Captan	16,000	4,000	25
	Daconil	20,000	10,000	50
	Picloram	20,000	10,000	50
	Amiben	20,000	20,000	100

[a]From Burchfield et al. (10).

[b]Group A, chlorinated hydrocarbon insecticides; Group B, organophosphate insecticides; Group C, phytocides.

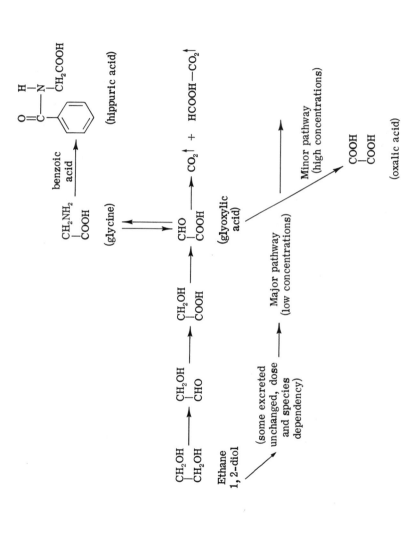

Fig. 1. Probable fate of ethylene glycol in the body (Influence of dose on metabolites formed and excreted). From Gessner et al. (12).

In other words, there is a bias in selecting the dose for a chemical or chemicals in favor of those with the highest acute toxicity. The disparity in such dose selections is shown in Table IV. Obviously, some of these chemicals in the same class would occur at comparable levels in the environment.

Weinhouse (11) has shown that glyoxylic acid conversion into oxalate in the rat liver homogenates is dependent upon dose. In Fig. 1, the probable fate of ethylene glycol in the body is shown as it reveals the influence of dose. These studies were reported on by Gessner and co-workers (12). Many more examples can be given on the influence of doses on metabolic conversions and pathways. All these factors are most relevant in making scientific interpretations on the results from any bioassay. "A substance with the potential of eluciting an adverse response at even an absurd level of human consumption is toxic" is a provocative statement from a recent report of the Federated American Societies of Experimental Biology (13). Taken literally, every existing substance is toxic.

G. *Statistical Considerations--Risk Assessment*

Undoubtedly, statistical considerations based on development of hypotheses and use of mathematical models are pre-eminent in the design of experiments, the evaluation of data, and the interpretation of findings. Certainly quantitative extrapolations from animal to man based even on dose-response data, not single-dose tests, may be of limited value. One can say that certain assumptions are extant in plotting the dose-response curve beyond certain points, namely, whether the shape of the curve is linear or curvilinear in the regions of low insult. If indeed there is a subcarcinogenic event or a threshold where there is a deviation from linearity, then a statistical treatment of data may overstate the risk.

The use of statistics has not solved the problems encountered in assessing the hazards of carcinogens (strong, weak, or equivocal) in the region of low dose. Statisticians are not to be faulted because most will agree that they build in some crude assumptions. Statistical procedures are directed toward quantification of results that assist in forecasting of potential hazards. Beyond the region of discernible response, mathematical models are applied; this approach has been labeled by Gehring and Blau (14) as scientific fiction.

Salzburg (15) indicates in his report on a critical evaluation of statistical procedures that no one who has used the tools of statistics ever expects to deal with probabilities

much below 1 in 500; however, some have proposed their utility
to predict probabilities of one in a million. In a more recent paper this same author, in discussing the use of examining lifetime studies in rodents to detect carcinogenicity,
states that in standard formulation of tests of hypothesis
where groups of 50 animals are used per group (standard bioassay) there is a 20-50% chance of having a false positive.
It is possible to define a "weak carcinogen" in terms of the
degree of effect that would produce a false negative less than
5% of the time (16). In essence, he questions whether hypothesis testing is a proper use of statistics in this context
and he proposed alternatives.

Another concept has been added recently by a statistician
(17) when Cornfield proclaims that "even if carcinogenesis is
an irreversible one-hit phenomenon between the ultimate carcinogen and DNA, the existence of a no-effect or threshold
level for the carcinogenic compound given is not precluded."
There is apparently no analytic basis for the sharp distinction drawn between the two classes of toxic reaction by present
safety evaluation procedures. He makes a clear point, long
shared by biochemists/toxicologists, that whether or not
thresholds do exist is dependent upon the protective mechanism
wherein the body becomes saturated by chemical doses. That
is, it may take a certain administered dose before the protective mechanisms become saturated or dismantled. One of
the best examples of this is the relationship between vinyl
chloride and glutathione, where sulfhydryl moieties may become saturated by the interaction with vinyl chloride. Thus,
Cornfield offers a new model for carcinogenic risk assessment.

H. Research Probes--Biochemical Intermediates

As previously stated, there is a need for reconsideration
of experimental designs in bioassay, especially the concept
and application of the maximum tolerated dose. Biochemical
intermediates that, by definition, include certain exogenous
chemicals such as the micronutrients and macronutrients
(vitamins, minerals, and fats, carbohydrates, and proteins--
amino acids) and some endogenous chemicals (metabolites of
intermediary metabolism) have been recently considered as
research probes. Some of these "presumed negative" chemicals
have revealed a tumorigenic response at high levels of dosing.
We have assembled data on such specific models because we
feel that this is instructive. If chemicals normally present
at the cell level as nutrients or intermediary metabolites are
capable of inducing neoplasia in animals in experimental
studies, then this should put the issue of the influence of

high dose administration into proper perspective. Some examples of biochemical intermediates that have been implicated in tumorigenesis are shown in Table V. The problems are well illustrated in the case of bladder cancer with diethylene glycol, xylitol, terephthalic acid, cyclamate, azotoluene, and Myrj 45. In these instances, these relatively inert and nontoxic materials--due to physiocochemical phenomena (supersaturation, crystallization, etc.) leading to nephricalcinosis and other irritant effects that are secondary--have produced tumors in the bladder under extreme conditions of dosage. The recent reported studies on xylitol are of great significance. Adrenal gland tumors and bladder tumors were induced at the massive doses of 20% for this polyhydroxy alcohol, which occurs normally in the body as a metabolite in the glucuronic acid pathway of glucose metabolism in man and animals.

Certainly, we need more data from extensive studies in this area. Research probes are a mechanism for eliciting more evidence on certain mechanisms and, hopefully, should place some of the current issues into better perspective in the design of experiments, especially the problem of dose selection.

I. Immunosuppression (Humoral Factors) and Hormonal Factors

Cell-mediated immunity and immunosuppression may play an important role in the development of spontaneous neoplasms and thus have a bearing on tumor incidence. That is to say, the immunosuppressant procedure or agent per se may not induce the tumors but may arise from transformed cells not necessarily associated with the chemical under test.

Similarly, the hormonal status has a bearing on a positive response when test chemicals are given. Here one is involved in a definition of carcinogenicity that really embraces co-carcinogenicity. Both of these factors, humoral and hormonal, should be considered in an approach to quantitative risk assessment.

J. Repair Mechanisms

One may assume that alterations in DNA may give rise to an initiated cell. But one cannot ignore the possibility of development from a structureless cell and that permanently altered cells may indeed be so-called quasi-permanently altered. Repair mechanisms thus can be operable. This effect may be dependent on the level of insult required to induce cancer, that is, in effect a threshold.

TABLE V

Biochemical Intermediates Implicated in Tumorigenesis

Intermediate	Physiological level required and biological function	Observed toxic dose for response	Adverse effect	Reference
Selenium	0.1–0.5 ppm Prevents erythrocyte damage	Rodent, 5 ppm	Liver tumors	(18)
Calcium	For man, 0.8 gm For bulls, 15–19 gm Bone growth, vitamin D metabolism	5 × rda (88 gm)	Ultimobranchial Tumors in bulls, nephricalcinosis, hypercalcitonin	(19)
Arsenic	(?) Hematopoietic system, growth function, phosphorylation mechanisms, controls selenium toxicity Deficiency effect on spleen, erythrocyte fragility	Man, drinking water, 2000 μg/liter occupational?	Skin cancer	(20)

Xylitol	Nonessential: in metabolism of pentose, xylulose, glucuronic acid pathway	Rodent, 20% in diet	Adrenal gland tumors, bladder tumors, bladder stones, bladder irritation	(21)
			Increase in urates decrease in serum phosphate	(23)
Sucrose	Provides energy, nonessential	Rodent, 20% in diet	Increase in renal tumors	(22)
σ-Amino-phenols (metabolites of tryptophane)	Metabolism of tryptophane in the urine	Mice, 9–11 mg in 4 parts of cholesterol in bladder implants[a]	Bladder tumors	(24)
Vitamin D_2	400 i.u., prevention of rickets, CA and P metabolism	Mice, 1 ppm in diet, C3H tumor virus, positive strain	Mammary tumors	(25)

[a]Significance of these implant studies questioned since feeding and injection experiments in other species were negative.

Perhaps the fact that some experimental animals and people experience a longer lifespan may be associated with such inherent repair mechanisms. The repair enzymes, if measurable, could indicate or detect and eliminate certain biochemical lesions as their levels in the DNA escalate. Assuming that human DNA has a high repair capability, risk assessment from the rodent to man may not be readily made.

IV. GENERAL COMMENTS ON LIMITING FACTORS

In the selection of the test chemical and the early design of a study, some of the limiting factors described previously come into play. Obviously, all of these factors should be seriously considered in evaluating and interpreting the data as to significance in the classification of carcinogenic activity. No attempt has been made to delineate all the limiting factors in this cursory review of some of the problems requiring clarification. A recapitulation of these limiting factors, which relate to dose-response data in animals relevant to the assessment of carcinogenicity and the process of risk assessment, is presented in Table VI.

TABLE VI
Limiting Factors in Relating Dose-Response Data in Animals in Assessment of Carcinogenicity and Probable Significance to Man (Risk Assessment)

1. *Inappropriate routes of administration*
2. *Improper test species and strains*
3. *Role of diet, nutrition, and diet contaminants*
4. *Contaminants in test agents (chemicals)*
5. *Body retention--role of lipophilic chemicals*
6. *Metabolic overload*
7. *Influence of dose on metabolic pathways*
8. *Validity of statistical procedures*
9. *Value of research probes--perspectives on dose administration and response*
10. *Humoral and hormonal factors*
11. *Prevalence of repair mechanisms*

V. SUMMARY

Neoplasia depends invariably on multiple factors acting concurrently or sequentially, and the impact could vary at different stages of the disease process. The dimensions of this process are first dependent upon the insult and response but are conditioned by the multiple factors mentioned above as well as a time factor.

In the discussion of the criteria and limiting factors, emphasis has been placed on those parameters associated with identification, characterization, and interpretation of data as to carcinogenicity of a chemical. To some, these specifications may appear as redundant but many of these qualifiers were overlooked in the decision-making process. To ignore these considerations may lead to loss of scientific excellence and credibility.

Emphasis has been placed on the significance of permissive factors that are active at the cell level to facilitate the development of neoplasia. Other factors that influence homeostasis, including genetic and physiological variants, control the mechanisms in a qualitative and quantitative manner. While responses, for example, may be noted at high-dose challenge, one can perceive the possibility that adaptive and protective mechanisms are operable in the low-dose area. There are increasing examples, such as the trace elements, wherein this situation prevails. Research probes with biochemical intermediates should elicit more information on such mechanistic possibilities.

Smyth (26) in his excellent treatise on "sufficient challenge" emphasizes the importance of pharmacological considerations in deliberations on dose-response and the concept of risk. He states that "any effects from high dose challenge which is statistically validated we accept. Any apparent benefit from our low doses which may be statistically validated we discount but it may not be dose-related." He challenges toxicologists with the manifestations of sufficient challenge to the extent that low-level insult may have apparent benefits. Nutritionists look for benefits at low-level insult. The lesson learned should be self-evident in the field of carcinogenesis in that the illustrations with selenium, calcium, arsenic, xylitol, and vitamin D_2 are provocative in this regard. Is there any reason to assume that the concept of sufficient challenge would not prevail for chloroform, carbon tetrachloride, trichloroethylene, and a host of other organic chemicals?

Unless we have explored the conditions and modifiers involved in various mechanisms that may prevail in experimental evaluation of carcinogenicity of a chemical, are we not really equating all events of equal significance with reference to characterization, interpretation, and proclamations issued on carcinogenic risk? We subconsciously classify certain chemicals in different risk categories.

An attempt, if not a plea, has been made to evaluate carefully the procedures and processes utilized, which lead to an indictment of a chemical as a carcinogen with ultimate banning by the force of law. The fact that renowned scientists have drawn quite opposite conclusions from the same pieces of data and information, while unsettling to the non-scientific community, certainly indicates the need for a flexibility in our thinking relevant to all procedures used in evaluation of carcinogenicity. This is not simply a matter of debate for the so-called advocates or antagonists. It is an issue of national consequence and an awesome responsibility for the scientific community at large. The view that one must accept a compromise between what is scientifically desirable in the performance of an experiment and what is fiscally possible is an oversimplification of a dilemma we all face. All studies must always be compatible with sound biochemical and pharmacological reasoning, and as Shubik (1) has emphasized, they should be consistent with current toxicological principles. This is a challenge we can ill afford to ignore on a societal basis if not in the interest of good science.

REFERENCES

1.   Shubik, P., Potential carcinogenicity of food additives and contaminants, *Cancer Res. 35,* 3475-3480 (1975).
2.   Kraybill, H. F., From mice to men--Predictability of observations in experimental systems (their significance to man), *Proc. Conf. Human Epidemiol. Animal Lab. Correlations in Chem. Carcinogenesis, June 1977, Mescalero, New Mexico* (1978).
3.   Roe, F. J. C., and Tucker, M. J., Recent developments in the design of carcinogenicity tests on laboratory animals, *Proc. Eur. Soc. Study Drug Toxicity 15,* 171-197 (1974).
4.   Grasso, P., Crampton, R. F., and Hooson, J., The mouse and carcinogenicity testing, *Br. Ind. Biol. Res. Assoc. (BIBRA) Rep. 01-643-4411,* Carshalton, Surrey, England (1977).

5. Tannenbaum, A., and Silverstone, H., The genesis and
    growth of tumors. IV. Effects of varying proportion of
    protein (casein in the diet), *Cancer Res. 9*, 162-173
    (1948).
6. Kraybill, H. F., Carcinogenesis associated with food,
    Clin. Pharmacol. Therapeutics 4(1), 73-87 (1963).
7. Kraybill, H. F., Helmes, T. C., and Sigman, C. C., Bio-
    medical aspects of biorefractories in water, *2nd Int.
    Symp. Aquatic Pollutants, Noordwijkerhout/Amsterdam, The
    Netherlands, September 26-28, 1977* (1978).
8. McCabe, L., Contaminants in chloroform, personal communi-
    cation to the author from Environmental Protection Agency
    (1978).
9. DHEW/PHS/National Cancer Institute, "Carcinogenesis Bio-
    assay of Chloroform," (NIH) 76-1279, March 1 (1976).
10. Burchfield, H., Storrs, E., and Kraybill, H. F.,
    "Lectures on Pesticides--Environmental Quality and Safety
    Book," Suppl. III, Thieme, Stuttgart, Germany (1975).
11. Weinhouse, S., *in* "Amino Acid Metabolism" (W. D. McElroy,
    and H. B. Glass, eds.), p. 637, Johns Hopkins Univ.
    Press, Baltimore, Maryland (1955).
12. Gessner, P. K., Parke, D. W., and Williams, R. T.,
    Studies on detoxification, *Biochem. J. 79*, 482-489
    (1961).
13. Federated American Societies of Experimental Biology
    (1977). FASEB report on health aspects of GRAS food
    ingredients--lessons learned and questions unanswered,
    Fed. Proc. 11 (1977).
14. Gehring, P. J., and Blau, G. E., Mechanisms of carcino-
    genesis: Dose response, *J. Environ. Pathol. Toxicol. 1*,
    163-179 (1977).
15. Salzburg, D. S., Mantel-Bryan--Its faults and alterna-
    tives available after thirteen more years of experimen-
    tation, *Food Drug Cosmetic Law J. 30*, 116-123 (1975).
16. Salzburg, D. S., The use of statistics when examining
    lifetime studies in rodents to detect carcinogenicity,
    J. Tox. Environ. Health 3(4), 611-628 (1977).
17. Cornfield, J., Carcinogenic risk assessment, *Science 198*,
    673-699 (1977).
18. Kraybill, H. F., Unintentional additives in food, *in*
    "Environmental Quality and Food Supply," (P. L. White and
    D. Robbins, eds.). Futura Publ. Co., New York (1974).
19. Krook, L., Lutwak, L., and McEntee, K., Dietary calcium,
    ultimobranchial tumors and osteoporosis in the bull,
    Am. J. Clin. Nutr. 22(2), 115-118 (1969).
20. Kraybill, H. F., Threshold levels in additive testing,
    *Proc. 19th Meeting Interagency Collaborative Group on
    Environ. Carcinogenesis, National Cancer Institute,
    Bethesda, Maryland, August 14, 1975,* 59-77 (1977).

21. Chemical and Engineering News, Xylitol causes some cancers in mice, *Chem. Eng. News* *55*(47), 121 (1977).

22. Hoffman LaRoche, Inc.--Finish Sugar Co., Huntingdon Research Centre (England), Unpublished report on toxicity of xylitol, sorbitol and sucrose to rodents (November, 1977).

23. Shills, M. E., *in* "Modern Nutrition in Health and Disease, Dietotherapy," (R. S. Goodhart and M. E. Shills, eds.). Lea and Febiger, Philadelphia (1973).

24. Allan, M. S., Boyland, E., Dukes, C. E., Horning, E. S., and Watson, J. G., Cancer of the urinary bladder induced in mice with metabolites of aromatic amino acid tryptophane, *Br. J. Cancer 2*, 222-228 (1957).

25. Gass, G. H., and Alaben, W. T., Preliminary report on the carcinogenic dose response curve to oral vitamin D_2, *IRCS Med. Sci.* *5*(10), 477 (1977).

26. Smyth, H. F., Sufficient challenge, *Food Cosmet. Toxicol.* *5*, 51-68 (1967).

ALTERNATIVES TO BE CONSIDERED
IN BENEFIT:RISK DECISION-MAKING

William J. Darby

The consumer, producer, politician, or scientist may not
similarly perceive the nature of benefit:risk. Their respective roles and responsibilities in society differ.
Kenneth Mees (1) wrote 30 years ago:

> A politician must understand the effect of emotional
> thought and must be prepared to utilize emotional appeal
> if he is to obtain popular support. A successful politi-
> cal leader must tend, therefore, either to believe his own
> emotional appeal or to become a cynic and to some extent
> a hypocrite if he exerts that appeal without belief. It
> is this difficulty that makes even the greatest democratic
> leaders seem insincere in many of their actions. The
> appeal to emotion is unavoidable if popular sanction is to
> be obtained, and yet their critics and often they them-
> selves in retrospect feel that appeal to be false and un-
> warranted. For this reason alone the political arena
> would seem to be unsuitable for the scientific man, and
> those who believe most fully in the value of the scienti-
> fic spirit should be prepared to understand and sympathize
> with leaders who must obtain general popular approval for
> their actions.

Francis Bacon defined the "real and legitimate goal of
the sciences" as "the endowment of human life with new inven-
tions and riches--not to make imperfect man perfect but to
make imperfect man comfortable, happy and healthy." The
scientist, concerned with unfulfilled social and economic
needs and wants, and dependent upon the continued support of
society, particularly that support made available through the
efforts of the politician, must maintain objectivity and com-
petence to command respect.

83

ISBN 0-12-192750-4

Both industrial and political leaders must espouse sound and truthful positions on issues of great public emotional concern, and these positions must be founded upon the best scientific evidence and judgment obtainable. The industrial leader must produce a quality product that is acceptable, and market and distribute it at a profit commensurate with the investment of his stockholders.

Responsible public officials must discriminate between the voices to whom they listen and react. They must discriminate emotion from logic, bias from objectivity, self-seeking from public-spirited motives, and the competent from the incompetent. Authority carries with it responsibility for such judgments and the silent public must be protected from risks inflicted and benefits denied through failure to weigh properly benefit as well as risk.

It is essential that the public be responsibly informed so that policy decisions are not influenced by ignorance, bias, or undue emotions on the part of an ill-informed public.

The nature of perceived benefits and risks varies (1) as cultural patterns vary; (2) as society moves from a subsistence state to affluence; (3) as cultural, socioeconomic and related factors change within a society.

Differing perceptions concerning benefit and risk often reflect the resources at the command of the society--or differing segments of it--at that moment.

CONCEPTS OF BENEFIT:RISK

Let us examine the concepts of benefit:risk in more detail.

The concept embraces *all* of the vital and nonvital, individual (private) and societal, voluntary and involuntary gains and losses resulting from a proposed action. Uncertainties cannot be eliminated and are part of most, if not all, individual and collective activities. Participation by the individual is either voluntary or involuntary. In the latter instance participation is imposed by the society in which the individual lives: imposed by law, regulation, social custom, or religion.

Societal acceptance of risk increases in a nonlinear manner in which the perceived benefits are derived from an activity; public acceptance of voluntary risks is estimated as some 1,000 times greater than involuntary risks. Societal policy for the acceptance of public risks associated with sociotechnical systems should be determined by the trade-offs between perceived social benefits and personal risks.

The individual in most cultures today strongly defends his right to enjoy the pleasures and perceived benefits, to fulfill his personal sense of need or want, and hence to decide what risks he voluntarily assumes. He resists undue or imposed curtailment of his "right" to these benefits, especially whenever he is unconvinced of the degree of risk involved or he feels that the risk is of such dimension in relation to his value judgment of benefit that he willingly accepts the risk when it makes him "comfortable, happy or healthy." This is the reason that prohibition failed in the U.S., why smoking is not abandoned, and why the public and the Congress prevented imposition of the proposed ban on saccharin.

THE QUESTION OF ABSOLUTES

A true conflict arises when it _s advocated that the food supply contain no detectable amount of any toxic substance, especially for which there may be some experimental evidence of tumorigenic effect. In 1969 this was clearly recognized by the Secretary's Commission on Pesticides. That report stated that to avoid all naturally occurring toxicants present in food would result in self-imposed starvation. To avoid all detectable traces of one man-made chemical, DDT, would proscribe "most food of animal origin including all meat, all dairy products (milk butter, ice cream, cheese, etc.), eggs, fowl and fish. These foods presently contain, and will continue to contain for years, traces of DDT despite any restrictions imposed on pesticides. Removal of these foods would present a far worse hazard to health than uncertain carcinogenic risk of these trace amounts" (2).

The relevance of such considerations to the task of meeting present and future food needs of the world is evident. Clearly, any demand for "absolute safety" of food, for food completely free of any potential toxic effect, is mistaken, misdirected, and in conflict with good nutrition.

Wider understanding must pertain of the concepts of hazard, toxicity, and safety. *Hazard* is the probability that injury will result from the use of a substance. *Toxicity* is the capacity of a substance to produce injury. *Safety*, as usually defined, is the virtual certainty that injury will not result from such use of a substance—that the hazard is minimal. Perfect safety is unattainable and not subject to experimental proof. Nature forever confronts us with risks, risks that for conventional foodstuffs present minimal hazard (and therefore are "safe," or acceptable).

THE VARIABLE NATURE OF RISK AND OF BENEFIT

A tolerated level of hazard or risk is that which is acceptable to the individual or to society in view of needs, wants, or demands. Society's needs, wants, and demands are determined by the lifestyle of the era. Toxicologists have evolved useful approaches to estimating biologically danger-ous or toxic levels of exposure of animals, and, with a re-markably good safety record, they have extrapolated such estimates to guide the setting of safe limits of exposure of man. But with the demand for "proof of safety" as recently was called for by the Commissioner of Food and Drugs in denying the petition to increase the fortification level of iron in flour and bread to a level that would provide the scientifically based recommended dietary allowance in the American diet, we are faced with a regulatory barrier to meeting normal nutrient requirements for maintaining health.

William O. Lowrance (3) summarized concepts of safety in a manner that I feel is especially useful (4):

> Nothing can be absolutely free of risk. The problem for the society is to compare the various risks and choose those it is willing to bear.
>
> Decisions about chemicals, as with all agents, are "relative" to the circumstances of use and exposure, to the availability and characteristics of alternatives, and to the society's value system.
>
> Artificial agents are not necessarily more trouble-some than natural agents. Technological development has indeed brought some undesirable effects, but many of these must be viewed as the expense of decreasing our vulner-ability to the hazards of the natural world; Mother nature is not always benevolent.
>
> It is sometimes easier to focus on the negative side of the decisional ledger and give attention to hazardous properties, to the neglect of the beneficial side. The properties of efficacy and benefit often need to be accorded fuller and more systematic appraisal.

It is my belief that much of the widespread confusion about the nature of safety decisions would be dispelled if the meaning of the term "safety" were clarified. One approach is to define *safety* as a judgment of the acceptability of risk, and *risk*, in turn, as a measure of the probability and sever-ity of adverse effects. Thus: *A thing is safe if its atten-dant risks are judged to be acceptable.*

This definition contrasts sharply with simplistic dictionary definitions that have "safe" meaning something like "free from risk." Nothing can be absolutely free of risk. One can't think of anything that isn't, under some circumstances, able to cause harm. Because nothing can be absolutely free of risk, nothing can be said to be absolutely safe. There are degrees of risk, and consequently there are degrees of safety.

Notice that this definition emphasizes the relativity and judgmental nature of the concept of safety. It also implies that two very different activities are required for determining how safe things are: *measuring risk,* an objective but probabilistic pursuit; and *judging the acceptability of that risk* (judging safety), a matter of personal and social value judgment. Although the difference between the two would seem obvious, it is all too often forgotten, ignored, or obscured. This failing is often the cause of disputes that hit the front pages of the newspapers—and land on the agenda of Congressional committees.

I advocate use of this particular definition for many reasons. It encompasses the other, more specialized, definitions. By employing the word "acceptable" it emphasizes that safety decisions are relativistic and judgmental. It immediately elicits the crucial questions, "Acceptable in whose view?" and "Acceptable in what terms?" and "Acceptable for whom?" Further, it avoids the implication that safety is an intrinsic, absolute, measurable property of things.

The business of measuring and estimating risks is an empirical, or at least a quasi-empirical, activity, usually led by members of the professional technical community; the findings are expressed as the probability and severity of adverse effects.

Just as risk can be measured scientifically, so can *efficacy,* the corresponding measure of the probability and intensity of beneficial effects. And just as risks are appraised on a social values scale in deciding how *safe* the agent in question is, the desirability of its degree of efficacy is appraised in determining *benefit.*

ALTERNATIVES

From such a perspective of safety and of risk and benefit,
what are the alternatives? To answer the question, one must
consider the impact of whatever action is contemplated.
Impact includes both primary and secondary effects--not only
the effect of permitting or withdrawing the agent in question,
but the result of such action on other safety considerations
such as the resultant change in use level of functionally
related substances, the disappearance from the marketplace
of a food or group of foodstuffs that may be nutritionally
similar or quite varied, but are affected by the action taken
because of similar processing needs. Whether there is a
functionally effective safe or safer alternative must be
decided.

As an example of impact considerations, one may reflect
on nitrate, which is traditionally used in curing and preser-
ving meat. The simplest benefit:risk to be weighed is the
balance between the prevention of a known risk of botulism
that well can occur when the meat is improperly stored in the
distribution system, particularly within the home and the
potential risk of carcinogenesis. The risk of carcinogenesis
would appear to be either negligible or its attendant risks
have long been judged acceptable. Currently this potential
risk is less than in the past, inasmuch as responsibly conducted collaborative studies conducted by the U.S. Department
of Agriculture, the Food and Drug Administration, and the meat
industry have defined reduced levels of nitrate that are
effective against botulism and appear to have no carcinogenic
potential under conditions of feeding to experimental animals.

In assessing risk of a given level of a substance in food
or foodstuffs, one must consider the total amount to which
individuals may be exposed through food, including the pattern
of exposure. This latter includes exposure from all sources
in the environment--water, atmospheric contact, industrial,
and so on. If the exposure of a substance such as nitrate
from meat sources constitutes but say 10% of the total exposure and a 50% reduction can be effected without increasing
the risk of botulism, the pertinent question then is whether
the elimination of the remaining 5% is of sufficient benefit
in view of continued total exposure burden to warrant the
risk produced by banning the substance.

In judging a functional alternative, it is important to
ascertain that it is effective, is at least as safe as judged
from both experimental and use experience. This is in keeping
with the "workable, ethical guidelines for acceptable risks"

stated by the Citizens' Commission on Science, Law and the
Food Supply (5): "When faced with alternatives, we should
choose that alternative whose worst outcome is better than
the worst outcome of any other alternative."

    I. Using this formula, the alternative of absolute
safety would usually be "worse" and the alternative of using
controlled chemical substances with their implied risks
properly evaluated and controlled.

    II. The phrase "worst outcome" in this general principle
should be construed as (a) a risk no greater following a
change than the risk of natural disease demonstrated to be
causally related to existing patterns of food use; (b) accum-
ulative net balance of risks and benefits in the population
as a whole, not the worst outcome in a single individual
case.

    Other risks or costs to be considered include those in-
direct ones of the possible decreased production of a food,
in this instance pork, with further indirect effects on grain
production, the price of cereals such as corn used as feed-
stuffs, and even less immediately obvious effects that would
result from a decrease in numbers of animals slaughtered.
The effects may reduce the availability of by-products that
are of great importance in medicine, by-products like insulin,
heparin, and ACTH.

ALTERNATIVES IN EXPERIMENTAL DESIGN

    Other alternatives that must be considered in assessing
risk are those involved in designing an experiment that will
properly test the "risk" in an appropriate use pattern.
This involves not only the question of level of dose or
levels to be tested, but also method of exposure, i.e.,
whether the substance will be consumed in practice once a
day, with each meal, at rare intervals, or with a variety of
foods, as a beverage, etc. One must design studies to include a series of levels of intake and they should permit or
include determination of changes in metabolic pattern of the
test substance at differing dose or intake levels. Even
essential nutrients are metabolized in one manner when given
in physiological amounts, and yet when presented in grossly
excessive quantities may be altered metabolically or excreted
differently.

When attempting to assess fetal exposure or the effect on the central nervous system, alternative test levels surely must be studied because the permeability of the placenta or of the blood brain barrier may prevent passage of a substance or metabolite until very large quantities overcome these barriers. The limited ability of the fetal organism to excrete accumulated materials must be reckoned with.

In testing *in vivo,* studies must be carried out to elucidate the metabolism or disposition of the compound in a series of organs or tissues. Metabolism *in vivo* may be quite different indeed from that in a single type of isolated cell.

Judgment of potential effects in human use rationally can be made only by considering evidence of alternative experimental design and level.

NUMERICAL ESTIMATES BASED ON ANIMAL MODELS

I am especially concerned with the recent trend toward use of animal data to calculate an estimate of the number of cases of illness or death that may occur in the population, i.e., the use of models as discussed by Peter Hutt. I recognize the reason for attempting to make such estimates, but these estimates invariably must be based on the assumption (and thereby carry the influence) that the human organism will be injured in the same manner that the most sensitive species tested is injured and that the test dose level administered to the animals would be comparable to the level to which the population would be exposed (despite the deliberately distorted high test dose used in such testing). These and other assumptions are rapidly passed over by the uninformed and one finds lawyers, legislators, the general public, and even scientists, subsequently claiming that "x number of cases of stomach cancer, bladder tumors, or other conditions or deaths within the population are attributable to a particular substance."

CONCLUSIONS

Such misunderstanding of the reliability of computer constructed models or of animal data reminds me of the quotation from William James:

Many persons nowadays seem to think that any conclusion must be very scientific if the arguments in favor of it are derived from twitchings of frogs' legs--especially if the frogs were decapitated--and that--on the other hand--any doctrine chiefly vouched for by the feelings of human beings--with heads on their shoulders--must be benighted and superstitious.

The need for judgment and reason in benefit:risk decision-making underscores generalizations made by Robert Ingersoll and Emile Duclaux, respectively: "Reason, Observation and Experience--The Holy Trinity of Science." "A series of judgments, revised without ceasing, goes to make up the incontestable progress of science."

Finally, the alternative to continued misunderstanding and ignorance on the part of scientists, officials, and the public concerning risk and benefit factors is education. Honest, informative information on labels, in advertising, and through the media must be comprehensible. Warnings that "eating may be dangerous to your health" merely create anxiety and do not inform. The growing plethora of such warnings on product after product results in seeing but not comprehending, hearing but not understanding. We who are knowledgable and concerned have a responsibility to speak informatively and clearly in a manner that will be heard and understood. C. P. Snow, in an essay entitled "The Great Issues of Conscience in Modern Medicine" wrote scientists:

It is not enough to make statements of the greatest possible truths; they must have the courage to carry those statements through because they alone know enough to be able to impress their authority upon a world which is anxious to hear, if only it can find voices which can speak with enough clarity and, I think I must say, enough noise.

REFERENCES

1. Mees, C. K., "The Path of Science," Wiley, New York (1946).
2. Report of the Secretary's Commission on Pesticides and their Relationship to Environmental Health, U.S. Department of Health, Education and Welfare, December (1969).
3. Lowrance, W. W., "Of Acceptable Risk: Science and the Determination of Safety," William Kaufmann, Inc., Los Altos, California (1976).

4. Lowrance, W. W., Testimony before the Subcommittee on
    Agricultural Research and General Legislation of the
    Committee on Agriculture, Nutrition, and Forestry,
    United States Senate, July 19, 1977, *in* "Food Safety and
    Quality: Regulation of Chemicals in Food and Agricul-
    ture," p. 48-51 USGPO, Washington, D.C. (1977).
5. Citizens' Commission on Science, Law and the Food Supply,
    A report on current ethical considerations in the deter-
    mination of acceptable risk with regard to food and food
    additives, March 25, 1974, *in* "Agriculture--Environmental
    and Consumer Protection Appropriations for 1975," Part 8,
    Study of the Delaney Clause and other anticancer clauses,
    p. 490. Hearings before a Subcommittee of the Committee
    on Appropriations, House of Representatives, USGPO,
    Washington, D.C. (1974).

REMARKS

DR. ERNST WYNDER: We have problems that no doubt require the
aid of Washington. As I listened to the previous speakers,
and all of us who have been involved in this very difficult
problem relative to the Delaney Clause, I feel like so many
of us do when we discuss vital political issues: it somehow
or other looks like we are facing a man in a white hat or in
a black hat and the issue continues to be emotionally charged.

Now it is our hope that in these deliberations we are
coming up with some kind of conclusions that the political
leaders of this country can and will understand, because I
feel convinced that those of us in science will not be able
to come up with a solution to this problem ourselves. Finally
as Dr. Darby says, it is really up to Congress, with our advice and counsel, to decide what is best for the people of
this country, not only for the present, but above all for a
future generation. In this regard we also want to make certain that whatever decision we make will be of benefit to the
people who follow us.

DECISION-MAKING ISSUES RELEVANT
TO CANCER-INDUCING SUBSTANCES

Albert C. Kolbye, Jr.

The decision-making issues described in this presentation
exist now and will continue to exist regardless of the Delaney
Clause. If the Delaney Clause did not exist, these issues
would still have to be resolved in order to improve the
effectiveness of any agency to regulate carcinogens, such as
FDA's under the general food safety provisions of the FD&C
Act. As you know, one of the basic precepts that motivated
passage of the Delaney Clause was that modern science was not
yet capable of evaluating carcinogenic risk with sufficient
precision to permit "safe" exposures to substances used as
food additives, if such substances were found to induce cancer in animals or man. Decision-making with or without the
Delaney Clause is complicated at best. Even if there were no
Delaney Clause, many complications would still exist and lead
to controversies concerning the reliability and significance
of scientific data, the interpretation of the scientific
evidence as to human risk from a regulatory and legal viewpoint, and the considerations relevant to where to draw the
line for specific applications of FDA's policy to resolve any
substantial doubt about safety in favor of the consumer.
With the Delaney Clause, some issues are simplified, but
others may be more complicated. For example, once a substance
has been determined to be a "carcinogen," then the regulatory
issues simplify in that no food additive uses shall be
deemed to be safe as a matter of law. This is essentially an
either/or equation, which tends to compress a spectrum of
considerations and to obliterate distinctions that may be important. Complications occur in determining what is a carcinogen vs. what is not; also whether certain substances arise
in food as food additives or not. There can be difficulty in
determining what the exact limitations are concerning the

93

ISBN 0-12-192750-4

legal boundaries between food additives and other ingredients
in food that may arise naturally or by inadvertent contamination of food itself or by environmental contamination related
to the production or processing of food.

Whenever a complex judgmental process is compressed into
an either/or framework such as safe/not safe, semantic confusion will be created in the minds of scientists and lay citizens alike, especially as to the degree of risk presented to
the public if a determination of safety is precluded by operation of law, and usage of a substance is deemed not to be
safe. There are important distinctions between "safe," which
is a value judgment at one end of the spectrum, and "dangerous," which is at the other end of the spectrum. Other terms
enter into the middle ground, such as "not shown to be safe,"
"not safe," "unsafe," "potentially harmful," each of which has
different connotations. Whenever preventive medicine and law
interact, problems of interpreting and applying either scientific evidence or legal phraseology emerge. And that is where
we are today. I should like to talk about some of the scientific problems and some of the legal problems, both of which
really involve the need for more explicit criteria and for
procedural safeguards to ensure fairness when applying general
criteria to specific problems.

It might be helpful to consider some of the changes that
have evolved over time since the food and color additives
legislation transformed the entire food safety field in the
late 1950s and early 1960s. One dramatic change has been
that analytical chemists can now detect a whole new galaxy of
low levels of substances in food. Previously the limits of
qualitative identification and quantitative measurement were
in the parts per thousand range: today parts per trillion are
not uncommon and in some instances routine. This represents
a millionfold increase in our ability to detect "chemicals"
in food. And while we can detect very low levels of some
compounds in food, we may not know whether the bioavailability
of such compounds at low levels is always similar to the
availability for absorption and potential toxicity when present in higher levels. Another dramatic change has been in
the way we test for the ability of a substance to induce cancer in animals. Chronic feeding tests have been expanded in
terms of numbers of animals used per sex per dose level,
dosing philosophy has changed somewhat with the adoption of a
new definition and philosophy of what constitutes a maximum
tolerated dose, and the criteria for interpreting a positive
result have broadened and at times fluctuate. Also, the
number of chemicals tested for carcinogenicity has increased
enormously since 1958. We know that in the majority of instances known human carcinogens have induced cancer in test

animals and as a society we operate on the theory that because this is so that therefore any substance given at any
dose that induces cancer in animals is a carcinogen for animals and therefore likely to be a potential carcinogen for
humans. We know we can influence the expression of tumors
in test animals by modifications of the basal diet and by
environmental circumstances unrelated to the test compound
itself. We suspect that some of the techniques that are
used to test animals may be imposing stresses that influence
the expression of cancer in ways we may not be adequately
able to isolate for purposes of controlling all relevant
variables.

The field of toxicology and related fields of pharmacology, endocrinology, and physiology have made important
advances since the early 1960s. There also has been a
dramatic increase in our awareness of molecular biology and
its application to toxicology in general and to carcinogenesis in particular. Very notable is our increase in knowledge
concerning some of the parameters of chemical carcinogenesis
and in our beginning ability to predict certain structure/
activity relationships for some classes of carcinogens. The
field of mutagenesis has had a revolutionary impact on the
fields of toxicology and carcinogenesis, with increasing
evidence that many chemical carcinogens are also mutagenically
active when tested appropriately.

Epidemiology has also increased in sophistication with
the advent of techniques to study the occurrence of chronic
diseases in humans in relation to particular exposures.
Although it usually is most practical to study the association
of past exposures to disease and develop some appreciation
of risk for humans, valuable insights are beginning to emerge,
at least in regard to substantial exposures such as those involved with cigarette smoking, therapeutic doses of some
hormones, to certain chemicals in occupational environments,
and to some major dietary exposures such as to aflatoxins or
dietary fat.

Our awareness of cancer also has increased largely due to
the fact that more people in the U.S. are dying of cancer today than ever before, although the incidence rates for many
types of cancer remain remarkably stable. In large part,
this is due to our increasing longevity gained by partial con-
quests over major killing diseases such as cardiovascular
disease and many infectious diseases.

One might ask, what has been the effect of all these
changes? We have greater insight into the relationships
between substantial exposures of either animals or humans to
certain carcinogens, but we still lack substantial insight
into the risk factors that may operate with regard to low-

level exposure. We still do not know whether the same parameters that apply to high-level exposures also necessarily
apply to low-level exposures that may occur singly or in
combination. We do know that the higher the dose, the higher
the risk. We have reason to believe that some exposures to
certain low-level carcinogens in combination may be additive,
while others may not be. We have similar questions concerning mutagens, but here one can test much more quickly to see
if multiple low-level exposures are additive or not.

One thing we can say, however: the world of chemical
carcinogenesis has expanded substantially in recent decades
and the issues have become more and more complex; consequently
differences of opinion abound. Today, honest differences of
opinion are too often labeled by one or another party-at-
interest as chauvinistic. The human mind prefers simplistic
problems and frequently prefers to think in terms of simplistic solutions, and thus polarization of viewpoints at a
meeting such as this usually becomes evident.

A regulatory agency such as FDA is frequently caught in
the middle of a "damned if you do, damned if you don't"
situation. On the one hand, no responsible person would
advocate permitting or continuing to permit dangerous exposures that might operate to substantially increase the risk
for cancer. To require absolutely complete evidence might
well delay appropriate regulatory action to protect public
health. On the other hand, responsible parties can and do
differ from time to time about the quality and the amount of
the evidence, the dose-response relationships involved, and
their estimates of the degree of danger potentially presented
by certain exposures to particular substances as compared to
other substances. Thus, when FDA takes regulatory action, it
is not infrequent that some advocates believe FDA did not move
fast enough while other parties at interest think FDA has
moved too fast or too far. One thing is evident: there is
no consensus either on the scientific side or the legal side.

Here is a brief review of some of the major points of
scientific controversy: What is the scientific definition of
a carcinogen? Is this coincidental with the same considerations involved with "inducing cancer"? Do we mean "cause,"
"potentially cause," "elicit," "lead into," "bring about,"
"effect," or "affect" cancer incidence? Said a different
way: does any exposure to any substance that can modify the
expression of tumors in test animals upwards make that substance a "carcinogen," or more specifically constitute sufficient evidence of "found to induce cancer"? Should we
differentiate factors that appear to influence the expression
of cancer when fed to animals at high dose levels but that do
not appear to be "initiators" of cancer in terms of current

experience from other agents more likely to act as tumor promoters? Does any factor that contributes to an increase in the incidence of cancer in animals mediated through a discernible toxicity pattern of intermediate phases of hyperplasia and metaplasia of epithelial or glandular cells constitute the same degree of risk at low-level exposures as another factor without apparent and predictable intervening toxicity? If the answer to that question is yes, then just about every hormone and hormonally active substance is likely to be deemed a carcinogen, because giving these substances at higher levels results in hyperstimulation and distortion of complex endocrine interrelationships, affecting tumor incidence.

What is a noncarcinogen? What level of proof is required to make a finding of carcinogenic vs. noncarcinogenic? What are the most appropriate ways of testing for carcinogenic potential? Is a particular strain or species of rodent always appropriate and relevant or are some more relevant than others as far as certain considerations are involved?

What dosage levels are most appropriate for which circumstances of anticipated human exposure? Since not only the levels and numbers of doses given, but also the intervals among graduated doses may affect the results, should all dosing follow a similar schedule or should we adjust doses according to specific criteria? If we use a maximum tolerated dose philosophy, how should we determine what the maximum tolerated dose should be under particular circumstances such as overt toxicity discernible by histopathological examination?

How much evidence should be required of a manufacturer to prove safety by demonstrating "noncarcinogenicity" for low-level uses where anticipated human exposures are very low and infrequent as compared to higher-level exposures likely to be substantial as far as some individuals are concerned? What criteria should be applied to determine whether or not the design and execution of the experiments were satisfactory? What criteria should be used to evaluate the significance of the experimental data? What constitutes a positive result? What constitutes a "clear negative" from a "suspect" outcome? What happens if the results of one experiment are at variance with other experiments on the same compound? How many positives or how many negatives constitute clear proof and how many negatives does it take to cancel out a positive result, if ever? Or is a substance always to remain "guilty" because of a positive outcome in a particular experiment? What about the categories between clear-cut positives and negatives? Often, the evidence is not clearly positive or negative; yet some positivity or suspected positivity can be interpreted

from some comparisons of selected parts of the data, such as
trends with marginal statistical significance, or statistically significant data that are fragile from the viewpoint
that a very small change in the numbers would completely
alter the results, or situations where there is some statistical increase in incidence as compared to particular controls
but not compared to historical controls, or where the findings lack biological coherence.

In animal experiments, some of the untreated control
animals usually will have tumors as a so-called spontaneous
phenomenon. One wonders why these particular animals are
susceptible to tumors under control conditions, but few satisfactory explanations are available. As far as treated
animals are concerned, the fact that some animals may have
tumors and others may not also points to factors influencing
the susceptibility of a particular species under specific
test conditions. In general, one can assume that the higher
the dose of a carcinogen, the higher the incidence of tumors
and in some instances the earlier the appearance of tumors.
Conversely, the lower the dose, the fewer the animals affected. As dose decreases, only the more susceptible animals will
be adversely affected. At some point, one could question
whether or not these animals who develop tumors in the lower
dose ranges do so because of the "triggering" effect of exposure to the test substance or in the alternative because
they are already hypersusceptible due to intrinsic factors
such as genetics and metabolism or extrinsic factors such as
exposure to toxic insults, nutritional deficiencies, or cofactors like viruses that act singly or in combination to decrease biological resistance and increase susceptibility.
One can also hypothesize that such factors may operate in the
etiology of many human cancers. My point here is that at
some point for each substance potentially capable of inducing
cancer, we must assess the vectorial strength of the exposure
to the "carcinogen" itself in comparison with the vectorial
strength of other independent factors (many of which are not
likely to be carcinogenic per se) that increase susceptibility of individual animals to cancer induction. In all of
the many bioassay-type experiments to detect carcinogens, one
rarely hears about meaningful efforts to compare susceptible
animals to resistant animals on a biological basis. Usually
one hears only about the gross and microscopic findings with
respect to tumor types, but rarely about any other comparisons
such as gastrointestinal microbiological flora, biochemical
parameters of metabolism, nutritional parameters, endocrine
status, enterohepatic interactions, liver metabolism, altera-
tion of enzymes, or excretion of biologically active metabolites that may or may not have mutagenic activity. I merely

suggest that if some of these parameters were explored more
systematically, we might learn much more about the cofactors
that may act more generally to influence animal and human
susceptibility to cancer. Some of these cofactors may be
stronger risk factors for cancer induction in humans than are
infrequent, low-level intermittent exposures to some of the
weaker initiators. We might recall that one of the most
important causes of cancer in humans--cigarette smoking--
involves very substantial exposures to a spectrum of substances that have a variety of toxic, initiating, and promoting capabilities. The toxic and promoting substances in
tobacco tar and gas appear to be very important factors in
the etiology of the smoking-related cancers in humans, as
well as the initiators.

In making a transition to some of the legal considerations
that will follow shortly, I think it important to remind this
audience that the questions posed earlier in this presentation
are very important regardless of the Delaney Clause because
we in FDA and EPA are faced with the problem of assessing the
degree of risk involved with human exposure to food already
contaminated with agents known or suspected to influence the
incidence of tumors in animals or in humans. For example,
we in FDA face this problem with respect to aflatoxins in
foods where the contamination can and frequently does occur
in corn while it is growing in the field, although certainly
such contamination can also occur under storage conditions.
We have also faced similar problems in relation to the massive
federal efforts to reduce exposures to PCBs, which most
recently have been considered to be tumor promoting agents by
some of the reviewers at the NCI Clearinghouse for Environmental Carcinogens.

In passing, it should be recalled that the Delaney Clause
does not apply to contaminants in food considered under
Section 406. If a Delaney Clause philosophy were to be
applied to environmental contaminants already in food, then
substantial portions of the food supply of the United States
would be deemed illegal by operation of law.

Some of the disparities of our societal approaches to
cancer prevention lead to confusion among lay people and
scientists alike. From my viewpoint, it appears to hinge on
the degree to which an individual can avoid exposures deemed
undesirable. As a society, we depend on the available food
supply in its original "raw" state (farm produce and food
animals) and can afford only a moderate degree of selection
when it comes to deciding which food per se is unsafe to eat.
If we act to ban food from consumption, there should be a good
reason: that the degree of risk is unacceptable. When we
come to the question of how food is processed, we have more

choices as a society. Here we can set up certain rules about
what substances will be considered as food additives and
whether or not such additives will be permitted, but my point
here is that Congress makes the law and we enforce it.
Enforcement may not be easy under some circumstances because
difficulty may occur when interpreting and applying statutes
to particular problems where several alternative interpretations might have some validity. That is the business of administrative law and why regulatory agencies exist. When
such regulatory problems simultaneously involve interpreting
the leading edges of science where scientific fact may be in
dispute, the problems compound themselves. It would seem to
me that the greatest problem we face in terms of regulating
potential carcinogens in the food supply is scientific and
not legal and until Congress and the scientific community
direct meaningful research to produce answers that can be
agreed to generally by the scientific community, the situation
will remain in a status quo of considerable scientific limbo.

Some clarifications could, however, be considered from
the administrative law viewpoint. For example, because the
statutory language has created an exception to the Delaney
Clause for additives and drugs used in food-producing animals,
FDA has published "Criteria and Procedures for Evaluating
Assays for Carcinogenic Residues" in the *Federal Register* of
February 22, 1977. While this document is in legal dispute,
it represents an important attempt to clarify the ground
rules involved in interpreting scientific and legal considerations.

It might be in order to consider publishing proposed
scientific and procedural criteria concerning interpretation
of the Delaney Clause for food and color additives in the
Federal Register also, the purpose being to clarify interpretation of statutory language more specifically in regard
to particular situations and to establish the scientific
parameters involved with respect to the evaluation of scientific evidence used for decision-making. For example: How
broad or narrow should the definition of "induce" be? What
scientific criteria should be applied to the definition of
cancer in terms of differentiating "noncancer" pathology and
how should statistical procedures be applied to the data?
Which tests when positive are deemed appropriate for fulfilling minimum conditions prerequisite to a finding that a substance induced cancer? What are the criteria for the minimum
conditions of such a finding? What constitutes primary evidence prerequisite to such a finding as compared to secondary
evidence that may be used to elaborate or substantiate a finding? How should "suspect" compounds be treated? What standard of proof should be required as a matter of administrative

law before action is initiated to protect humans from undesirable potential exposures: some substantial evidence, a preponderance of the scientific evidence in perspective, the most recent scientific evidence, or evidence beyond a reasonable doubt? Which scientific groups should be empowered to evaluate the data from a regulatory viewpoint and decide whether the scientific evidence is sufficient to support a regulatory finding of cancer induction? What happens if another august body of "experts" renders an opposite opinion? Who is most qualified to judge--a court of law? What rights do concerned parties at interest have to submit additional evidence, seek clarifications, and dispute conclusions at the review group level? What rights of appeal exist before administrative action is initiated by a regulatory agency? What about "pretrial publicity" when a nonregulatory agency issues certain findings and the public media get into the act? How can objectivity of decision-making be maintained under the variety of pressures generated when interested parties and advocates interact if the general milieu is unduly alarmist?

Having participated in decision-making and also in preparing criteria for decision-making, I can only say that considerable effort would be required to develop comprehensive written procedural and scientific criteria for interpretation of the Delaney Clause, and any proposal would immediately be challenged by some party at interest. Arriving at a societal consensus in the midst of many uncertainties would be difficult to attain at best, but maybe the attempt should be made.

There are those that suggest that a risk-benefit approach be taken with respect to additives to food when such additives have been shown to be potentially carcinogenic. While toxicologists know that decreases of dose facilitate biological tolerance to many substances, and cancer experts agree that risk increases with increases of dose, very few people are willing to say minimal exposures are associated with negligible risk as far as carcinogens are concerned. There is argument over what constitutes a "minimal" exposure or a "negligible" risk. Concern arises because of the identification of some powerful carcinogens such as aflatoxin that are effective when fed to animals in the low parts per billion range, a growing linkage between electrophilic carcinogens and mutagenicity, and the kinetics of radiation-induced cancer. Mathematical models have been generated that assume that the biological mechanisms involved with observable tumor induction in animals at relatively high daily doses over a lifetime are similar to those operating at much lower doses of lesser frequency and shorter duration. Estimates of risk

generated by some of these statistical models project that
fewer cases will occur in a population as dose is decreased,
but at no time will they go so far as to predict a zero
incidence.

Because the public at large has great difficulty in
understanding comparative risks in perspective, the fact that
some theoretical increase in risk for cancer can be described
in finite numbers from extrapolation models mitigates against
agreement on an acceptable level of risk that would be fully
supported by a societal consensus. Benefits are similarly
difficult to quantitate at times, and some advocates are
likely to present them to the uninitiated in terms of dollars
of profit vs. some theoretical deaths from cancer. Thus a
societal consensus about appropriate parameters for a risk-
benefit equation would also be difficult to attain. Many of
these issues are presented to the public in oversimplified
form, and the public is not likely to appreciate that the
real issues go way beyond simplistic descriptions. The real
issues involve how society allocates resources to feed itself
and to live and work, while minimizing undue risks for cancer
or any other health hazard. Our resources of food, person
power, and money have limitations from an individual, family,
or societal viewpoint. If we overreact, we pay one price,
and if we underreact, we may pay another. Ultimately, the
choice should be public's.

We should concentrate our efforts to define the issues
more descriptively from both a scientific and a legal viewpoint in terms that are understandable by the public at large
and let them decide. There is a good management rule of
avoiding unnecessary surprises by anticipating problems and
coordinating communications and remedial efforts. There is
a societal need to improve how we manage our national resources to ensure our national health and welfare.

Hopefully there is a way to subject new developments to
a rational procedural and scientific analysis that will ensure improved awareness and understanding of the issues,
risks, and options involved so that we as a society can react
to "surprises" more constructively and make appropriate
decisions about remedial efforts with greater deliberation
of the issues without undue haste or delay.

GENERAL DISCUSSION

DR. WYNDER: I notice that one group has been missed. I do not see anywhere on the program the contribution that the epidemiologist could make to this debate. Now I recognize that the Delaney Clause specifically does not talk about human evidence, but it would seem to me that there are a number of agents that have been around long enough, so that the epidemiologist can use his ability and art to determine whether we have the experience, looking retrospectively, that a Delaney Clause decision 30 or 40 years ago could have, in fact, led to a benefit for man. Because let's face it, whatever we do, finally, it must be an epidemiological interpretation to determine whether our decisions have benefitted man or not.

DR. COULSTON: Dr. Kraybill, I think someone else mentioned the need for looking at multiple factors and presumably multiple exposures and synergism. I wonder to what extent has true synergism been demonstrated among carcinogens and has it been sufficiently frequent to suggest a serious problem?

DR. KRAYBILL: I believe there are a lot of other people here who are more competent to speak about the issue of promoters, particularly Dr. Clayson and Dr. Lijinsky, but I recall in identifying some of the water contaminants, we characterized some 27 promoters in the aquatic system. They are hydro-xyaromatic compounds and I do not have a list of those here, but we have them in our publication.

The other examples that come to mind are perhaps benzo-pyrene and sulfur dioxide. I refer to the work of Shabad with asbestos coated with benzopyrene. I guess in this context one might consider asbestos and cigarette smoke, the two together causing cancer in man. Also, Dr. Kolbye mentioned a while ago something about the arochlors being potential promoters.

Some people have even thought that perhaps DDT or DDE might be considered in that context a promoter because it was a carcinogen in the mouse, not the rat, but what is the effect of contaminants in the diet in the presence of DDT?

Perhaps the problem of dose dependency with certain estrogens might be put in this context. I think this was alluded to by Dr. Kolbye. Are estrogens at the appropriate level in the milieu acting in a homeostatic mechanism as a hazard, but at higher doses can estrogens act as promoters in combination with other chemicals?

ISBN 0-12-192750-4

Last, but not least, I think we have a whole array of situations to consider where we think of a semisynthetic diet vs. a lab chow, because we get apparently a different tumor incidence. We see this amplified considerably and one might ask the question how much do the contaminants in our lab chow have to do with their impact on the particular chemical that we are testing?

DR. LIJINSKY: There have been unfortunately very few tests of synergism by a group of carcinogens. We have enough trouble, as Dr. Kraybill verified, testing individual compounds. The logistics of testing combinations of compounds at low levels are very, very great. I do not think there is enough money or pathological expertise to do very much of this kind of research, but what evidence we have shows that there can be synergism and there can be inhibition. The problem we are faced with when evaluating the risk in human exposure to a compound that is identified as a carcinogen by animal experiments against a background of other carcinogens is that we do not know what factor to put on it. And, I think that in the interest of common sense, we have to assume that there will be synergism. That is a positive effect rather than a negative effect. You cannot assume a negative effect and claim we are protecting the public health.

DR. CLAYSON: I take a slightly more optimistic view than Dr. Lijinsky. I prefer not just to talk about synergism between carcinogens but interactions between environmental factors other than carcinogens. We have, as far as our literature is concerned, I think very much more frequently turned out negative associations. These factors have been inhibitory, rather than positive associations. Although like Dr. Lijinsky, we have got to keep our eyes and ears very wide open, if we are not to miss something very dreadful. But I think this question in fact turns very strongly on one that Dr. Kolbye asked us and that is what we really mean by a carcinogen. Do we mean any agent that increases the tumors in a population? Would we therefore say that an excess fat diet as shown by Al Tannenbaum many years ago is a carcinogen? Or do we want to have a more restrictive viewpoint? To summarize my feelings, I think as we find out more about the mechanisms of carcinogenesis, we shall find it much, much easier to classify these things and possibly to classify them so that we can say when the regulations appear, we do not risk anything with that chemical. It may be that the element of risk is far, far different.

DR. PARKE: I'd like to comment on Dr. Kraybill's remarks
about the need to adhere to accepted toxicological principles
and consider metabolism and kinetics.   Dr. Korte referred to
this partly this morning when he said that, on repeated dosage
or following enzyme induction, different pathways of metabolism could be affected to different extents.   And, finally,
Dr. Kolbye's remarks that the validity of scientific evidence
was very important in decision-making.   It is now generally
realized that the enzyme system in all living organisms,
which protects us against toxic chemicals, detoxicates acutely
toxic chemicals and is the very same enzyme system that activates carcinogens.

This is one of the most profound anachronisms of toxicology.   To put it simply, we have an enzyme system that if
it acts one way detoxicates or protects us and if it acts
another way, activates carcinogens.   Now this enzyme system
is about 10-20 times more active in the mouse and the rat,
the rodent.   So, when we are validating acutely toxic chemicals, the mouse is metabolizing, deactivating these about
10-20 times the rate that man or subhuman primates do.
Therefore, in order to get data that are really valid for
man, we have got to give these small rodents at least 10
times the dose.

On the other hand when we are evaluating carcinogens,
exactly the converse applies.   If we give them the same dose
as man, we are actually getting ten times the carcinogenic
effect.   So, we are using a telescope in toxicology: we look
through it one way when we are looking at the acute toxicity
of chemicals, and we look through it the other way when we
are looking at carcinogenicity.   I think very few of us realize we are doing this.   Certainly there are a number of
friends and colleagues like Dr. Coulston, who a few weeks ago
with me looked through an enormous amount of data on a very
large number of chemicals.   These data came from all over the
world and I looked through most of them to see if any company
or National Institutes of Cancer were validating their toxicology this way.   Not one of them was.

I think this is an enormous anachronism that most of us
have not realized.

DR. WYNDER: You really have, it seems to me, four issues.
One relates to what Dr. Parke just said, the metabolic difference between one species and another.   Another one to which
you referred is the type of tumor.   Does indeed hepatoma in a
mouse relate to squamous cell cancer in man?   Another one relates to site of application.   We, of course, tend to make it
easy on ourselves, but is it true indeed that ingestion of a
compound relates to topical application, which is perhaps one

of the more common ways in which we expose ourselves? Of course a final and major issue to which you referred relates to the dose. Does a high-dose, short-term time exposure relate to a low-dose, long-term time exposure? It seems to me this is the crux. And I like to think we can probably argue these points to doomsday and not be better off than we are now, because this is not the first conference that I've attended on this subject.

Therefore, I had hoped there would be some epidemiologists here who could debate with us whether some of these long-term, long-time exposures are safe or not. We have been exposed, as you know, to some sweeteners, hair dyes, and decaffeinated coffee for many, many years. And, it would seem to me that we ought to have the epidemiologic ability to determine whether indeed we have been at an increased risk.

DR. ED EGAN (NIOSH): I think this is a very interesting aspect of carcinogenesis, because when we talk about carcinogenesis in man we are talking about carcinogens applied to what we believe are normal men or women under normal circumstances. We have a project underway that will be of interest to you, conducted by Dr. Plotnik in our Cincinnati branch in which he subjected a group of rats to the inhalation of EDB. Ethylene dibromide is a normal additive to gasoline and he produced very few tumors at the acceptable threshold level. Then he added disulfiram, which some of you will recognize as anti-alcohol, a common treatment for alcohol. He virtually got an epidemic of tumors, so much so that we have put out a mini-alert to all people who are producing EDB and advising them to tell their workers who are on disulfiram either to come off the drug or give up their work.

DR. COULSTON: Just a comment. First of all I am very suspicious of people, when they start using this word synergism. In my area of expertise, I do not know of a single example pharmacologically or toxicologically of an unexpected tremendous increase of some kind of activity, which we would call synergism. Perhaps, you mean potentiation or additive effects, which is quite understandable. But the term synergism is tricky. I do not want to see it creep into our lexicon here.

Now, I just want to say this about Dr. Kraybill's statements. I do not know whether the audience really realizes and understands what he said. He said some of the most amazing and important things that I think will come out of this entire meeting. He pointed out that the current testing procedure, not only by the National Cancer Institute, but that followed by almost all regulatory agencies in this country, is done now by almost everybody, even the group at Lyon. This test uses the maximum tolerated dose and half of the tolerated dose, two dose levels and a control group. He showed a slide that really is amazing, in which one of these tests done by the NCI used the equivalent of the total amount of that chemical manufactured in one year to do one test on mice to see whether it was a carcinogen. My goodness, this shows how safe that compound was. When you get up to 10,000 ppm, how ridiculous is such a kind of procedure. If that is what has to be done as a test procedure, we had better evaluate it in terms of a reasonable dose relation to the intended use of the chemical.

Now I want to make this point very clearly, his data showed that we need a reconsideration of the testing procedure. We have heard from many other people that there are better ways to do carcinogenic tests than to take this heroic dose, a maximum tolerated dose, which is probably an LSD whatever way you look at it. A dose should be established that does not change the normal physiology of that animal as a low dose, and then use a higher dose to establish the safety factor. Three dose levels would be best. Anyway by using almost 15-20% in the animal diet, nutritional deficiencies are certain to occur.

DR. KENSLER (Cambridge): I think we should point out that there is a dose/response relationship in terms of cocarcinogenesis as well.

But on the other side, I would like to support what Dr. Clayson said that there have been quite a number of examples of inhibitory effects. One of the early classic ones was the work of Richardson, where he found that the polynuclear aromatic hydrocarbons prevented the carcinogenic action of azo dyes on the liver. And this, in fact, opened up the whole microsomal enzyme area. Another example would be the effect of BHT on inhibiting acetylaminofluorene carcinogenesis. So this is not just a one-way street, and there are fortunately apparently things in our environment that are acting in our favor.

DR. KRAYBILL: Well I appreciate the laudatory remarks made by Dr. Coulston but I would hasten to add that toxicologists, I believe, have improved in their methodological approaches in recent years. Some of the data that I showed were gleaned from the earlier days when we did a 6-week range finding. We found in many instances that this will not work, and we have now gone to 3 months' study, at least as I indicated, and sometimes we have to go to 6 months of dose ranging.

This so-called subacute or subchronic testing now has been buttressed by metabolic studies getting at the body burden, if possible, the pharmacokinetics. Some of the chemicals are now tested at NIEHS to get a good handle on their pharmacodynamics and pharmacokinetics. I think we have a better handle at least on what the dose will be, for example, for a dioxin or some organic chloride compound, so that we can start the animals on that test dose and hope that we get what I call good performance: so that those animals can be carried through for 2 years and prove our neoplasia without having the interfacing of overtoxicity.

DR. KENSLER: I wanted to thank Dr. Kraybill for leaving something for me to talk about tomorrow but I also meant to mention that, with respect to the clearing house and PCBs where Henry Peto was pushing to call them cocarcinogens, there is great disagreement as to whether this was legitimate. And the compromise motion that was passed said that essentially it is possible that it is a cocarcinogen for man. Since that meeting, which I think was last November, there has been a paper published showing that arochlor inhibits aflatoxin carcinogenesis in the trout, for example.

DR. WYNDER: Going on to the second paper, cost:benefit. As Dr. Darby said, whose cost and whose benefit, and we could argue this forever. But clearly, when we look at a substance like aflatoxin, where we have evidence that at one part per billion it will produce hepatomas in a certain species, and if we were that strict, there are certain foods in particular peanuts that we would be unlikely to eat and I suppose we would ban peanuts. We would hear from certain segments of the city. Clearly, we continue to constantly make cost:benefit decisions.

It is obviously an area that we could discuss for weeks, but I wonder whether any of you have some constructive questions for Dr. Darby, and perhaps whether anybody has a secret formula where you could equate a certain number for cost, a certain number for benefit, and come up with an answer that society could live with. Would anyone like to take on this very difficult subject matter?

DR. MRAK: How serious is this modeling system you are talking about? This is something new to me. I was astounded that the computer is putting the population out of business.

DR. DARBY: I do not know how serious it is, but I think only
time will tell us.   I am concerned, but let us take this
subject that Dr. Wynder just raised: aflatoxin.   Well if you
are talking about aflatoxin in peanuts that are eaten at a
cocktail party that's one thing.   But when the recognition of
the aflatoxin problem arose I happened to have been chairman
of the protein advisory group for WHO/FAO UNICEF.   Now immediately a question came up.   What do we do about the use of
peanuts in the formulations that are being prepared and distributed in the countries where babies die, 50% of the babies
born die by the age of 4 or 5 years of age of malnutrition
and feeding of peanuts is a life-saving measure?   There is no
computer that can do this.   I think these are human judgments,
and we cannot put in a formula what is going to tell us the
answer to these things.   That is why I maintain you have to
consider the culture, the socioeconomic situation, the stage
of cultural and human development, the alternatives.   There
was not any alternative at that moment that you could do anything about in a place like Nigeria.

    Even if you prepared something else, suppose you made
fishmeal or fish flour, it would not have been eaten.   So you
have to understand what use you are making of these foods,
and then make your decision as to the benefit and the risks
and not just cost.   I object to using cost rather than risk,
in this instance, because cost implies only the economic cost
to people and it's not just that.   It is the usefulness of
the product to mankind.

DR. WYNDER: We are still on the subject of risk:benefit.   It
is really the kind of thing that our legislators have to come
up to grips with.   What kind of advice do we want to give
them?   Do we merely want them to depend on their own consti-
tuency or what other formula can we give to them?

DR. UPHOLT: I would say that in EPA the agency has been considering cost:benefit, not as a formula or a mathematical
equation, but rather simply as a method, as you might say, a
checklist to make sure that we have in fact considered all of
the benefits to be derived from the pesticide or whatever else
the substance is as well as the risks involved.   We try to do
this in as much of a quantitative fashion, not a numerical
fashion--I hope you get the distinction between quantitative
and numerical--that permits actual balancing of dollars or
human lives against something of this sort, but rather to
provide to the decision-maker, a reasonably complete and
understandable statement of what the cost to society will be,
if he reduces further the level of exposure to a certain
carcinogen, in order hopefully to reduce the risk of carcinogenicity.

DR. DARBY: I think that that is the objective. While it is
true that they regard this as an index, if you will, as soon
as one says that if we make these assumptions based upon animal data, the maximum number of cancer cases that occur in
the human population is so and so, then this is immediately
translated to the public as a figure because the public cannot
understand that this is an index. One of our problems, and
that is why I was quoting C. P. Snow, is that we must try to
get the public to understand not the quantitative aspects of
this but the qualitative aspects: what the values are and
what these mean to them. Why, with what effects, and for
what benefits, particular risks are assumed. As Sam Stumpf
has said and is pointing out in a rather scholarly article
that will appear in the March *Bioscience*, if we start at the
most basic of the values that we are concerned with, the first
one is life itself; then survival; then we move up through
the prevention or cure of malnutrition; and we eventually get
to a point of the regular availability of foods: the erasure,
if you will, of seasonable products, and the appeal of those
foods through color and other properties that may make the
food acceptable where it otherwise would not be, something
that is highly desirable on the part of the individual although it may have no particular health benefit whatsoever.
However, as I cited earlier, we have good evidence that cigarette smoking is not beneficial, and yet people assume this
risk because of the pleasure, the quality of life that it
gives them from their own sense of value judgment.

So this is one of the problems that we have I think with
the model system. I was before one of two or three committees
on the saccharin question, and I heard quotations of how many
bladder tumors would be prevented in the United States by removing saccharin from the diet. I do not think anybody can
possibly make that prediction with any assurance whatsoever.
That is the maximum amount of tumors that could occur based
upon certain assumptions. In fact, I feel uncomfortable about
making such assumptions.

DR. HOLLIS: Is it possible, in consideration of the use of
some useful chemicals like pesticides in the production of
food to consider, and I say consider, their use on the basis
that food is the only indispensable product produced by man,
and so is unique in this regard? Should we consider instead of risk:benefit the risk vs. risk? In the case of alfa-
toxin the risk of using an insecticide that will control the
insect that opens the pericarp that allow the fungal spores
to get into the seed and produce the toxin. There are many
instances here where we might be able to overcome some of the
problems that occur naturally with a man-made product, even
though it has a risk attached to it.

The risk vs. risk seems to put apples and apples and oranges and oranges together better than risk vs. benefit does.

DR. WYNDER: This is a good point and we now go on to the last of the subjects; namely, that presented by Dr. Kolbye. A regulatory agency does not make laws, but can only interpret existing laws and regulations. We must not forget that. The EPA and the FDA probably cannot do better than whatever we in our wisdom suggest they can do with existing laws and regulations.

DR. COULSTON: I was very intrigued of course with Dr. Kolbye's paper and we could spend indeed many hours discussing it, but one thing he said hits me so hard as a toxicologist: You can do negative test after negative test, you can do 20 of them. Then one scientist does a test that is a positive and right away the regulatory agencies will react and say that a risk exists. Even if you did 100 negative tests there is still always the question that there is a risk, and sooner or later someone will design a positive test. What is the point then of establishing safety? I sat not so many years ago on a committee discussing the safety of NTA, where at least 18 tests were done to show the negativity of teratogenesis for NTA. And then one test was done, which happened to be by a governmental agency, where they showed it was a teratogen. On that basis, I believe alone, the EPA never allowed NTA to be used in this country. How safe is safety! NTA is used in Canada very successfully; it is used in other areas of the world. I would like to hear Dr. Kolbye discuss how you get this problem turned around, or do you ever?

DR. KOLBYE: I am reminded of a conversation with a few scientists, a few years ago on the mathematical extrapolation models, modified Mantel Bryan, where the suggestion was made that negative data ought to be fed into that model because we scientists are not sure, that we failed to detect a true positive and that we should develop a model for the negatives and then apply some huge mathematical factor to come out with permitted levels. That carries some absurdities, at least to my way of thinking, even one step beyond where it tends to be today. I don't know the answer to your question Dr. Coulston. I was asking the same question. I think it is up to society to start developing some tentative answers in a way that will have effectiveness in terms of influencing how we make decisions in this country.

Simply as a footnote, I would suggest that the various parties at interest, be they scientists, companies, or consumer advocates, ought somehow to practice getting together and coming up with some criteria. You have to remember you cannot have the criteria both ways. You cannot have criteria that say that there "ain't" a problem, to use the vernacular, when you don't want a problem, but that there is a problem when you feel, that it does not gore my ox, so therefore we will let it be a problem. That takes effort, precision of language, precision of thinking, and a determination to assist regulatory agencies in decision-making.

If you people do not do it to help us, we are left with doing it ourselves and do not complain. I realize that sounds like a harsh phrase, but within FDA we have very limited manpower and I am not very confident that, even if the Commissioner were to support an effort to develop criteria that would impart or address your contention, Dr. Coulston, the result would be a high quality product that we could live with as a society. We need interaction among scientists here.

DR. MRAK: This morning I was inspired to make notes of the factors that might be involved in making a new policy so I started writing them down. I did not get them all but I got to 71 different points and issues that probably should be considered: 71. So the next thing I wanted to say is Dr. Kolbye, I did not hear you say anything about epidemiology, or did I?

DR. KOLBYE: Well, I use epidemiology to try to create a perspective for certain decision-making in FDA and I might simply remind us of a very interesting epidemiological experiment that is going to get underway in the not too distant future in cooperation with the National Cancer Institute with respect to saccharin. NCI will probably be relating to a fair number of outsiders on that, Dr. Mrak, and I am sure Dr. Wynder will be carrying some of his work forward and that other investigators will be carrying their work forward. That is going to be an interesting problem, because we are going to see whether or not saccharin potentially can increase the human risk for bladder cancer. Hopefully we shall see.

DR. LIJINSKY: To supplement you, Dr. Kolbye, I hope that the search for epidemiological evidence on the carcinogenic effect of saccharin will not be restricted to the bladder. There is so much species variation in response to carcinogens, I think it would be a gross error to only look at the incidence of bladder tumors in man. If saccharin, for example, has the colon as the target organ it would make the epidemiologist's job very much harder, but it is nevertheless worth doing.

DR. KOLBYE: I am aware of that sentiment. However, at the
moment, I think the emphasis is going to be on the main
question raised thus far. I am aware of the evidence in
general that there is tissue specificity depending on the
enzyme inducibility that varies species to species with respect to a particular carcinogen.

DR. WYNDER: May I say on this that we at the American Health
Foundation are looking at the saccharin question for all
major types of human cancer and I, Dr. Kolbye, offer you our
continued assistance in answering this important epidemiological question.   Incidentally, I read for the first time in
The New York Times that the FDA was doing a study in this
regard. I would hope, since I have been involved in this for
many, many months, that in the planning of this we could
certainly come to you with the ongoing study and perhaps we
could gain from one another.

DR. BLAIR: I would like to address one aspect on this regulatory agency activity and that is the problems associated
with one agency vs. another. As we go into policy-making, as
we relate data into all types of public hearings and scientific bodies, I would like to discuss very briefly benzene,
where there might be some concern as to whether it is or is
not a leukemic agent. When OSHA first gets around to attempting to establish a regulation under an emergency standard,
because something all of a sudden seems to be apparent, then
the rule-making really comes in and the level is set at one
part per million in a chemical plant down to a bulkhead storage station, but at a gasoline pump there is no rule at all.
    I think that these are the problems that society is
looking at. As we talk about scientific data and regulations,
as they are brought out into the public arena, there really
is no reason why OSHA decided one way or the other, other than
maybe the fact that they felt they could not get by with it.

DR. WYNDER: Are there any other comments? If not may I say
that in this study on saccharin that the FDA and NCI are doing, we are spending 1.3 million dollars in public funds to
try to answer this particular question as to cancer.
    As I look at this audience, clearly no one here and no
one in the country wants to see one single cancer occurring
that we know to be preventable. The question that we ask
ourselves is: At what price to society? And particularly,
and this is of great concern to me, with what assurance that
whatever decisions we make today will we be thanked by the
coming generations? Therefore, I would like to come back to

the point that we must utilize more the science of epidemiology, particularly in those areas where an answer could be given, because a given agent has been around for many years.

And I think we also need to consider the amount of spending that we have on these subjects in terms of expenditures of scientific funding and in terms of personnel that gets involved in these particular questions. I really, often, remind myself when I attend meetings, and I am as interested in the Delaney Clause and the occupational carcinogens as anyone here, if I were going to hold a meeting here on the question of tobacco carcinogenesis, and all of us have known for many years that tobacco contributes to more cancer death in man than any other factor that we know of, I would be lucky to have eight or ten people here.

And I ask myself whether we as scientists should not put more of our attention on those factors that really contribute to most of the cancers. Obviously we have got to be of some help to the regulatory agencies, which are practically swamped with one positive animal test after another. Sometimes I ask myself whether we do not spend too much of our regulatory activity on agents that perhaps do not produce cancer at all!

I was riding in a cab the other day. Much of my knowledge that I have in science I learn from the local cab drivers in New York City, who incidentally work without a government grant. And, as this fellow was puffing away on a cigarette, I was listening to one of the more recent pronouncements from the EPA. I do not know whether the radio station was correct, but this local announcement just said that we had done another really important public health service to the American people, because we had just removed, really by new regulation, all carcinogens from the water supply by forcing all cities of over 75,000 people to put in a carbon filtering system at the cost of about $460 million plus or minus. The cab driver was saying, "You see my government protects me, and, furthermore, we are really living in a sea of carcinogens, so I might as well continue to smoke."

We are also putting a tremendous amount of pressure on the Congress. The Congress obviously has to come to grips with these various environmental carcinogens to which we expose them. I do not have to tell you that Congressmen are elected for two years and in these two years we want them to learn about energy controls, air pollution, the Delaney Clause, and the Food, Drug and Cosmetic Act. It is a very tough thing to do!

Finally, of course we have the subject of the public press. Now, I was recently interviewed by one of the major television stations and they wanted to know whether or not they should have a program on smoking and health. When we were all finished, they said we should not have a program because after all this is not news.

What would be news is another dye or something found to be carcinogenic. This, indeed, is news! Indeed, if I were the public press, I would also be confused by what an environmental carcinogen really is, because there are new ones appearing all the time. I say again it is finally up to the epidemiologist to clear up this particular debacle for the American people.

Now I hope when we leave here, we and the organizers of this conference will come up with one or two specific recommendations so if I were a regulatory agency, if I were the head of OSHA or NIOSH or the NCI, I could take that back with me. Because just another conference will, I think, not clear up the air. I have heard many of you talk about this thing, and you are convinced in your particular area. It seems to me that what a democracy is all about is that we can convince the majority within the Congress to make the right kind of decisions, those which are best for the majority of the American people. I hope we can come up with some clear suggestions, not a whole book full, but two or three that our Congress can understand, that our Congress can implement. I am looking forward to these continued deliberations, because, as all of you know, we have felt for many many years that the science of epidemiology makes it clear that most human cancers relate to environmental causes, and it seems to me that no environmental cause identified to be linked to cancer should remain in our environment. I am convinced that if we are ever going to make headway in the cancer field, it will only be made through primary prevention.

SESSION II

INTRODUCTORY REMARKS
(SESSION II, MORNING)

DR. COULSTON: The first part of this meeting should be not
only stimulating to the group, but should be a lot of fun.
It was in June last year that the *Chemical Engineering News*
decided to have a forum on the Delaney Clause. I think I
mentioned they invited Dr. Lijinsky, Congressman Martin, Dr.
Sidney Wolfe, and myself to participate. We each wrote a
statement without knowing what the other was writing and then
we saw what each other wrote. I must say it was all in good
taste, even though we wrote some strong words, from time to
time. But it was an opportunity for each of us to say something that we have always wanted to say, to put our thoughts
in print with no holds barred. I think for the first time in
my career, I was able to put down all the things I wanted to
say.

Our rebuttals were kind of interesting too, because we
took little cracks at each other. I said some things such as
that Dr. Lijinsky should learn to be a toxicologist, and he
said a few things about me, but the key point is that the
three people represented here this morning are good friends.
There is no animosity, but good camaraderie between us and I
speak very highly of Dr. Lijinsky and Congressman Martin.
The fact that Dr. Lijinsky is here speaks for itself. He had
the courage of his convictions. And that's the key to good
scientific discussion: not to take it personally and not to
react emotionally, but to present the facts as we know and
believe them and then get on with it. That is what you are
going to hear in this session.

Now we have heard a lot about the Delaney Clause and we
have heard from Mr. Hutt that it's meaningless. Mr. Hutt
said very carefully yesterday that the Delaney Clause does
not really matter, the FDA has all the authority it needs
without the Delaney Clause. As a matter of fact, I said the
exact same thing in the *Chemical Engineering News* statement.
The Commissioner and his people did not have to use the
Delaney Clause. All they had to do was delist saccharin as a
generally recognized as safe (GRAS) substance, the same as
they did with cyclamate, etc.

So the point is the Delaney Clause, by the admission of
almost everybody, is meaningless in its use by the regulatory
agency (FDA), but they can use it when they wish. They have
used it very rarely. I said that in my statement in *Chemical
Engineering News*, and again you heard Peter Hutt say the same
thing. In fact it has only been used two or three times for

119

ISBN 0-12-192750-4

minor purposes.   It is the intent and philosophy of the
Delaney Clause that is the proper discussion, and this un-
leashes, if you will, the whole concept of chemical carcinogenesis, the testing of chemicals, the prediction from
animals to man.   That is what we are really here to talk
about.

THE DELANEY CLAUSE: AN IDEA WHOSE TIME
HAS COME FOR A CHANGE

J. G. Martin

The Delaney Clause provides, and I quote the relevant
section, "that no additive shall be deemed to be safe if it
is found to induce cancer when ingested by man or animal."

No additive shall be deemed to be safe and thereby permitted as a food additive, if it is found to induce cancer
when ingested by man or animal. In this day and time when
regulatory agencies are pressed to discharge their responsibilities to a vulnerable public, some 1000 of whom each day
will die of cancer in the United States, faced with a variety
of less certain alternatives for controlling carcinogen
hazards in the food and water supply, in the air we breathe,
in the workplace environment, and in the consumer products
we handle every day, the Delaney Clause sets forth one fundamental principle to the exclusion of all others, it sets for
us an absolute zero tolerance level for any proven carcinogen
that operates upon ingestion--ban it.

Diregard conflicting evidence, it tells us. Disregard
degrees of risk, absolute or relative. Disregard all counter-
vailing benefits, including any risk associated with the removal of the substance from society. Those after all are
merely scientific refinements and the Delaney Clause is policy,
not science, we are told.

So we have a pivotal approach to public policy; it is
direct in its absolutism. The public must not be exposed to
any detectable trace of any food additive that, even if only
upon overwhelmingly massive daily lifetime overdosage, causes
a significant increase in cancer in even one hypersensitive
test animal.

Analytical chemists know that their continuing assault
on the threshold of detection sensitivity reduces that dimin-
ishing tolerance level to anachronistic absurdity. I submit
to you that such a policy is far from trivial and irrelevant,
although I recognize the element of irony in such a characterization as employed by yesterday's keynote speaker.

ISBN 0-12-192750-4

I submit that it is far more pervasive of influence than just a merely "convenient symbol for protagonists." Right now as we here question the validity of this policy for the Food and Drug Administration, other regulatory agencies are considering incorporating this same absolute prohibition into their jurisdiction, and the Occupational Safety and Health Agency proposes to adopt what may well turn out to be just such a regulatory standard.

Furthermore, while it is true that FDA has banned few food additives solely upon the Delaney Clause, and claims it could ban saccharin via other general safety provisions of the food law, it is also unquestionably true that as long as the Delaney absolutism remains legislatively secure, its mandate guides the FDA in its interpretation of its own responsibilities under other provisions of law.

Thus if Congress indeed be the ultimate authority on policy, and if Congress were to repeal or modify this Delaney proviso, surely that change would not be lost on the FDA proceeding under the general safety statutes.

And I might add that Congress of course would be negligent in dealing with this policy if it did not suitably and harmoniously also amend the general safety section as well, which would solve that problem.

After yesterday's discussion it's clear that we must not only adopt a more reasonable, more scientifically valid, more modern carcinogen policy for the Food, Drug and Cosmetic Act, but must extend this same new policy to other regulatory jurisdictions.

Now for historic reference let me describe for you Bill HR-5166, which I have drafted and introduced with 201 cosponsors in the House of Representatives. It begins with the Delaney Clause resting intact with its one current exception for animal feed additives and then proposes to add an additional exception.

If the Secretary of Health, Education and Welfare makes and publishes a finding that based on all the evidence, the benefit to the general public from permitting the use of the additive outweighs any risk to human health that permitting such use may present, then a second exception is made and the additive is allowed.

Public health and nutrition benefits would be weighed against public risks. In evaluating this balance the Secretary would consider a report that he would commission from an advisory committee required in this bill, a committee including toxicologists, nutritionists, economists, lawyers, and representatives of consumers and the food additives industry affected.

The Secretary must (1) evaluate the intake level at which the food additive causes cancer in animals in relation to the reasonably expected intake level of the additive by humans. He must (2) evaluate the quality of any test data and the validity of any tests that may have been performed on the food additive. He must (3) factor in the human epidemiological and exposure data respecting the food additive and statistical data on human consumption of it. He must (4) evaluate any known biological mechanism of carcinogen effect of the food additive. He must (5) evaluate the means available to minimize human exposure to the risks presented by the food additive, and the adequacy of the data available on such means. Finally, the Secretary must evaluate the probable effects of prohibiting the use of the food additive, and evaluate the probable effects of permitting its use, such evaluations to be made in accordance with the following priorities.

First, he must consider health risks and benefits. Second, he must consider nutritional needs and benefits and the effects on the nutritional value, cost, availability and acceptability of food. Third, come environmental effects and, fourth, the other interests of the general public.

With the exception of a few (what I would characterize as anti-intellectual) legislative proposals to ban rat tests and disregard any evidence based on massive dose, which after all may be the only early warning screening test that we have, this bill that I've just described, HR-5166, is the first serious attempt in 20 years to modernize the Delaney Clause.

There's an important lesson for us today in the consideration of the lack of an earlier serious legislative challenge. It's not that no member of Congress ever questioned the Delaney totem. Questions were raised, but only to be quickly silenced by the inevitable counterattack in which the latent heretic was reminded that his constituents "certainly deserved better than a representative who seemed to be in favor of a little bit of cancer."

For all practical purposes, that modest bit of terrorism soon dismissed the need for any further open inquiries. As courageous as I felt in the personal belief upon arriving in Congress that I would one day have to deal with this absolutist dogma, yet even as late as one year ago, I saw no political virtue in introducing any modification of Delaney. Such a solo flight of fancy would have been premature and short-lived to say the least.

In fact, how many of you five years ago recognized that for one reason or another the Delaney Clause needed to be revoked or changed so as to allow value judgments? And then let me ask how many urged your Congressmen to do so?

Unquestionably there is a great need for scientists to
speak up on relevant public policy issues such as this. But
then on March 9, the historic opening arrived as the FDA
announced its intention to ban saccharin, the last of the nonnutritive sweeteners permitted in this country.

Now let's don't kid ourselves; by no means was this the
century's most profound challenge to the liberties of free
citizens, now was saccharin the most satisfying focus for
the debate that at last can no longer be avoided.

Let me say in that regard that a much more preferred
cause de guerre would have been for the FDA finally to leap
the canyon of consistency and apply their absolutist principles not just to food additives but as well to all the hosts
of natural carcinogens present in natural foods.

How mischievously delicious if their test case had been
tryptophan, for which massive overdosing produces some evidence of a risk of approximately the same level as found with
saccharin.

What happens when you ban an essential amino acid? And
to whom will it happen? Or, consider banning the calcium of
mother's milk. The fortuitous political advantage of
saccharin was that there were in the land no fewer than 50
million diet-ridden Americans who knew from personal experience that without an artificial sweetener they would in all
probability smoothly substitute sugar beverages and desserts,
lose control of their diets, and suffer great harm.

In theory, they could all forego any sweets, but the
earlier interdiction of cyclamates had left compelling historic evidence that human nature typically lacks that iron
will that FDA would recommend.

Led by the American Diabetes Association, the more mili-
tant Juvenile Diabetes Foundation, joined by an array of
weight control associations and their bariatric physicians
to which later were added state medical and nutritional societies and the American Institute of Chemists, then joined by
the industry's calorie control council, urged these and all
other real consumers--I say real consumers--to exercise the
age-old privilege of writing Congressmen.

These real consumers spoke for themselves with an implied
vengeance. The resulting flood of over a million letters
fixed the attention of Congress upon a growing unrelenting
truth that a saccharin ban would do far more harm than good.

So for the time being, we have resoundingly rejected the
proposed ban during an 18-month moratorium.

Meanwhile the National Academy of Sciences is to assess
the potentiality of weighing risks against benefits, as
proposed in my Bill HR-5166. Not just for saccharin but for
food additives generally.

Saccharin, in its own sweet way, has thus set the stage for an unprecedented critique of the scientific and legislative basis of the 1958 Delaney Clause. Saccharin, for all the smug assertions that it is merely a crutch for its users, has thus finally and uniquely strengthened a heretofore timorous and politically weak-kneed legislative assembly to stand up to the advocates of an omnibus safety dictatorship.

Now the burden is on the absolutists to justify the extra incidence and/or aggravation of diabetes and hypertension and cardiovascular disease and hypoglycemia and obesity and arthritis and dental caries that will result when millions of dieters are denied saccharin and smoothly switch to sugar, as they did when cyclamate was removed in 1970.

Now the burden is on the absolutists to justify the risks by banning preservative nitrites. Now the burden is on the absolutists to justify the risks of banning BHA and BHT, which are suspect because of massive overdose experiments, while the normal use of these antioxidants appears at least historically to correlate with a historic decline in stomach cancer.

Now the burden is on the absolutists who at last must be careful not to accuse others of being for a little bit of cancer in the face of increasing evidence that obesity predisposes to cancer, knowing as even they do that cheating on a bland diet predisposes to obesity.

Now the burden is on the absolutists, who deserve it.

THE PREDICTIVE VALUE OF ANIMAL
TESTS FOR CARCINOGENICITY

William Lijinsky

The history of our progress in medicine is largely a
history of the prevention of disease. The great scourges
that lasted until the end of the last century, cholera,
typhoid, smallpox, and tuberculosis, have become much less
common causes of death because improved hygiene and better
nutrition have led to their diminution in the industrialized
world. It seems likely that a similar approach, that of prevention, will reduce the appalling incidence of cancer.
Prevention, however, implies the removal of causes, and it
is the identification of causes of cancer that is occupying
the attention of so many scientists and governmental bodies.

It has been variously claimed that 60-90% of cancers are
related to exposure to chemicals, but this is impossible to
verify. Nevertheless, it is certain that chemicals make a
substantial contribution to the overall cancer risk in the
industrialized world, and it is exposure to these that is of
concern. It does not matter substantially whether the
chemicals are industrial products, to which the major exposure is the workplace, or waste products of factories, or
components of food, or everyday components of our civilized
environment, including medicines and food additives. If any
of these carries with its use the possibility of an increased
risk of cancer, that should be at least known, and preferably eliminated. This is to say nothing of new chemicals
introduced into the environment, some of which might provide
an increased risk of cancer, and which should, therefore,
be used with caution.

Cancer is not a rare disease, but a very common one,
killing about 1,000 Americans every day, approximately one-
sixth of all deaths in this country. Neither is it a new
disease, having been recognized in antiquity. The new thing
about cancer is that the organs in which cancer is commonly
found in industrialized countries of the West are not those
in which cancer commonly occurred in the past, nor those in

127

ISBN 0-12-192750-4

which cancer is generally found in less-developed countries.
This makes our particular environment suspect as increasing
the risk of cancer of the lung, colon, pancreas, and bladder.
Many chemical carcinogens induce in laboratory animals just
those types of cancer that are common in Western countries.

Most of our knowledge about environmental causes of cancer is due to observations by alert physicians, going back
200 years to Percival Pott, who observed an uncommon skin
cancer, cancer of the scrotum, in young men who had been or
were chimney sweeps. He attributed their cancer to contact
with coal soot. Only a small proportion of chimney sweeps
developed this cancer. Similarly, exposures to several
environmental agents have been shown to be associated with
certain types of cancer. In 1876, Volkmann reported skin
cancer in coal tar workers, and in 1895, Rehn found bladder
cancer, which he called "aniline cancer," among dye workers
in the Ruhr. Between 1900 and 1920, a whole range of agents
--mineral oils, shale oil, creosote--was shown to be causa-
tive agents of skin cancer in many of the workers who made
contact with them. In 1932, lung cancer was observed in
people who had been engaged in nickel refining, again in only
a small proportion of people exposed. In the 1950s, Hammond
and Wynder, Doll, and Hill showed the connection between lung
cancer in man and cigarette smoking. And more recently there
have been cases of cancer in the vagina in young women who
have been exposed in the womb to diethylstilbestrol, which
their mothers took to prevent abortion. Finally, vinyl
chloride has been shown to produce a rare cancer of the liver
in some of the plastic workers who were exposed to this compound.

So we see that contact with environmental agents, mostly
in the workplace, but also by habit, can be responsible for
some types of cancer. Most people exposed to these agents,
even cigarette smoke, did not develop that particular cancer.
The variation in response, that is, the fact that some people
respond with cancer and some don't, might be genetic. This
difference in susceptibility might also be because of concurrent exposure to other carcinogens, which also increase
the risk of cancer.

Research workers became interested in the possibility
that the cancers of man associated with exposure to chemicals
could be induced in experimental animals by the same chemicals,
making the process of carcinogenesis more amenable to laboratory investigation. And so it was, starting with the pioneer
studies of Yamagiwa and Ichikawa, who induced skin cancer by
painting coal tar on the ears of rabbits. A few years later,
it was found that mice would do just as well, and mice and
rabbits were used to follow the isolation of the fluorescent

carcinogenic compound in coal tar, benzopyrene. Subsequently, it was found that rats (because they responded better), hamsters, or less frequently dogs were more suitable than mice for testing of certain carcinogens. Indeed, there are carcinogens that are inactive in mice, but induce tumors in rats or hamsters, and vice versa. Aflatoxins are good examples.

Overall, there is very good agreement between carcinogenicity of a substance in man and the induction of cancer by the substances in experimental animals (even though only a small number of cancers have been connected with specific chemical exposures, because their rarity in the general population made them easy to notice). The only exception is certain derivatives of arsenic, which have been connected with cancer in miners and metal smelters. Furthermore, there are now several instances in which tests of a compound in animals gave rise to cancer and the same chemical was later shown to increase the risk of cancer in people exposed to it.

In what some say were less enlightened times, the finding that the azo dye butter yellow, which was to be used as a colorant of margarine, gave rise to liver cancer when fed to rats, caused withdrawal of the substance from that use. Butter yellow does not induce cancer in mice. Similarly, more than 30 years ago, acetylaminofluorene, an effective insecticide, was never approved for use because it was found to be a potent carcinogen in both rats and mice, giving rise to tumors of many organs.

On the other hand, although diethylstilbestrol was known to induce tumors in mice more than 20 years ago, it was used as a drug to prevent miscarriages and was incorporated in cattle feed to increase weight (and I believe still is). The consequences of using it in cattle feed we don't know, but its use in women had tragic consequences, since several of the female children born to these women developed a cancer of the vagina, again detected only because it is rare. Had the warning of the animal test been heeded, this would not have happened. Even those babies that did not develop that particular cancer have an increased risk of cancer in general because of exposure to that carcinogen. Likewise the case of vinyl chloride, the monomer from which polyvinyl chloride is made and to which hundreds of thousands of workers have been exposed. Six or seven years ago (in experiments conducted in Italy), this was shown to induce cancer in mice and rats. No action was taken until a few workers in vinyl chloride plants died from a rare cancer of the liver, angiosarcoma. When our government ordered the exposure levels to this compound be reduced drastically to protect the workers, complaints were made that this was so costly that the factories would have to close. In the event this was a smokescreen, since few if any

factories did close. In the meantime, the delay in taking
the action indicated by the positive animal test has led to
years of unnecessary exposure of people to vinyl chloride,
with an undoubted increase in their risk of cancer.

Our tests of chemical carcinogens in animals show us that
the characteristics of chemical carcinogenesis in animals
and in man are identical, even though one chemical might produce different cancers in different species. Low doses given
continuously are the most effective in inducing tumors.
Large single doses are usually ineffective because most of
the compound is excreted unchanged. In fact, when we overload the body with a foreign compound we usually ensure that
most of it will be excreted unchanged because the enzymes
can't handle it. We know very little about the mechanism of
action of carcinogens. It is probable that the routes of
metabolism of carcinogens that we readily observe are not related to carcinogenesis and that carcinogenesis is in every
case a minor pathway.

Comparing pathways of metabolism in man and animals in
relation to carcinogenesis is risky because we are likely to
be misled. In response to higher dose rate more of the test
animals develop tumors and sooner. Most carcinogens require
metabolism and are not directly acting agents. There is a
long latent period between the initiation of the treatment
and the appearance of tumors. Younger animals and fetuses
are usually more susceptible than adults. Finally, there are
large differences in susceptibility from one animal to another;
the most susceptible develop tumors earlier. This is strikingly evident in some of my own experiments.

A group of 30 rats was given identical treatment with a
carcinogen, nitrosomorpholine, dissolved in drinking water.
The treatment was for 30 weeks. During the treatment one rat
died with a liver cancer. Other rats died at intervals with
the same cancer during the next 1-1/2 years, and some died
without that cancer. Two rats were alive at the end of two
years and both had liver cancer. So a treatment that was
sufficient to cause liver cancer in 4 months in one rat, induced the same cancer in other rats over their whole lifespan
and some did not develop that cancer at all. And these rats
were much more closely related than most members of a human
population.

I think that all would agree that since some chemicals
cause cancer in man, there must be a way of finding what they
are and reducing peoples' exposure to them. No one is able
to determine without testing whether a compound is a carcinogen. How then, do we devise a test that will give reasonable
certainty that a substance is free of carcinogenic risk?

There is no absolute certainty, but indications are that only a small number of chemicals are carcinogenic, this being a rare and special property.

We must select at least two animal species, since, as I have indicated there are many cases in which one species is refractory, and there is no reason to believe that man would be refractory to that particular compound. Then we must include both sexes, since there is some indication that one sex might be more resistant than the other in a few cases. We must choose a dose that is not so acutely toxic that the animals will die before sufficient time has elapsed for tumors to develop. So we do a pre-test in which the animals are treated with a series of doses for 6 months. From the result, we choose the dose that leads to an insignificant difference in weight gain between the test animals and a group of untreated controls. For the chronic test, we give one group of animals that selected maximally tolerated dose and another group one-half that dose, in case animals at the higher dose should die prematurely in spite of our care. Finally, we must have a group size that is not unwieldy, but yet adequate to give assurance of significance of the result of the test. So our test group usually consists of 50 males and 50 female rats and mice given either of two doses of the chemical.

The size of the test group is small and yet has to represent a large exposed human population. Let us assume that the substance, at the dose normally used by people would induce a 0.1% incidence of cancer, or 1 in a 1000 or 20,000 in 20 million, which might be the number of people ingesting a food additive. Let us assume that this exposure would continue for 50 years to produce that effect. This 50-year exposure has to be represented by a 2-year exposure to rats (which don't live much longer than 2 years). This also presupposes that man is only as susceptible as the rat; he might be more susceptible. Therefore, we must increase the dose to the rats by 25 times to compensate. (There is no evidence that the time of appearance of tumors is related to lifespan of the animal; instead it seems related to dose and dose rate). If we gave the rats this 25-fold dose and these conditions pertained, we would expect to see not a single tumor in the 50 rats. Instead, to observe the same incidence we are trying to detect in man, we would need a group of 1000 rats to observe even one tumor. To aggravate the problem, we would not consider one animal with a tumor a significant finding, even if that particular tumor were never seen in the untreated rats. Instead, we need five animals with that tumor in a group of any size before we can claim a positive cancer incidence; this would require a test group of 5000 rats, clearly an unwieldy size, unless we had only one chemical to test instead of hundreds. Remembering that incidence of tumors increases and the time for tumor development decreases as the dose is increased, we compensate for the moderate size of our group of test animals by increasing the dose. In practice, to make the test as sensitive as possible, we increase the dose as much as we can, to the maximally tolerated dose.

If, after a lifetime treatment, we find that there are fewer than five animals with any tumor that does not appear in a concurrent group of untreated control rats, we say that the compound tested is not carcinogenic and is safe. We say this even though we have only established that the carcinogenic risk in exposure to the substance is such that only one in 1000 (or 1 in 10,000) people might get cancer from it. On the other hand, if there is a significant incidence of tumors (usually five animals with a tumor that does not appear in untreated controls) induced in the test group of rats, we must say that the compound is a carcinogen. This carries with it the strong probability of an increased carcinogen risk for man. This is particularly important when considered together with possibly inevitable exposures to other carcinogens, the effect of which might be synergistic with the compound we tested. It would be irresponsible for a regulatory agency faced with the positive results of such a test not to consider the substance unsafe, particularly for chronic exposure to people. Likewise, while we cannot estimate numerically the risk to humans from exposure to the substance based on the results of such a test, no objective person could conclude that the substance is safe compared with a compound that does not induce tumors in the test. This is true even if the tumors induced are benign.

If our aim is to protect the most sensitive members of society from carcinogenic risks, we must reduce exposure to any chemical found to induce tumors in a properly conducted animal test. The most sensitive individuals include children and babies in the womb, neither of whom can claim to have freedom of choice. Therefore, because of the potential chronic exposures at a vulnerable age, no carcinogen detected in an animal test can be allowed as a food additive, which is the meaning of the Delaney Clause. Even if we knew that a threshold existed for a carcinogenic effect, we have no means of establishing what it is for any substance. If we make an estimate, and it is too high, that amount of the one carcinogen will add its effect to the threshold doses of all of the other carcinogens that we encounter and increase the risk of cancer. Unfortunately, mistakes we make now might not be revealed for 20-40 years.

GENERAL DISCUSSION

DR. COULSTON: Some things have been said here about vinyl chloride, about factories, and cancer and I think one would be very remiss to allow those statements to go without some kind of explanation. This business about the women treated with DES (diethylstilbestrol) with the daughters that got cancer is still not a proven fact. There are some states in the United States where you cannot show this phenomenon, like the State of California. After all women make estrogens all the time and some women make more estrogens than others.

I am not advocating that stilbestrol be used in women or cattle; it is not used in Canada and there are alternate chemicals. There are other nonsteroidal estrogens that can be used and as far as we know they are safe for their intended use.

DR. LIJINSKY: My point about stilbestrol is that it might not in inducing cancer be behaving as an estrogen. It might be a consequence of its not being a natural estrogen. Does it have a side product? That's why I talked about metabolism and not knowing enough about it.

DR. COULSTON: This question of the long latent period to produce cancer in man or animals leaves me cold. We know very well that the real carcinogens, true alkylating agents like mustards and so on, produce cancer quickly and you do not have to wait the lifespan of the animal to get a neoplasm. You do not even have to wait the lifespan with nitrosamine. You will hear more about this from Dr. Adamson. Just stop and think what we do with these animals. We take an amount that approaches the maximum tolerated dose and give it for the life of that animal and every day we pump that dose into the animal. Many of these compounds are fat-soluble compounds, they store in the fat, they store in the liver, the cells enlarge. The poor liver cells, I do not know how they can take it as long as they do. They literally burst and die with all this smooth endoplasm reticulum that is being produced. Yet every day we pump the chemical into the cells at these high doses. We have an unphysiologic situation in these animals; they are no longer normal. I don't know why we persist in this myth that this is a safety test. It is not! I would far rather see a probit-type test dose-response curve produced. However, if a chemical produces cancer in more than one species on a reasonable dose in the mouse, rat, dog, cat, rabbit, monkey, I would never allow this chemical to be used in man. I would declare it a carcinogen right away and ban it.

133

ISBN 0-12-192750-4

But, when you have the situation we have today where only certain inbred strains of mice show a cancer and rats and monkeys do not, I would not say the chemical is a carcinogen to man. We just had a high-level international meeting where we set ADIs for pesticides such as dieldrin, chlordane, and lindane. We said we will disregard the mouse data, because these chemicals cause no cancer in any other species, as shown even by our own National Cancer Institute. We decided that if tumors only occurred in the mouse at tremendous doses over long periods of time we could discount those data.

CONGRESSMAN MARTIN: First let me deal with the rhetorical question that my friend Bill Lijinsky asked me early in his remarks and that is what if saccharin were as potent as cigarettes. I would say in response that if that were the case then I would want to put a warning label on the package of saccharin. I would want to add a tax, as the Secretary of HEW has proposed, and I would want to educate the public about the potential harm from saccharin.

The fact is it is not as potent. Second, I was interested in the distinction that was made to the effect that the Delaney Clause should only apply to food additive regulation and not to other areas. I think this is a welcome differentiation insofar as the interests of all other kinds of regulatory jurisdictions are concerned. But it seems to me that if the principle of the Delaney Clause is valid, if the absolute zero tolerance level principle is a valid one, in the interest of the public, if the risk of a drug or food additives exceeds its benefits, then ban it. Conversely, if the benefit of a food additive indeed exceeds the risk of that food additive, then what public purpose is served by banning it?

Third, it is true that cancer incidence is slightly increasing. I am told that you could defend a fairly surprising postulate to the effect that the increased incidence of cancer correlates very well with the progress of modern medicine, principally because of the increased life expectancy. The fact is that while there are a thousand cancer deaths per day in this country that would also be true whether or not you banned saccharin.

Finally, let me suggest as an addendum to my remarks the necessity for four discreet steps to be taken in any regulatory policy. First of all there must be a determination whether a substance based on scientific evidence is carcinogenic. If it is not then you go on to other considerations. If it is then you proceed to the next three steps rather than stopping at that point as the Delaney Clause requires of the FDA. The second step would be to determine the potency of that carcinogen. If it is a potential carcinogen, how potential is it? This would be done on the basis of epidemiological data, if such are available and if they reveal a measurable potency. If not, then rely on dose:response data or some other procedure using test animals to derive some estimate of potency. Admittedly as many have said this will not be precise, but it is certainly better than the overly rigid precision of the Delaney Clause principle in my opinion. Third, once you determine the potency, you should then determine what the actual risk is to the exposed population. If you combine a potency factor and an exposure factor, taking into account how many people actually use the chemical, then you can derive a risk.

I would say, for example, that if you made all three of those steps for the animal test data on saccharin and assume that humans are as sensitive at low doses as the test animals are at the high doses, you would conclude that there would be some 26 additional cases a year in the United States. That comes out to 13 cases per 100 million and that risk is about 1/4 the risk of taking one transcontinental airplane flight per year. That is the risk of getting cancer from the cosmic radiation of that flight. It is about 1/40 the risk of accidental electrocution. It is about 1/80 the risk of average U.S. exposure to medical x rays. It is about 1/100 the risk of drowning while fishing. It is about 1/1000 the risk of a fatal falling accident. And I say make that determination as a third step. Do not disregard that. And, then as the fourth step, compare that estimate of risk with the best estimate of benefits that you have available and if the risk level is trivial, ignore it. If the risk level is substantial, but not a great hazard, then compare the benefits. And if the risk is enormous then, of course, flatly prohibit the substance without regard to other benefits on the grounds that banning it will lead technology to find some substitute in that case.

DR. COULSTON: The real issue in all of this on a scientific basis is the methods of testing that are employed today, particularly by the National Cancer Institute. I have the feeling and I have some reason to feel comfortable that this thinking, this philosophy is changing. There will be more than two doses used in the future; there will be more of the concepts that we have heard from Dr. Kraybill brought into this testing program.

I am sure that the testers will look for no-effect levels in the future even for lifespans, and there is a no-effect level even for cyanide. No-effect levels exist by whatever criteria you want to look at. If you cannot find anything wrong with the animals, the no-effect level is found.

Now the real question is what is a low dose? The hang-up,
Dr. Lijinsky, is that the analytical chemist will find a
molecule or a part of a molecule today with many chemicals.
In the meanwhile, the tester for cancer at the highest dose
says a cancer exists. The chemist finds one molecule, and if
you follow the concept that there is no threshold you have to
either restrict or ban the chemical.

DR. LIJINSKY: I was very interested in your comments, Mr.
Martin. I don't think you addressed the problem that I was
really dealing with, which is do you trust an animal test or
do you have a better way of assessing this matter? I am
particularly concerned about this, because I am at least as
worried about the chemicals that are now being introduced, for
which there cannot be any epidemiologic evidence, and these
are often useful chemicals. What do you do about a chemical
that produces cancer even at high doses in animals and which
is proposed for use as a food additive? This is the real
crux of this question. The second point that bothers me a
little is that all measures of potency and estimates of risk
have very big areas of error attached to them. I don't think
any estimate of risk is worth the words in which it is phrased,
because we have no basis on which to make a risk estimate.
And estimating the number of deaths from cancer from a certain
exposure to a chemical is not a very rewarding or fruitful
exercise. Because it ignores the possibility that each exposure to a carcinogen increases the risk of that person who
is exposed getting cancer by some amount, which we cannot at
the moment measure, but which is finite. It is the sum of all
these increased risks that is probably the reason that we
have so much cancer in the human population, and that is what
bothers me. To project the number of deaths from cancer due
to that particular cause is not what we are really concerned
with. I think it is misleading. And I'm interested in rais-
ing the total exposure to carcinogens, which eventuates in
the incidence of cancer that we see, which is a lot, not a
small amount.

DR. COULSTON: Dr. Lijinsky, I shall give you a minute re-
buttal in just a second, but you keep saying that the incidence of cancer is increasing. I don't know of any such evidence. Let's ask Dr. Wynder.

DR. LIJINSKY: The risk.

DR. COULSTON: The risk is increasing. Dr. Wynder you have a
message here that we can use?

DR. WYNDER: Yes, I would like to again speak up for epidemiology, which somehow gets lost here. As all of you know, our
institute has just as many mice, rats, and hamsters as any
other institute, but at the same time we are keenly concerned
about disease in man. And this is really what this debate
is all about. Just as Dr. Lijinsky, I am a great friend of
history; I have in my own library the early volumes of Potts,
Folkman, Canaway, and Lindsey, and what I really learned from
all of this, of course, is that there is a good correlation
between animal tumors and man, particularly when you consider
the dose. As you know, it goes all the way back to Paracelsus,
who said in paraphrase that it is the dose that makes a poison.
But, it so happens that there are a number of agents in our
environment where we can use the art of epidemiology to show
retrospectively whether our decisions, indeed, would have
benefitted man. This certainly should involve DDT, which has
been around a long time; it certainly should involve saccharin,
which has been around a long time, and we can certainly make
such determinations on hair dyes.

It is particularly true when an agent has been used by a
significant number of populations. Just the other day, we
looked at our favorite computer (I must say years ago I did
all my work by flowsheets; today, we are using computers,
because we have many thousands of cases and many variables to
study). Looking at the controls and the last sample of 700
cases we find that some 21% of males and 24% of females use
saccharin. In other words, we have plenty of people to study.
Among this group 11% of the males used at least 7 drops or
tablets a day and 90% of the women; 8% of the men had used
saccharin for more than 20 years. The point I'm making is
that we have a considerable number of people who have used
saccharin in tablet or drop form for more than 20 years.
Sixteen percent of our males and 19% of our females use
sweetened drinks. Thirteen percent of these men have used
from 15 to 21 bottles a week and 9% for more than 15 years.

The point I am making is that we have the epidemiologic
capacity through appropriate case control studies to determine
whether sweeteners represent a risk in man. In correlation
studies, we have shown that sex ratio, education, profession,
smoking, coffee, weight, religion, and diabetes all affect
sweetener use. So what we really would expect, if saccharin
is a factor in a given population and is more commonly used,
is that population is at higher risk. Now my colleague Sir
Richard Doll has made a particular point of this for diabetes:
the use of saccharin is at least three times greater than in
the general population. You would expect a greater risk of
bladder cancer in them, when properly standardized for smoking
habits.

I certainly agree with you that if we study saccharin we should look at all cancers, which indeed we are doing. Another way of looking at it, during World War II there were certain countries in Europe where sugar was not available and they had tremendous use of sweeteners. One would like to see whether then or 20 years thereafter there was an increase in certain types of tumors. All of this is being done.

There is one possible problem, Dr. Kolbye, that both you doing the study and we must look at. In recent years, there has been so much attention given to saccharin and sweeteners as a possible cause of bladder cancer that we must make very certain that we do not have an interviewer bias. This is something that perhaps your group and I can look into, because I believe if I go today to the bedside of a patient with bladder cancer and ask about sweetener use it is likely, if they have ever read the *Washington Post* or the *New York Times* or some other one of my favorite medical journals, they will have a different reaction to this question than if I would interview a neighborhood control sitting at home watching his favorite television station over a beer. So I'm just indicating this to you that we have the capacity to study the epidemiology of this problem and hair dyes. I think this conference, as I said earlier, should have been more heavily focused on epidemiology. It is a science and after all must make the final evaluation whether the decisions that we make here are correct or not.

I would like to point out to Congressman Martin that having been in these debates for a long, long time and having several of my good friends on the other side of the aisle, who I wish were here, I would expect that this debate would have certainly been hotter, if Sidney Wolfe and the environmental group had all been here. I would like to believe that we ourselves cannot really come to grips with this question, because those of us who believe in zero tolerance and no threshold will always believe it.

Finally, I would believe that if we are going to solve this problem, it must be done by you people in Congress, because you had the wisdom to give us certain regulatory laws and many years have passed and perhaps the time has come to modify these. But we can only advise you. I think it's up to our Congress to evaluate what is in the best interest of the public. As I said yesterday we can sit here another week or two or two years. We will never make this evaluation by ourselves.

DR. COULSTON: Dr. Wynder, thank you. You know we did invite Dr. Selikoff, we did invite Dr. Wolfe. Dr. Selikoff had one reason, Sidney Wolfe had still another reason not to come. We did invite these people. At the last meeting of this International Academy we had many good epidemiologists and many good discussions on similar problems.

We appreciate very much what you say and you know I am a great believer in epidemiology, but there is also the other side when Dr. Lijinsky points out that this is fine, a lot of the older epidemiological studies are retrospective. He's worried about new compounds.

CONGRESSMAN MARTIN: Let me say first with regard to the question that was put to me by Dr. Lijinsky asking whether I trust the animal test, let me reassure him that, yes, implicitly I do. Therefore, I make no criticism of your major thesis of the importance of continuing researches. Let me say, as I did say in my speech I am not one of those who wants to ban animal tests, who wants to ban evidence based on massive dosage, and I have never taken that position even in the early stages of this debate. Rather it has been my purpose to focus on what I regard to be a flawed policy that says that you must ban saccharin at every drop of a rat. Now in the second case is it important to estimate the projected number of deaths or is it true as you say that such estimates are not profitable? Certainly it seems to be profitable from the point of view of the FDA to project an estimate of some 1200 additional bladder cancer cases a year when in fact, if they take the actual test data and transpose that on a straight line, in a proportional way, to the incidence that would be expected if they were to consume the amount of saccharin that is actually consumed in this country, you would get not an upper limit of 1200 bladder cancer cases a year but 26. Therefore, it seems to me to be profitable to undo whatever profit was assumed by the FDA in projecting that kind of estimate in the first place. Let me say that although this correction has been published and although it has been the subject of some debate, the Office of Technology Assessment in its review saw fit to simply take the 1200 bladder cancer estimate of the FDA for its own. However, it is simply not valid on any mathematical basis, if you take into account the actual exposure of the public.

Third, and finally, I would say that while Congress does have the responsibility to set policy, Congress does need your help and does need the input of all of you on this and every other scientific question that comes up. This was the reason that I wanted you to know earlier in the meeting of the formation of the Science Forum of members of Congress

working with the American Academy for the Advancement of
Science, and seeking to involve all of the learned societies
such as the American Chemical Society and the Society of
Toxicology working with Congress, congressional committees,
and our Forum. These groups will improve the quality of
public discussion of these questions.

DR. FRAWLEY: A couple of comments. I never like the record
to stand indicating that 90% of the cancers are caused by
chemicals. Dr. Lijinsky backed off from 90% and went down to
80, 50, and 40%. I think it has to be clarified for the
record that these cancers really are not all chemical; many
are environmental, include the cancers from cigarettes, from
alcohol, skin cancers from sunlight exposure, etc. I think
we're really talking about a much smaller percentage of cancers that we're suffering with, being caused by chemicals
that man has put in the environment.

Another point I have to take issue with is that the maximum tolerated dose bioassay is a safety test. It was never
really intended to be a safety test. It was a screening
test. And that is the way NCI designed it. It has been up-
graded over the years so that it's getting closer to a safety
test, but it was a screening test. It pulls out the phony
carcinogens as well as the real carcinogens and it is unfortunate, indeed, that the regulatory agencies have to react when
they find a phony carcinogen as well as a real carcinogen.
It should indeed be considered a screening test and an indication that in-depth definitive studies be undertaken on the
compound.

DR. FURMAN: Dr. Coulston, to reinforce what you said, I
would like to make a few observations relative to the remarks
of Dr. Lijinsky and the presumed carcinogenic role that *in
utero* DES exposure played in the genesis of clear cell adenocarcinoma. Implicit in his remarks, I think, is a fairly serious indictment of the scientists and physicians and members
of the United States Pharmaceutical Industry who marketed
and sold DES for the purpose of chronic and threatened
abortion several decades ago. As you all know, there is
maintained at the University of Chicago and at Harvard University, a clear cell adenocarcinoma registry of these unfortunate women who developed clear cell adenoma of the vagina.
The number now is roughly 350 such cases give or take a few,
and among these 350-odd cases there is a good medical history
in approximately 2/3. Of these 2/3, there is *no* medical evidence whatsoever of exposure to DES or other estrogens, synthetic or steroidal, on the part of the mother. The registry
interestingly enough includes a girl who is one of a set of

monozygotic twins born of a mother who was treated with DES.
One girl has had clear cell adenocarcinoma, fortunately
successfully remedied by surgery and radiation, the other
girl is perfectly healthy except for the teratologic effect
of adenosis which so far, fortunately, appears to be a benign
lesion.

Dr. Lijinsky said in effect that had we paid attention to
the animal experiments of 30 years or so ago, DES would not
have been administered for this purpose. There were, as far
as I'm aware of, no animal experiments that demonstrate that
DES or any other estrogen will induce carcinoma of the genitourinary tract in female offspring. DES and many other
estrogens, years and years ago and for many years, have been
known to induce breast carcinoma and cancer of the uterus in
experimental animals. The use of DES was based on experiments
from the Smiths at Harvard, which, although their technique
and technology subsequently was shown to be incorrect, nevertheless established on the basis of technology of the day that
DES increased the progesterone availability. There was and
still remains today a consensus that in certain cases of
threatened abortion, replacement or addition of progesterone
or progestins will be therapeutically useful.

DES gradually fell into disuse as progestins came on the
market and they replaced DES for this purpose. Dr. Priscilla
White along with a good many other physicians, particularly
at the Joslin Clinic in Boston, was convinced that the use of
DES brought to term many diabetic mothers who in her experience would not have completed pregnancy. She remains even
today convinced of the efficacy of DES in this situation, and
I think would continue to use it if she were not prohibited
from doing so by that regulation.

Finally, the offspring of DES-treated mothers, when the
use of the drug gained widespread clinical acceptance, were
examined obviously very carefully in order to detect whether
or not there were abnormalities of the offspring. There was
no evidence whatsoever of any increased incidence of any
structural or other abnormality in these children. I think
that I want the record to show the basis for use of DES and to
make it clear that there are many cases of clear cell adenocarcinoma in the registry for which there is *no* evidence of
DES exposure.

DR. COULSTON: I'm delighted that you put this on the record
because I've been listening to this story all around the
world. I know better, and I try in my own way to say it, but
people tend to disregard it for an obvious reason: I am not
an expert in that area. But from you, Dr. Furman, we appreciate having this in the record. It is not a problem that has

been played up in the press because of what you said; the FDA
has never banned DES from use for specific purposes in human
medicine. Whether it should be used in cattle or not is another matter and I do not want to get into that story. DES
is a useful compound that has saved hundreds of babies in
this world and it should still be, in my opinion, used in
competent medical practice.

Dr. Etcyl Blair I hope you'll talk a little about vinyl
chloride and do away with some of the myths we have had on
this subject. Particularly, I want you to emphasize if
possible that Dr. Lijinsky has always given, in good faith
and correctly, examples of industrial accidents, one way or
the other. A man goes down a vat to clean out a vinyl chloride tank and is not protected properly and gets sick. Why
should he not? If he went down a tank with benzene he would
also get sick or die. The point is the vinyl chloride cases
in this world, which now number some 50, are all directly
related to companies that, I emphasize this, did not protect
their workers sufficiently. A clear-cut accidental type of
exposure due to the lack of knowledge.

DR. BLAIR: It was following World War II, in the mid-1940s I
guess, that industry was faced with rebuilding itself and
supplying goods to a market, to citizens who had been long
associated in the conflict overseas. During that period of
time, there were some studies going on in toxicology, but
certainly nothing on cancer in animals. At the Dow Chemical
Company, we were very concerned about all chlorinated hydrocarbons, because of known problems with liver damage. Our
studies on vinyl chloride in those early days in the 1950s
were conducted with the idea of seeing if there was harm that
could be done to livers. Indeed, from those studies recommendations came out, and they were published in the literature,
that a work place level could be set at some 50 parts per
million or lower. Certain elements within the government
thought that was too low. It did not become a national
standard; it became a standard for our organization. A number
of years later, another segment of the industry identified
angiosarcoma in the work place but prior to that there were
concerns about vinyl chloride. The industry in the United
States working with the industry in Europe sponsored studies
with Maltone at Bologna, Italy, to do cancer research. So
again, through misunderstanding, we in industry frequently
are criticized for not doing work. All of that early work
was sponsored by both American and European chemical industries. Maltone found that vinyl chloride could indeed induce
cancer in animals. Dow Chemical Company studies showed the

same thing. We went to a series of public hearings. The levels were lowered to emergency standards at 10 parts per million and the final standard was at one part per million.

Dr. Lijinsky made some statements to the fact that industry could not operate or would shut down. The jury is still out on that. There will be a number of companies that will eventually fail in this endeavor because as we pushed toward one part per million and as we push toward maybe even lower levels, if that can be done, there will be industries that will shut down.

Time will only tell. However, I do want to make another comment about this particular activity, because as we continue to pool our resources, say on vinyl chloride, we are depriving ourselves of studies on things that we do not know about. I think there is an enormous amount of effort being put in on this repetitive challenge that we are faced with. Trying to deal with these issues, we are not getting on to discover new things that we need to know. The problems associated with these become not just FDA problems, they become EPA problems, OSHA problems, and they become the so-called problems of industry and of society.

Also on this matter of disease in the workplace, actually the level of disease from vinyl chloride with the workers I think is more like 60 ppm than it is 30. I have heard Dr. Selikoff talk about the iceberg effect. Dr. Lijinsky talked about it and if there was an iceberg effect we should begin to see more and more cancer each year. Actually the cancer coming from vinyl chloride each year is less and less. We have never observed it in our manufacturing plant at Dow. We have been manufacturing vinyl chloride for close to 30 years. Those exposure levels have been at 50 parts per million or less most of that time. I might add also that while the level in the United States is one part per million, it ranges all the way from 10 parts per million in Europe to five and three in Germany, and it may be one part per million in Sweden. All are looking at the same data and I submit that the Europeans do not have less information than we have, yet they have come up with a higher level of permissible exposure that they think is not harmful.

Again as Dr. Frawley stated, this statement that one always hears, 90% of cancers are environmentally-induced, ends up being chemical substances. I would like to support Dr. Frawley in this. Really, we are talking about tobacco, alcohol, our social factors, and many of these as discussed here today, are really the big contributors to cancer. While we are unhappy with the percentages that we do have occasionally from the chemical industry, rest assured we are working mightily to try to make sure that that figure is reduced.

The more effort that we have to put into say vinyl chloride to go from a one part per million to a tenth of a part per million, means those resources and those laboratories are not available to be dealing with the issues we should be dealing with.

DR. HARRISON: The group here I am sure understands the vocabulary we are using and the connotations involved but the public does not. We are using the term chemicals, meaning synthetic substances that are produced by industry. Included in your environmental exposure are the chemicals that are parts of natural systems. I think, it seems to me, that one of the great confusions in the public is about a certain compound that Dow or any other company may produce. It may even be the same compound. If we can use our vocabulary to make the distinction between chemicals in natural systems that are a part of our environment, and those introduced due to the technology as synthetic substances I think we may dilute some emotion.

DR. COULSTON: Your point is very sound. Let me just leave this with you. There are now over 2000 chemicals that are on suspect cancer lists of various agencies of our government. However, the facts are that there are only around 20 cases of distinct chemicals that have ever been shown to be carcinogenic in man. You would have to include many natural substances like estrogens (estrone or estriol or estradiol); these are all cancer-producing substances in man. There are the alkylating agents, there are many cancer-producing chemicals that are used to treat cancer. It's an enigma but it's a risk:benefit relationship. One of the best known of all cancer drugs today is a very potent carcinogen to man and yet if it weren't for that drug today many women would be in very great trouble with mammary carcinoma. This chemical is widely used and is a great boon to mankind. Yet one dose given to an animal like a mouse produces cancer. One dose, or more, it is that potent!

So you see, everything is relative and what we are talking about is that we have to consider all points of view as has been said by everyone. You have to consider the benefit, you have to consider the intended use of the compound. I do not want to use an alkylating agent to treat a simple headache, but I would use it to treat cancer. Why should I use an agent that may cause cancer in its own right? On the other hand if I have mild pain or fever I will use aspirin. And it is so safe, I will let the public decide whether to take 2, 3, or 4 tablets. Put the warning labels on the bottle and with a safe drug, let the consumer decide the dose and even the brand.

All these factors come into balance. What I want to see happen some day is a reasonable approach in the U.S.A., in a similar way to what goes on in the rest of the world. Is that too much to hope for? What do we know that Canada does not yet know? Canada does not ban cyclamates and saccharin. What do we know different than England or Germany or India and Japan where these chemicals are still used. Where have we come in our thinking that we have to go so far? You have just heard that even with vinyl chloride, all the countries using vinyl chloride are more understanding of the socio-economic risk:benefit relationships than we seem to be and set more realistic exposure levels than we do here in the U.S.

THE DESIGN CRITERIA AND APPLICATION
OF DOSE-RESPONSE RELATIONSHIPS
TO INTERPRETATION OF CARCINOGENESIS BIOASSAY

Morris F. Cranmer and Neil Littlefield

I. INTRODUCTION

There are several mistaken impressions operational in the
confidence placed on current bioassays. Theoretical discussion is available elsewhere (1,2). The examination of the
rationale behind and data generated by a large dose-response
experiment (3) support the following general statements.

We are misleading the public when we indicate that we
have the ability with any great certainty to demonstrate in
animal bioassay that chemicals do or do not cause cancer at
low doses in large population.

We are only capable of detecting the carcinogenic potential of chemicals when the difference of nonneoplastic to neoplastic toxicity is large.

We are technically capricious when we agree to divide by
100 one day and extrapolate linearly to risks approaching zero
the next.

We are deluding the public and ourselves if we think that
we are applying any measure of consistency to "positive and
negative" determinations of carcinogenicity.

We currently do not exercise the toxicological capability
of separating carcinogens and potentiating or accelerating
agents or nonphysiological conditions.

We are encouraging bad or limited experimentation by our
strategy of two standards. Too little attention is paid to
the quality of data.

ISBN 0-12-192750-4

II. STATEMENT OF PURPOSE AND NEED

A. *General*

   The ever-growing knowledge and impact of chemicals on
human health has greatly increased the need for a scientific
basis to determine what, if any, would be the largest quantity of a chemical that could be absorbed for a more or less
indefinite period without producing adverse health effects.
For this purpose, the toxicologist has long relied on dose-
response relationships and extrapolation downward on the curve
for the purpose of estimating a "safe" dose. Very few informed toxicologists find it is possible to deny that the extent of induced adverse effects is proportional to the extent
of exposure and absorption of a chemical. An exception to
this established toxicological concept, however, has evolved
in the area of chemical carcinogenesis to the degree that
chemicals are classified as carcinogens, noncarcinogens, or
procarcinogens, regardless of dosage involved. Since this is
not consistent with classical toxicological concepts, a
rigorous evaluation must be made of testing procedures for
chemically induced cancer from a toxicological viewpoint.

B. *Existing Deficiencies*

   Most of the deficiencies in current carcinogenic testing
result mainly from the concept that this testing involves
only the determination as to whether or not a compound can be
made to produce a neoplastic tumor. However, it is now recognized that carcinogenic testing must, of necessity, consider
both qualitative and quantitative factors. The main deficiencies in past studies of these factors involve primarily
two areas: experimental design and definition of endpoints.

   1. Experimental Design
      a. *Statistics.* The bulk of technical literature
         reflects the lack of statistically valid experi-
         mental design including adequate numbers of ani-
         mals at low levels of carcinogenic response.
      b. *Dose-Response.* The prediction of risk at a given
         exposure level requires dose-response information.
         Only limited use of dose-response studies are re-
         ported in the technical literature for the purpose
         of determining tumor incidence and time-to-tumor
         in terms of dose-rate and total dose.

    c.   *Low-Dose Studies*. Little information is available
on dose-response studies at low levels of exposure
and response. At low levels of exposure environmental factors may alter extensively the quantitative aspects of a response.

    d.   *Mathematical Models*. There is only limited mathe-
matical definition of the dose-response curves at
low levels of exposure in terms of variables
affecting a chemical carcinogenic response.

    e.   *Life-Shortening*. There is limited use of experi-
mental designs that permit proper observation and
evaluation of life-shortening effects of chemical
carcinogens in relation to dosage.

    f.   *Age Sensitivity*. There are only limited studies
available on the influence of age on the sensitivity of an animal to a chemical carcinogen. The
hazards involved in exposure to a chemical carcinogen depend not only on the nature of the chemical
itself, the route or exposure, and the extent of
exposure in terms of amount and time, but also on
susceptibility of the animal at the time of exposure.

    g.   *Recovery*. There is a lack of evaluation of the
possible regression or progression of pretumorous
lesions such as hyperplasia in relation to dosage.

    h.   *Tumor Growth Rate*. The technical literature shows
an impressive lack of study of the possible dependency of tumor growth rate on dosage.

    i.   *Reproducibility of Results*. There is an extensive
lack of evaluation of the quantitative reproducibility of chemical carcinogenic testing.

2. Endpoints

    a.   *Tumorigenesis*. There can be no doubt from a sur-
vey of the technical literature that benign neoplasms are often precursors of malignancies. Hyperplasia and number, type, grade, and individual
distribution of tumors must all be carefully used
as endpoints in the evaluation of chemical carcinogenesis.

    b.   *Time-to-Tumors*. Time-to-tumor may be a very
sensitive endpoint permitting estimation of
"acceptable exposure levels" from dose time to
tumor curves.

    c.   *Life Shortening*. As indicated earlier, there is
limited use of experimental designs that permit
proper observations and evaluation of life-shortening effects of chemical carcinogens in relation
to dosage.

  d. *Pathology.* It is of the utmost importance that a
      complete and accurate pathological examination be
      conducted on all animals used in carcinogenesis
      studies. All lesions, including precancerous
      lesions such as hyperplasia, must be described.
  e. *Biochemistry.* Each compound must be evaluated
      individually on the nature of its absorption,
      distribution, metabolism, retention, and excre-
      tion.
  f. *Pharmacokinetics.* In order to provide a firmer
      basis for evaluation of results obtained in the
      large chronic low-dose carcinogenic bioassay, it
      will be essential to develop a correlation between
      dietary level of the carcinogen, total and/or
      daily intake of chemical, incorporation of chemi-
      cal into the target site, and the incidence of
      tumors as a function of duration and level of ex-
      posure. Involved also in this correlation is the
      need to evaluate the role of blood levels (total
      as well as unbound) and urinary excretion patterns
      of the chemical and/or its metabolites.

C. Extrapolation Problem

    Of utmost importance to a toxicological approach to car-
cinogenic testing is the capability of estimating "safe"
levels of exposure in a given experimental species by extrapolating downward from positive results obtained at exposure
levels above the actual use levels. The Mrak Technical Panel
on Carcinogenesis pinpointed the major problem:

    The basic problem is that extrapolation outside the range
    of observation must be based on a generally unverifiable
    assumption about the mathematical nature of the dose-
    response relationship near zero dosage.

This is precisely one of the main objectives of the NCTR
Chronic Studies; that is, to establish the mathematical nature
of the dose-response curve at low levels of dosage in order to
facilitate this type of extrapolation. The lack of information about the mathematical nature of the dose-response relationship near zero dosage for a carcinogen and associated uncertainties in the extrapolation from higher to lower dosages
have led the FDA Subcommittee on Carcinogenesis to the following conclusion:

Although it is possible in principle to estimate "safe" levels of a carcinogen, uncertainties involved in "downward extrapolation" from test levels will usually result in permissible levels that are the practical equivalent of zero.

In other words, the uncertainties in downward extrapolation must first be eliminated if one is to consider a safe level of a carcinogen other than zero. Several mathematical models, such as the probit curve, logistic curve, and the one-particle curve have been offered to predict the incidence of effect at low levels of insult. These models all agree with data observed in experiments utilizing high relative incidence of response, but differ widely in predicting incidence at low levels of chemical exposure.

D. Definition of the Problem

As early as 1969 a National Center for Safety Evaluation had been called for by the Mrak Report. The NCTR was established in 1971 because of the lack of scientific data that would support the accurate determination of a small risk to an annual population from long-term, low-dosage exposure and the inability to extrapolate with confidence the existing data to the human experience. Recognizing these inadequacies, the NCTR set out to design an experiment and to establish a protocol to answer these problems. Specifically, the ED_{01} experiment set about to answer these problems and to help fulfill the mission of the center by (1) determining the adverse health effects from a long-term, low-incidence, low-dose, chronic study of a known chemical carcinogen, (2) by mathematically defining the previously experimentally undefined lower limits of the dose-response relationship, and (3) by the integration of experimentation in comparative pharmacology, metabolism, and pathology that will support the appropriateness of the extrapolation.

III. RESULTS

The data (4) presented in this manuscript represent findings using both pathologist-verified data as well as data that have been screened by the parapathologist. There have been 22,273 animals examined by both groups with a total of 1919 mice remaining. All the remaining mice are from the

life-span portion of the study. Since the data are not
completed at this point, an in-depth analysis has not been
attempted; however, some of the relevant findings are presented. These initial findings prompt many questions that
will be answered as a more in-depth analysis is done.

Part of the experimental design is shown in Table I.
There are seven dose levels plus a control, ranging from 30
to 150 ppm of 2-acetylaminofluorene (2-AAF). The total seven
doses are separated only by a factor of 5, which represents
a very compact design. There are eight sacrifice intervals,
starting at 9 months and ending at 24 months. Most of the
sacrifices are located between 12 and 18 months. There is a
group of mice allocated as a life span study. These will remain on the study until they are dead or are in a moribund
condition. The study was conducted in six rooms (each considered a replication) under SPF/DF barrier conditions.
There are a total of 81 treatment groups represented in the
entire study. A total of 24,192 female BALB/c mice were
allocated to the study over a 9-month time period. The numbers in the table represent the number of animals allocated.

Table I.
2-AAF ED^{01} Study: Serial Sacrifice

Time (months)	9	12	14	15	16	17	18	24	Life
150 ppm	144	144	120	96	96	72	144	336	144
100 ppm	144	144	120	96	96	72	144	336	144
75 ppm	144	144	120	96	192	144	288	600	288
60 ppm	288	288	240	192	288	216	288	792	288
45 ppm					288	288	432	864	432
35 ppm						432	864	1296	864
30 ppm							1728	1728	1728
0 ppm	144	144	120	96	192	144	432	744	432

Body weight changes are probably the most sensitive indicator of toxicity. Figures 1 through 4 illustrate some of
the reactions that are typical of the doses and replications.
Figure 1 shows the body weight of all animals from the control
group of the lifespan animals from each of the replications.
The graph is plotted for average animal weight in grams/mouse/
week. The numbers represent replications 1 through 6. At the
peak, the replications range from 27 to 29 gm. Figure 2 represents the same parameters except the animals were dosed at
30 ppm. Again, at the peak, the ranges were between approximately 27.5 and 28.5 gm. The total graph showed little vari-

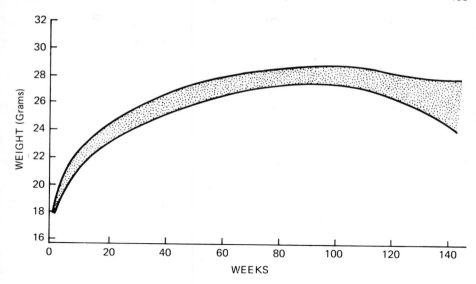

Fig. 1.   Average animal weight (grams/mouse/week), dose = control.

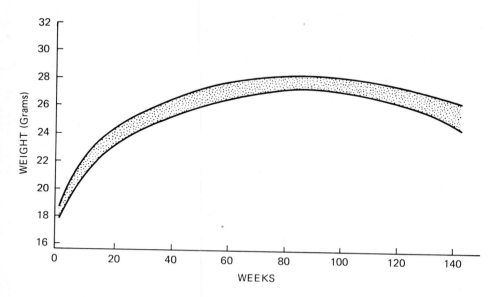

Fig. 2.   Average animal weight (grams/mouse/week), dose = 30 ppm.

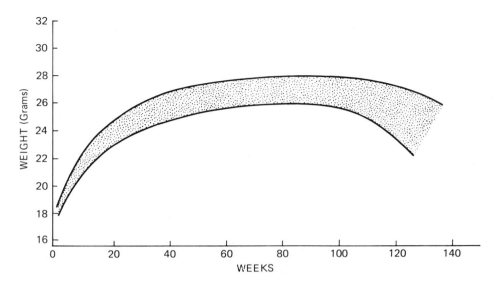

*Fig. 3.   Average animal weight (grams/mouse/week), dose =
150 ppm.*

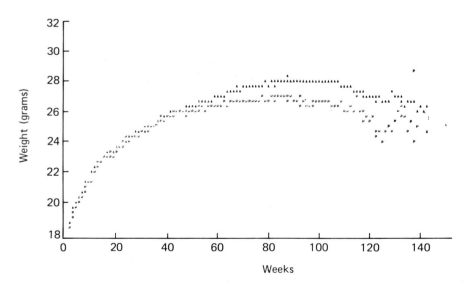

*Fig. 4.   Average animal weight (grams/mouse/weeks), A =
control, P = 150 ppm.*

ation between the different replications. Figure 3 used the same parameters except the dose was at 150 ppm 2-AAF. At the peak, there was a range from 26 to 28.5 gm, showing more variation than the other dose levels. In addition, there was a more pronounced weight decrease after 2 years on the study. Figure 4 represents the body weights when plotted in relation to dose. Doses 0 and 150 ppm were plotted to illustrate the effects of dose level. The control dose peaks at about 28 gm, however, the high dose peaks at about 26.5 gm, a 1.5 gm difference in the average body weight. The other dose levels were all located at varying levels between these two curves.

The next three figures deal with mortality of mice during the study. The graphs have been corrected for sacrifices. Figure 5 represents all animals at 150 ppm. The plotted numbers represent replications. Mortality is plotted as time (days) vs. percentage response. The figure shows that 60% of the animals were dead at 720 days. There is approximately an 8-week range between replications in the time to death at most time intervals. Figure 6 shows mortality plotted against the same parameters except the doses from only one replication are plotted. Each plotted number represents a dose with 1 equal to control and 7 equal to 150 ppm, with the other doses corresponding to the respective intermediate numbers. Of significance in this graph is that mice on dose 7 (150 ppm 2-AAF) exhibit a higher mortality when compared to the other doses. Figure 7 is plotted similarly to Fig. 5, except the dose of 30 ppm is used. Since the plotted numbers represent replications, it can be seen that replications 1, 2, and 3 have a slightly higher mortality rate than replications 4, 5, and 6. This may be due to the fact that after about 2 years on the study, replications 1, 2, and 3 were adjusted to include a 24-month sacrifice while replications 4, 5, and 6 were not changed and contained a terminal sacrifice at only 18 months. This manifestation will be examined in more depth. It is thought that this may be an exhibition of the relative sensitivity of this study where large numbers of animals are used.

Figure 8 represents the average dose (micrograms/kilogram/mouse/week) that was eaten by each dose group of mice. When plotted over the time course of the study, the doses remain essentially parallel and there is no crossing. The higher doses demonstrate an increase in 2-AAF consumption after about 2 years, but this can be correlated with the body weight data, which show some weight loss at the same general time period.

The occurrence of bladder neoplasms in sacrificed animals is plotted by time in months in Fig. 9. Neoplasms occurred in the 75-, 100-, and 150-ppm doses. In the 150-ppm dose

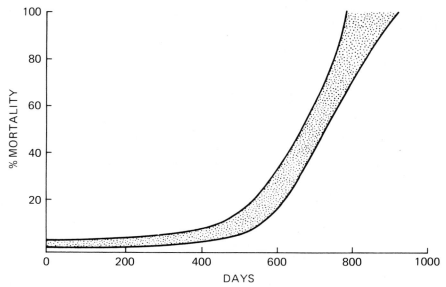

Fig. 5. Mortality dose = 150 ppm.

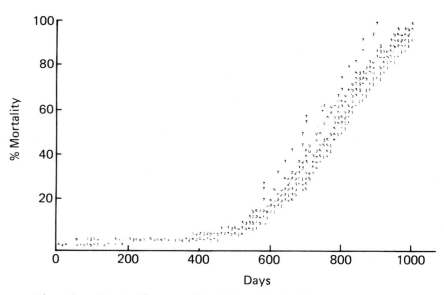

Fig. 6. Mortality replication #4 by dose.

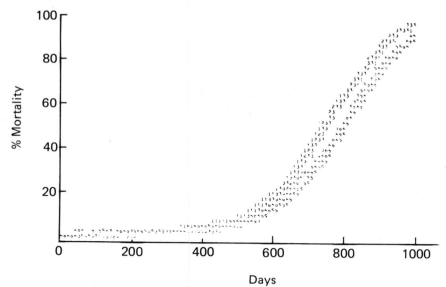

Fig. 7.   Mortality to ppm by replication.   30 ppm by replication.

level, the tumors occurred at low incidences by 9 months and increased rapidly between 12 and 18 months. The sacrifice intervals were placed close together during this time inter- val since a preliminary study with 2-AAF indicated there would be rapid growth of neoplasms during this time interval. At 100 and 75 ppm, the growth and induction of the tumors was much lower and later in time. At the 18-month sacrifice, there were still approximately 90% of the mice left on the study. Since the lower doses of 2-AAF were not causing neo- plastic growth at this time interval, it was decided to post- pone the remaining sacrifices on three of the six replications and schedule them at 24 months. Since death rates were low, there was a large number of mice available for this transfer.

Figure 10 has the same data as that shown in Fig. 9, but is plotted in relation to dose, using the 18- and 24-month sacrifices. This figure shows the apparent effect of extend- ing the sacrifices by 18 months. There are substantial changes at the 100- and 150-ppm dose levels; however, the 6- month extension did not appear to induce changes at 75 ppm or below. Plotting these data in this respect shows an apparent minimal effect level.

The data shown in Figs. 9 and 10 were from mice that were alive at their respective scheduled sacrifice dates. It be- came apparent that this study will provide data to analyze the problem of whether mice that are allocated to a scheduled

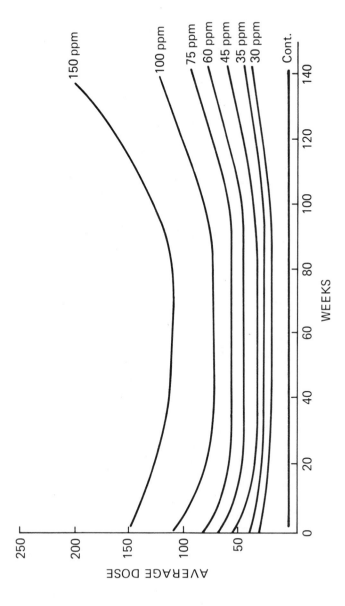

Fig. 8. Average dose (microgram/kilogram/mouse/week).

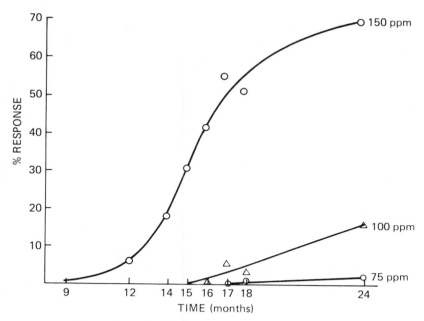

Fig. 9. Bladder neoplasms.

sacrifice group and die before they reach their sacrifice date should or should not be included in the sacrifice data. Figure 11 shows the bladder neoplasm data plotted in both respects. The sacrificed animals include only those animals that were actually alive at the sacrifice times, while the total animals includes all animals that were allocated to the specific treatment group, irrespective of whether they survived to the sacrifice time. The data show no apparent differences up to 18 months, but a decrease in the percentage response of the total group is apparent at 24 months. This may be a direct function of the fact that only about 10% of the mice had died by the 18-month time interval, but the number dying had increased to approximately 50% by 24 months.

The occurrence of liver neoplasms in sacrificed animals is plotted by time in months in Fig. 12. The data show a definite dose-response effect. Of great significance here is that this study, like many carcinogenic studies, was originally scheduled to terminate at 18 months. If only 18-month data were available, this study may have concluded that there was minimal or no effect in the liver tumor development. Figure 13 has the same data as that shown in Fig. 12, but is plotted in relation to dose, using the 18- and 24-month sacrifices. These results are different than the bladder neo-

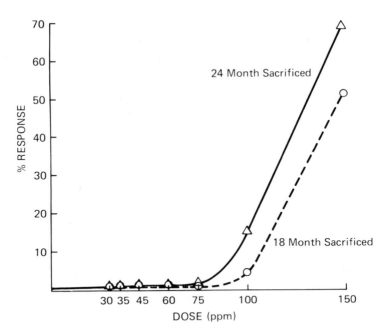

Fig. 10. Bladder neoplasm.

plasm results in that there does not appear to be any minimal
effect level at either sacrifice group. The data are very
linear and appear to extrapolate directly to zero in both
sacrifice groups. Figure 14 has the same data, but the liver
neoplasms for the dead and moribund group are included as a
comparison of the effects of different reasons for removal
from the study.

A comparison is made in Fig. 15 of the bladder and liver
neoplasms at 150 ppm. As pointed out earlier, the bladder
tumors appear early and grow rapidly, while the liver tumors
appear later in the study.

Hyperplasia of the bladder epithelium is thought to be a
precursor of bladder neoplasms. In Fig. 16, the bladder hyperplasia occurrence in sacrificed mice is plotted in relation
to time. Hyperplasia develops early and remains at a constant
level throughout the study period. As in development of
bladder neoplasms, hyperplasia was noted in the mice in the
75-, 100-, and 150-ppm dose levels. The mice at 100 and 75
ppm exhibit a specific level of response that remains at
essentially the same level throughout the sacrifice periods,
which would indicate that the formation of hyperplasia is
more a function of dose rate rather than total dose.

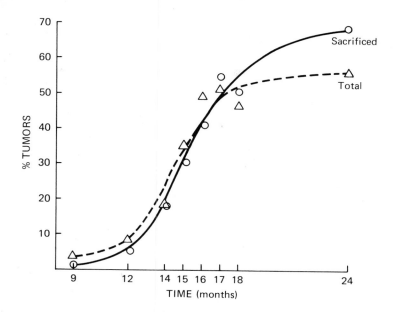

Fig. 11. Bladder neoplasms a 150 ppm. Sacrifice vs. Total
    (sacrifice + dead and Moribund).

The remainder of the data presented have to do with the
mice in the serial treatment groups. Table II shows the remaining ED^{01} experimental design delineating the serial
treatment protocol. There are four dose levels (60, 75, 100,
and 150 ppm) with two sacrifice intervals (18 and 24 months)
and three treatment intervals (9, 12, and 15 months). The
mice were fed 2-AAF for 9, 12, or 15 months and then sacrificed at either 18 or 24 months. All other conditions are
the same as the rest of the study.
    The effects on bladder hyperplasia when treatments are
discontinued are shown in Figs. 17, 18, and 19. Results are
plotted in relation to time and are compared to the mice in
which the 2-AAF dosing was not discontinued. At each time
interval wherein the treatment was discontinued, bladder hyperplasia appears to be readily reversible when the carcinogen is removed. At the 9-, 12-, and 15-month intervals, the
percentage response decreases in each group as well as each
dose level.
    The effects of discontinuing 2-AAF in the feed on bladder
neoplasms is shown in Figs. 20, 21, 22, 23, and 24. Figure
20 shows that after 9 months on 2-AAF at 150 ppm, the tumor
incidence continues to increase to 18 and 24 months, but the
rate is much slower than if the treatment is continued to

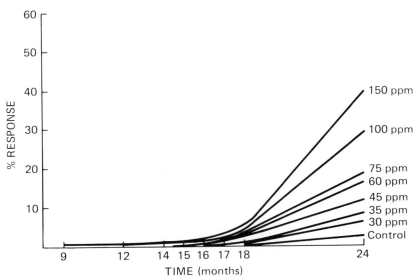

Fig. 12.   Liver neoplasms — sacrificed animals.

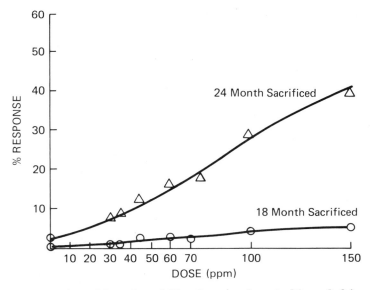

*Fig. 13.   Sacrificed animals at 18 and 24
months.   Liver neoplasms.*

Table II.
2-AAF ED^{01} Study
Serial Treatment

Months of sacrifice:	18	24	18	24	18	24
Months of 2-AAF:	15	15	12	12	9	9
150 ppm	72	72	72	72	72	72
100 ppm	72	72	72	72	72	72
75 ppm	144	144	144	144	144	144
60 ppm	216	216	216	216	216	216

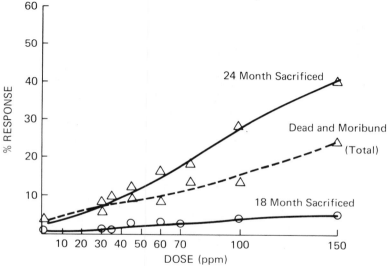

Fig. 14. Sacrificed animals at 18 and 24 months.
Liver neoplasms.

the sacrifice times. This would suggest that induction of
bladder tumors is not completed after 9 months on 2-AAF.
Figure 21 exhibits essentially the same response, that is,
tumor induction has occurred, but at a much lower rate than
in those animals in which treatment is continued to the 18-
and 24-month groups. These same data were tested using the
sacrificed plus the dead and moribund mice in one grouping
(Fig. 22). Using these data, the percentage response at the
24-month sacrifice was not less than that noted in the 18-
month sacrifice, as was noted when only sacrificed animals
were used. Figure 23 shows the results from the animals that
were treated to 15 months and sacrificed at 18 and 24 months.
The percentage response does not materially change during the

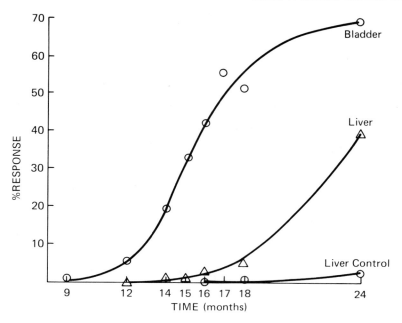

Fig. 15. *Liver neoplasms vs. bladder neoplasms at 150
ppm (sacrificed animals).*

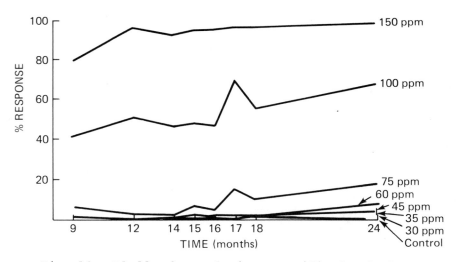

Fig. 16. *Bladder hyperplasia − sacrificed animals.*

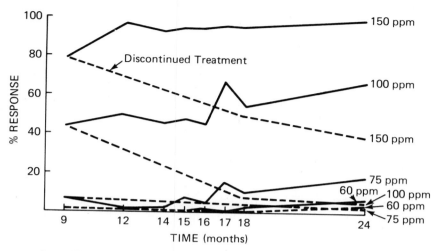

Fig. 17. Bladder hyperplasia - serial vs. continuous treatment discontinued at 9 months.

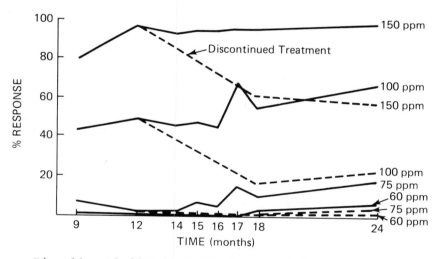

Fig. 18. Bladder hyperplasia - social vs. continuous treatment discontinued at 12 months.

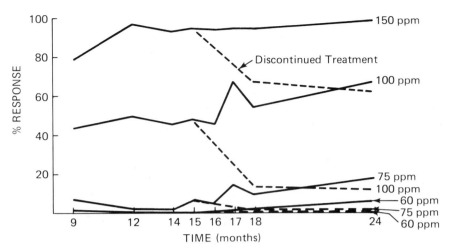

Fig. 19. Bladder hyperplasia - serial vs. continuous treatment discontinued at 15 months.

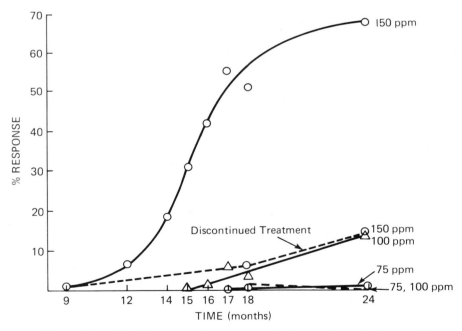

Fig. 20. Bladder neoplasms, serial treatment discontinued at 9 months.

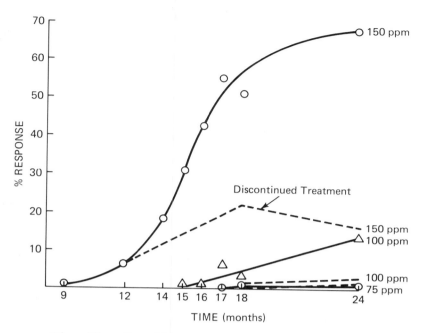

Fig. 21. Sacrificed only.

three sampling periods, indicating that no further tumor induction took place following removal of the carcinogen from
the feed. Figure 24 uses the same data combined with the
dead and moribund mice (as was Fig. 22) to show that essentially no variation results when compared to use of sacrificed mice alone. A slight response was, however, noted in
this group at the 100-ppm group that was not noted in the
sacrificed mice. Figure 25 summarizes the results to show
them at 150 ppm from the three time treatments.

Figures 26 and 27 show the effects of discontinuing 2-AAF
in the feed on the development of liver neoplasms. Figure
26 shows that after 9 months on 2-AAF, the tumor incidence
continues to increase with time at all four dose levels. The
rates of increase are, in most cases, only slightly lower
than in those mice that were treated for the full 18 or 24
months. This would indicate that induction of liver tumors
appears to take place by 9 months to a high degree. The data
from mice fed to 12 and 15 months show essentially the same
results. Comparing the liver and bladder tumor development
shows that bladder tumors become apparent much sooner than
liver tumors, but need the presence of the carcinogen if the
incidence is to continue. Liver tumors, on the other hand,

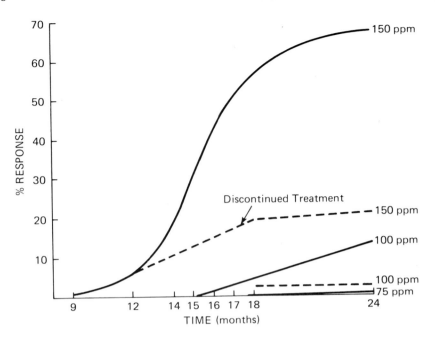

Fig. 22. Bladder neoplasms - serial treatment discontin-ued at 12 months.

become apparent at a much later time in the life of the mice, but do not need the continued presence of the carcinogen for the incidence to continue to increase.

The data covered in this presentation represent only a small percentage of the subjects that will be analyzed from this study. As data continue to become available, some of the toxicologically pertinent topics that will be evaluated will be

1. more in-depth analysis of all the data shown today;
2. comparison of total doses when given at different times and at different levels;
3. effects of serial sacrifice, terminal sacrifice, and lifespan data when applied to different statistical and toxicological models;
4. application of the different statistical models to determine their significance in determination of risk estimation;
5. effect of barrier conditions in relation to development of toxicological signs, neoplasms, lifespan, etc.;
6. biological variability between replications;

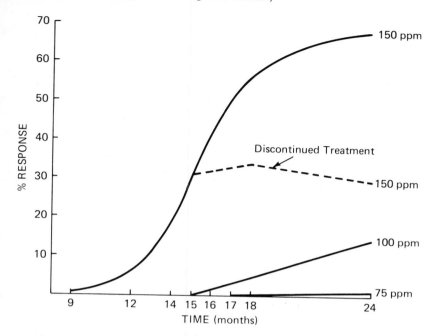

Fig. 23. Bladder neoplasms - serial treatment dis-continued at 15 months.

7. evaluation of toxicological and statistical method-ology; and

8. slope functions and risk estimation procedures.

In addition, other areas such as information gathering, storage and retrieval, animal husbandry, pathology baseline data, and microbiological surveillance techniques will un-doubtedly be able to use the data of the ED^{01} study to report significant contributions in their respective areas.

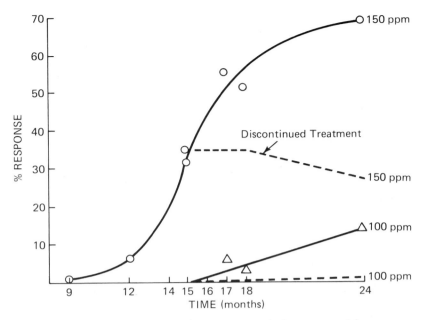

Fig. 24. Bladder neoplasms — serial vs. continuous, discontinued at 15 months, sacrificed plus dead and moribund. Sac + D + M.

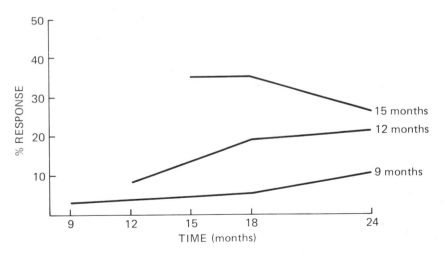

Fig. 25. Bladder neoplasms — serial treatment at 150 ppm, sacrifice + dead and moribund.

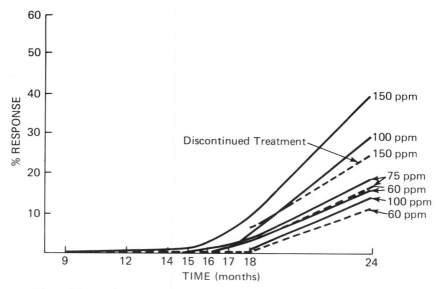

Fig. 26. Liver neoplasms – serial treatment discontinued at 9 months.

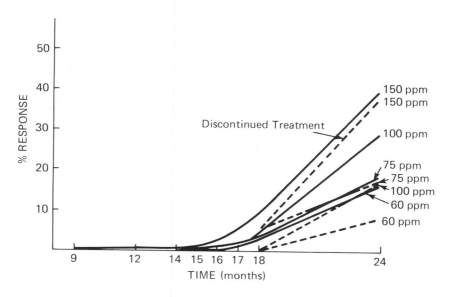

Fig. 27. Liver neoplasms – serial treatment discontinued at 12 months.

REFERENCES

1. Cranmer, M. F., Estimation of risks due to environmental
    carcinogenesis, *Med. Pediatr. Oncol.* *3*(2), 169-198,
    (197X).
2. Cranmer, M. F., Hazards of pesticide development and
    mammalian toxicity: Carcinogenicity, teratogenicity,
    and mutagenicity, *Proc. XVth Int. Congr. Entomol.*,
    719-736, (1977).
3. Cranmer, M. F., Reflections in toxicology, *J. Nat. Acad.*
    Sci. USA 64(2), 158-179, (1974).
4. Meeting of the NCTR Science Advisory Board, October
    (1977).

THE ROLE OF THE ENDOPLASMIC
RETICULUM IN CARCINOGENESIS

Dennis V. Parke

Most adult mammalian cancer arises in epithelial tissues namely, the skin, the gastrointestinal, respiratory, and genitourinary tracts, the breasts, and the liver. These epithelial tissues are characterized by their synthesis of extracellular proteins. The cells of the epithelial tissues frequently exhibit a transformation in morphology and function that precedes this synthesis of extracellular proteins, and they contain a well-developed endoplasmic reticulum where this protein synthesis occurs. Nevertheless, the endoplasmic reticulum has other functions perhaps more widely known to the pharmacologist and toxicologist namely, the metabolism of hormones, drugs, and environmental chemicals.

DETOXIFICATION AND ACTIVATION OF CHEMICALS BY MICROSOMAL
ENZYMES

The role of the enzymes of the hepatic endoplasmic reticulum in the metabolism of drugs and the detoxication of chemicals is well known. More recently, attention has been directed toward the converse effect namely, the *activation* of toxic chemicals by these enzymes and, more especially, to the activation of carcinogens. The microsomal enzymes involved in the detoxication and activation of chemicals are mixed function oxidases, mediated by the terminal oxygenase, cytochrome *P*-450. Other microsomal enzymes, concerned primarily with detoxication, and not activation, are epoxide hydrase and glucuronyltransferase, which further metabolize the oxygenated metabolites, converting arene oxides into dihydrodiols, and conjugating the diols, phenols, etc. to form the corresponding glucuronides. Microsomal enzymes are not confined to the liver, and mixed function oxidase, epoxide

ISBN 0-12-192750-4

hydrase, and glucuronyltransferase activities have been demonstrated in the lung, kidney, intestinal tract, and many other tissues (Lake *et al.*, 1973).

One of the characteristics of the microsomal enzymes, particularly the mixed-function oxidases, is their substrate-mediated induction (Parke, 1975). Three major prototypes of enzyme-inducing agents are known; namely, (a) *drugs,* which result in an increase of cytochrome *P*-450 and the microsomal reductase, i.e., NADPH-cytochrome *c* reductase, (b) *carcino-genic polycyclic hydrocarbons,* which increase cytochrome *P*-448, a modification of cytochrome *P*-450, but do not increase the reductase, and (c) *steroids,* which increase the reductase but not cytochrome *P*-450. These increased enzyme activities are attributed to *de novo* synthesis of the enzyme proteins, and are believed to involve genomal derepression with conse-quent increases in transcription and translation. Enzyme induction by drugs generally reaches a maximum at about 24 hours after administration of the inducing agent. However, with carcinogenic polycyclic hydrocarbons, and possibly other carcinogens, a much earlier increase in cytochrome *P*-448 has been observed, usually within 2-6 hours of administration of the carcinogen (Alvares *et al.*, 1973) and this is independent of protein synthesis and may result from a conformational change of cytochrome *P*-450 (Imai and Siekevitz, 1971; Kahl *et al.*, 1977; Parke, 1977c).

The carcinogenic polycylic hydrocarbons, such as benzo(*a*)-pyrene and benz(*a*)anthracene, are characterized by the pres-ence of the phenanthrene nucleus within the chemical structure, and Jerina (Jerina and Daly, 1977; Jerina and Lehr, 1977) has proposed that activation of polycyclic hydrocarbons to the proximate or ultimate carcinogens necessitates "bay-region" epoxidation of this phenanthrene nucleus. Other workers (Sims *et al.*, 1974; Swaisland *et al.*, 1974) have found that the diol epoxide is the active carcinogen, with the initial polycyclic hydrocarbon epoxide being converted into the cor-responding dihydrodiol, by epoxide hydrase, then further oxygenated to yield the diol epoxide (see Fig. 1). These diol epoxides do not appear to be suitable substrates for epoxide hydrase, and form stabilized internal ion-pair com-pounds that react as carbonium ions with DNA, even in the presence of glutathione and proteins, so that no threshold concentration of the diol epoxide need exist for DNA aryla-tion and consequent damage of DNA and carcinogenesis (Hulbert, 1975). Microsomal oxygenation is also known to be involved in the activation of carcinogenic aromatic amines, such as 2-naphthylamine (Radomski and Brill, 1970), carcinogenic amides, such as 2-acetamidofluorene (Weisburger *et al.*, 1972), and carcinogenic nitrosamines (Magee and Barnes, 1967).

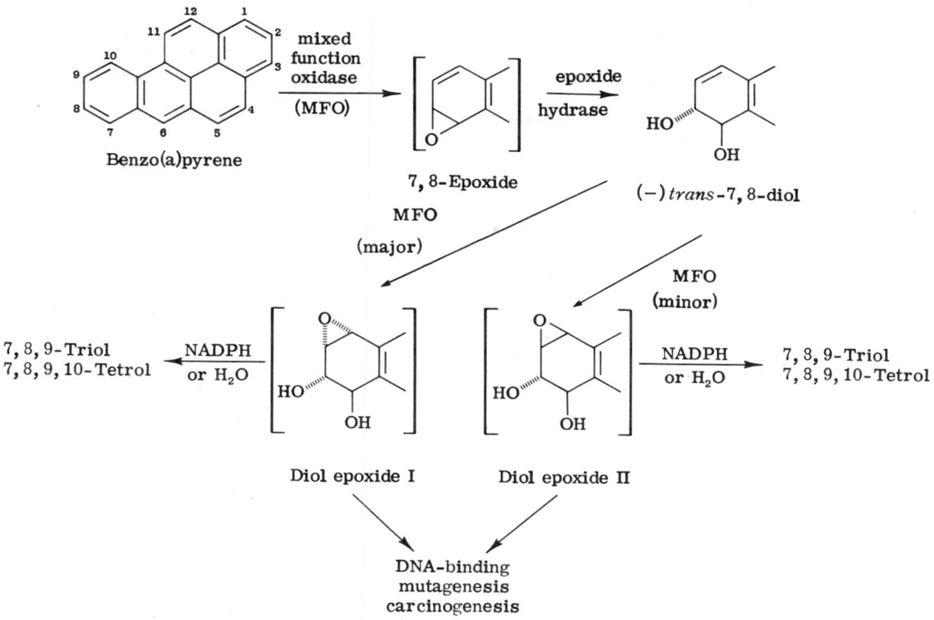

Fig. 1. Activation of benzo(a)pyrene by epoxidation.

Microsomal oxygenation of the model substrate biphenyl yields two major products, namely, 2-hydroxybiphenyl and 4-hydroxybiphenyl. This is of interest as both the 2- and 4-hydroxylations of biphenyl are catalyzed by cytochrome P-448, while 4-hydroxylation only is catalyzed by cytochrome P-450 (Burke and Bridges, 1975; Atlas and Nebert, 1976). This specificity of cytochrome P-448 for the 2-hydroxylation, and the known formation of cytochrome P-448 by carcinogens has suggested a correlation between the effect of carcinogens on the endoplasmic reticulum and an increase in the enzymatic 2-hydroxylation of biphenyl (Creaven and Parke, 1966; McPherson et al., 1976). Furthermore, the irreversible displacement of ribosomes from the endoplasmic reticulum (degranulation) by carcinogens (Williams and Rabin, 1971; Delaunay and Schapira, 1974) has suggested a further relationship between degranulation of the endoplasmic reticulum and the expression of the enzyme biphenyl 2-hydroxylase (Parke, 1977a). The structure of biphenyl contains the spatial arrangement of the phenanthrene nucleus, and the 2-hydroxylation, but not 4-hydroxylation, may be considered as bay-region oxygenation.

The interaction of activated carcinogens, namely, the electrophilic or carbonium ion ultimate carcinogens, with DNA, to form alkylated or arylated derivatives, seems to be fundamental to mutagenesis and chemical carcinogenesis (Parke, 1977b). However, not all mutagens are carcinogenic, and it would appear that some biological damage additional to DNA alkylation is requisite for the expression of carcinogenesis. The essential requirement is for an increased rate of DNA synthesis, increased mitosis, and decreased immune surveillance to ensure the replication of the damaged DNA and the establishment of a clone of malignantly transformed cells before DNA repair can be effected (see Fig. 2) (Parke, 1977b).

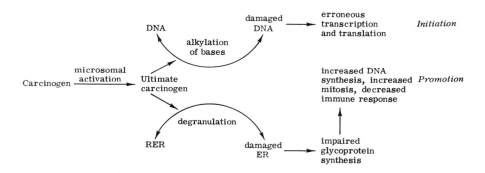

Fig. 2. Mechanism of chemical carcinogenesis.

THE ENDOPLASMIC RETICULUM, GLYCOPROTEIN SYNTHESIS, AND MALIGNANCY

Another essential biological function of the endoplasmic reticulum is the synthesis of glycoproteins. Indeed, the initiation of glycoprotein synthesis appears to be exclusive to membrane-bound ribosomes (Hallinan et al., 1968) and the glycosyltransferases are found only in the membranes of the endoplasmic reticulum, Golgi complex, and plasma membrane (Schachter, 1974). Glycoproteins play a vital role in the overall function and economy of the cell, and in the regulation of the whole eukaryotic organism. Glycoproteins secreted into the plasma membrane glycocalyx regulate cell division, and glycoproteins secreted from the cell, such as the immunoglobulins, play a major role in immune surveillance. The nature of the carbohydrate moieties of these glucoproteins appears to be the critical feature determining the rate of mitosis, the immune characteristics of the cell, and the immune surveillance within the living organism. However, the mechanism by which the various glycosyltransferases within the endoplasmic reticulum are regulated is as yet unknown (Schachter, 1974).

The binding of the ribosomes to the endoplasmic reticulum, which is necessary for the formation of the glycoprotein polypeptide backbone, appears to directly involve cytochrome P-450. An increase in cytochrome P-450 produced by phenobaritone or methylcholanthrene leads to enhancement of polysome binding, whereas loss of cytochrome P-450, by administration of cobalt salts, diminishes binding. Purified cytochromes P-450 and P-448 also bind polysomes, so that it is likely that it is the destruction of this cytochrome, or the loss of the activity of this enzyme, that leads to degranulation (Takagi, 1977).

The degranulation of the endoplasmic reticulum that accompanies carcinogenesis (Delaunay and Schapira, 1974) must of consequence be accompanied by interference with glycoprotein synthesis (Parke, 1977b). Indeed, in malignantly transformed cells, one of the earliest changes observed is the loss of a high molecular weight glycoprotein from the cell surface (Warren et al., 1974) and changes in the nature of the terminal sugar moieties of the glycoproteins of the glycocalyx (Van Beek et al., 1973). Elevated levels of certain glycosyltransferases have been observed in the plasma of cancer patients, and high levels of fucosyltransferase, with correspondingly low levels of sialyltransferase and galactosyltransferase were demonstrated in various human metastasizing tumors (Kessel et al., 1977). We have also shown a marked inhibition of glycosyltransferase in the synthesis of gastric mucus in patients with gastric carcinoma (Parke and Symons, 1977). It would therefore appear that both the quantitative and qualitative aspects of glycoprotein synthesis are profoundly affected by malignant transformation, and that the changes occurring in the glycoprotein of the cell surface are associated with many of the characteristics of the malignant cell, namely, loss of adhesiveness, accelerated cell division and altered antigenicity (Warren et al., 1974). Furthermore, degranulation of the endoplasmic reticulum might lead to increased synthesis of intracellular proteins by the cytoplasmic ribosomes, which, as in the neonate, would lead to the growth of the cell and to accelerated cell division. Hence, degranulation would result in a switchover of the economy of the cell from a predominance of synthesis

of glycoproteins for export (glycocalyx, immunoglobulins,
mucus) to the fetal/neonatal state of predominantly intracellular protein synthesis, with accelerated cell growth and
mitosis.

CARBENE FORMATION WITH CYTOCHROME P-450, MEMBRANE
DEGRANULATION, AND PROMOTION OF CARCINOGENESIS

    The weak hepatocarcinogen safrole, when administered to
rats, results in induction of the microsomal mixed function
oxidases, degranulation of the liver endoplasmic reticulum,
increased hepatic cytochrome P-450, increased biphenyl 2-
hydroxylase activity, but a progressive loss of other mixed-
function oxidase activity.(Parke and Gray, 1978). These
pathobiological changes are associated with the hepatic
metabolism of safrole to form an active intermediate that
binds immediately to cytochrome P-450 to form a carbene
complex (see Fig. 3) (Ullrich, 1977). In consequence there
is a marked induction of the microsomal enzymes, including
cytochrome P-450, but since the P-450 reacts immediately with
more activated safrole to form more carbene complex, there
is no increase in the P-450-mediated mixed-function oxidase
activity. However, as the result of damage to the endoplasmic reticulum, cytochrome P-448 is formed and biphenyl
2-hydroxylase activity, but not the 4-hydroxylase activity,
is expressed.

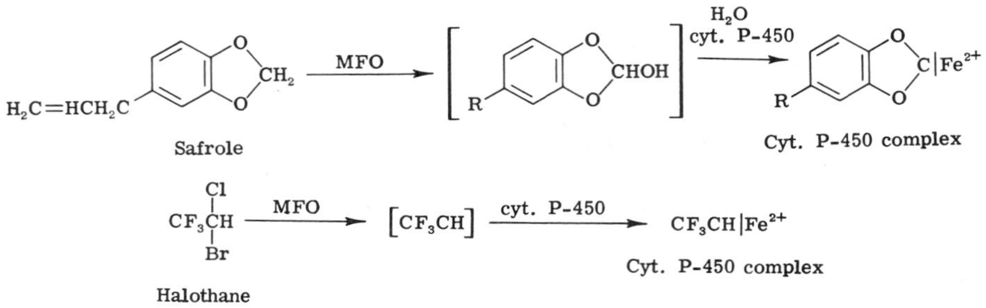

*Fig. 3. Formation of carbene complexes of cytochrome
P-450.*

Similar complexes of cytochrome P-450 have been shown to be formed by carbon tetrachloride, chloroform (Wolf et al., 1977), halothane, and other halogenated anesthetics (see Fig. 3) (Takahashi et al., 1974) and, indeed, probably arise from reductive dehalogenation of many organohalogen compounds. Certain other chemicals, such as methylene dioxyaryl compounds and amphetamine derivatives, are also able to form complexes with cytochrome P-450. Moreover, not only is cytochrome P-450 of the liver so affected, but so also is the cytochrome P-450 of lung, intestine, and probably other extrahepatic tissues (Buening and Franklin, 1976).

Using ^{14}C-labeled carbon tetrachloride, trichlorofluoromethane (Freon 11), chloroform, and halothane, Uehleke et al. (1977) have shown covalent binding of radioactivity to microsomal protein and lipid, dependent on microsomal metabolism, with each of these organohalogen compounds (see Table I). However, none of these compounds showed any mutagenic activity with tester strains of Salmonella typhimurium, and are therefore unlikely to bind to DNA or to initiate carcinogenesis.

Similar effects on the hepatic microsomal mixed-function oxidases, with selective inhibition of the terminal oxygenase, cytochrome P-450, have been observed with a number of chlorophenol pesticides, chlorodiphenyl ethers (predioxins), and dioxins (Arrhenius et al., 1977). These compounds were found to stimulate the microsomal flavoprotein reductase, which is the situation that occurs following carbene formation; namely, the induction of both the terminal cytochrome and the reductase but with the activity of the former inhibited by the binding of an active metabolite. This state of affairs results in the disruption of the normal coupled electron transfer to cytochrome P-450 from the reductase, with a corresponding change in the predominant mode of hydroxylation from C-oxygenation to N-oxygenation (Arrhenius et al., 1977) and possible lipid peroxidation and damage of the endoplasmic reticulum (Suarez et al., 1972). There is a close chemical and biological correlation between N-oxygenation of aromatic amines, epoxidation of polycyclic hydrocarbons, uncoupled mixed-function oxygenase, cytochrome P-448, and chemical carcinogenicity, on the one hand, and C-oxygenation, dihydrodiol formation, cytochrome P-450, and detoxication, on the other.

Carcinogenesis, particularly in mouse-skin tumorogenesis, has been considered as a two-stage process (Chu et al., 1977) consisting of a primary stage of initiation, which probably involves DNA damage, and a secondary stage of promotion, which leads to the rapid expression of this information by DNA replication, protein synthesis, and inhibition of glycoprotein synthesis (Parke, 1977b). The potential of chemicals to initiate carcinogenesis would be related to their chemical

Table I

Covalent Binding of Carbon Tetrachloride, Freon 11,
Chloroform, and Halothane to Rabbit Liver
Microsomal Protein, in vitro[a]

[14]C-Labeled substrate	Amount of radioactivity bound in (nmole/mg protein)	
	10 minutes	1 hour
Carbon tetrachloride	7	20
Freon 11	5	7
Chloroform	2	5
Halothane	1	3

[a]Data from Uehleke, et al. (1977).

reactivity, or that of their activated metabolites, and their
ability to alkylate DNA. The potential of chemicals to *pro-
mote* carcinogenesis would depend on their ability to damage
the endoplasmic reticulum and modify or inhibit microsomal
cytochrome *P*-450, with uncoupling of the electron-transport
chain and loss of polysomes. The former group of chemicals,
the *initiators,* would be mutagens detectable by the Ames
test, while the latter group, the *promoters,* would be substances that by various means (damage to the microsomal membranes, carbene complex formation, uncoupling of electron
transport) would lead to the conversion of cytochrome *P*-450
to *P*-448, which is readily monitored by the biphenyl test and
the increased level of biphenyl 2-hydroxylase activity
(McPherson *et al.*, 1974). Obviously, most carcinogens mani-
fest both properties and are both initiators and promoters,
which means that they are *complete carcinogens.* However,
many weak carcinogens are compounds of the second class and
are only promoters, like chloroform and the other halogeno-
hydrocarbons, which Uehleke *et al.* (1977) showed did not give
a positive Ames test. Nevertheless, exposure to these promoters would be likely to contribute to the overall incidence
of carcinogenesis, although it is likely that they would produce a significant effect on the endoplasmic reticulum only
at relatively high dosage, since both enzyme induction and
degranulation of the endoplasmic reticulum are functions of
substrate concentration, that is, of dosage.

SPECIES DIFFERENCES IN MICROSOMAL ENZYME ACTIVITY AND IN
CARCINOGENESIS

Recent studies into the safety evaluation of industrial
and environmental chemicals has shown that many compounds are
carcinogenic only in the mouse. These chemicals, which, on
prolonged feeding at high dosage, lead to the formation of
hepatocellular carcinoma in mouse, fail to produce any malignancy in other animal species. Many of these chemicals are
organohalogens and most of them give negative Ames tests for
mutagenicity, which would suggest that they are functioning
as carcinogenesis promoters, and lead to potentiation of
endogenous oncogenic viruses or of subthreshold levels of
initiating carcinogens.
Malignant cell transformation has been shown to result
from the combined effects of chemicals and tumor viruses
under conditions in which either agent alone was ineffective
(Price et al., 1972). Dexamethasone, which gives a strongly
positive response in the biphenyl test for endoplasmic reticulum damage, stimulates by 20-fold the expression of murine
mammary tumor virus, in mouse cell livers (Parke et al.,
1974). This induction of cancer by combinations of viruses
and chemicals has been extensively reviewed by Roe and Rowson
(1968) and the reviewers' proposed mechanisms for this synergism include interference with detoxication, changes in
hormonal regulation of cellular metabolism, inhibition of
immunological defense, and induction of hyperplasia to
facilitate the establishment of clones of transformed cells.
The reason why carcinogenesis by these promoting agents
is so often seen only in mouse liver is probably because of
some fundamental species abnormality in microsomal metabolism.
A number of such abnormalities are seen with the mouse, which
therefore make this animal a most unsuitable species for
monitoring carcinogenic potential, especially of promoters.
The first of these is the high rate of hepatic microsomal
metabolism in the mouse and Brown et al. (1974) have shown
that chloroform, which is a hepatocarcinogen in the mouse, is
metabolized--and hence metabolically activated--at a much
faster rate in the mouse than in the rat or squirrel monkey
(see Table II). The conversion of $^{14}CHCl_3$ to $^{14}CO_2$, an index
of overall metabolism, is about 80% of the dose in the
mouse, 70% in the rat, and only 20% in the squirrel monkey.
Hence in the mouse there will be greater generation of the
active metabolite, and hence greater damage to the endoplasmic
reticulum, and greater potentiation of carcinogenesis, for a
given dose of promoter. Other abnormalities include the
greater predominance of the rough to smooth endoplasmic re-

Table II

Species Difference in Rates of Metabolism of Chloroform (60 mg/kg)[a]

Species	Body wt. (g)	Dose (%) excreted in 2 days as		Total recovered
		$^{14}CO_2$	$^{14}CHCl_3$	
Mouse				
CBA	20–30	76 ± 5	7 ± 2	83
C57	20–30	79	5	84
CF/LP		76	6	82
Rat (Sprague–Dawley)	200–300	66 ± 4	20 ± 5	86
Monkey (Squirrel)	500–1000	18 ± 2	79 ± 3	97

[a] Data from, Brown et al. (1974).

ticulum in the liver of mouse than in rat, guinea pig, or monkey liver (Gram et al., 1971), the higher activities of mixed-function oxygenase in the rough endoplasmic reticulum of mouse than other species (Gram et al., 1971), and the high ratio of high spin/low spin of mouse hepatic cytochrome P-450 (Levin et al., 1973). These features suggest that the extent of degranulation of the endoplasmic reticulum for a given amount of promoter substrate would be much greater in the mouse than in other species.

The mouse also has a high potential for the formation of other active metabolites leading to toxicity, e.g., in the hepatic necrosis resulting from the N-hydroxylation of paracetamol. At a dose of 500 mg paracetamol/kg body weight, the mouse exhibits an incidence of 75% of hepatic necrosis, the hamster (because of the high normal levels of liver cytochrome P-448) shows 100% incidence, whereas the rat does not experience any hepatic necrosis at all at this dose and even at 1000 mg/kg shows an incidence of only 2% (Gillette, 1977). Moreover, whereas enzyme induction by phenobarbitone leads to a decrease in the covalent binding of paracetamol metabolites and to a decrease in liver necrosis in the hamster, both binding and necrosis are increased still further in the mouse by this treatment (Gillette, 1977).

A further factor that may make the mouse more vulnerable to the action of promoters of carcinogenesis is the possibly greater level of malignant initiation activity. The high rate of microsomal metabolism, consequent degranulation, and hence carcinogen activation by bay region epoxidation, would potentiate the effect of environmental mutagens so that a given dose would likely lead to greater DNA damage than occurs in other larger species. Furthermore, there is evidence that there is a much higher level of oncogenic viruses present in the mouse than in most other species. All of these factors may explain the anomalous behavior of the mouse in chemical carcinogenesis and indicate that the formation of hepatocellular carcinoma in this species, in consequence of the administration of a number of organohalogen compounds and other environmental chemicals, is a spurious index of the true carcinogenic potential of these chemicals in other animal species, such as man. Since so many of these organohalogen chemicals are of great medical, social, and industrial value, such as the volatile anesthetics, pesticides, and organic solvents, and indeed many are produced naturally by the chlorination of water supplies (Tardiff et al., 1975), their safety evaluation is of paramount concern. The molecular mechanisms herein reviewed hopefully may form a basis for the

true evaluation of the carcinogenic potential of these compounds, which should be undertaken in an appropriate animal
species and with due regard to the biotransformations and
chemobiokinetics of the chemicals concerned.

References

Alvares, A. P., Parli, C. J., and Mannering, G. J. (1973).
    Induction of drug metabolism. VI. Effects of phenobarbital
    and 3-methylcholanthrene administration on N-demethylating
    enzyme systems of rough and smooth hepatic microsomes.
    Biochem. Pharmacol. *22*, 1037-1045.
Arrhenius, E., Renberg, L., Johansson, L., and Zetterqvist,
    M. A. (1977). Disturbance of microsomal detoxication
    mechanisms in liver by chlorophenol pesticides. *Chem.-
    Biol. Interact.* *18*, 35-46.
Atlas, S. A., and Nebert, D. W. (1976). Genetic association
    of increases in naphthalene, acetanilide and biphenyl
    hydroxylations with inducible aryl hydrocarbon hydroxylase
    in mice. *Arch. Biochem. Biophys.* *175*, 495-500.
Brown, D. M., Langley, P. F., Smith, D., and Taylor, D. C.
    (1974). Metabolism of chloroform. I. The metabolism of
    (^{14}C) chloroform by different species. *Xenobiotica 4*,
    151-163.
Buening, M. K., and Franklin, M. R. (1976). The formation of
    cytochrome P-450 metabolic intermediate complexes in
    microsomal fractions from extrahepatic tissues of the
    rabbit. *Drug Metab. Disp.* *4*, 556-561.
Burke, M. D., and Bridges, J. W. (1975). Biphenyl hydroxyla-
    tions and spectrally apparent interactions with liver
    microsomes from hamsters pretreated with phenobarbitone
    and 3-methylcholanthrene. *Xenobiotica 5*, 357-376.
Creaven, P. J., and Parke, D. V. (1966). The stimulation of
    hydroxylation by carcinogenic and non-carcinogenic com-
    pounds. *Biochem. Pharmacol.* *15*, 7-16.
Chu, E. H. Y., Trosko, J. E., and Chang, C. C. (1977).
    Mutational approaches to the study of carcinogenesis.
    J. Tox. Environ. Hlth. *2*, 1317-1334.
Delaunay, J., and Schapira, G. (1974). Ribosomes and cancer.
    Biomedicine 20, 327-332.
Gillette, J. R. (1977). The phenomenon of species variations;
    problems and opportunities. *In* "Drug Metabolism--From
    Microbe to Man," (D. V. Parke and R. L. Smith, eds.), pp.
    147-168. Taylor and Francis, London.

Gram, T. E., Schroeder, D. H., Davis, D. C., Reagan, R. L.,
    and Guarino, A. M. (1971). Enzymic and biochemical comp-
    osition of smooth and rough microsomal membranes from
    monkey, guinea pig and mouse liver. *Biochem. Pharmacol.*
    20, 1371-1381.
Hallinan, T., Murty, C. N., and Grant, J. H. (1968). The
    exclusive function of reticulum bound ribosomes in glyco-
    protein synthesis. *Life Sci. 7,* 225-232.
Hulbert, P. B. (1975). Carbonium ion as ultimate carcinogen
    of polycyclic aromatic hydrocarbons. *Nature (London) 256,*
    146-148.
Imai, Y., and Siekevitz, P. (1971). A comparison of some
    properties of microsomal cytochrome P-450 from normal,
    methylcholanthrene-, and phenobarbital-treated rats.
    Arch. Biochem. Biophys. 144, 143-159.
Jerina, D. M., and Daly, J. W. (1977). Oxidation at carbon.
    In "Drug Metabolism--From Microbe to Man." (D. V. Parke
    and R. L. Smith, eds.), pp. 13-32. Taylor and Francis,
    London.
Jerina, D. M., and Lehr, R. E. (1977). The bay-region theory:
    A quantum mechanical approach to aromatic hydrocarbon-
    induced carcinogenicity. *In* "Microsomes and Drug Oxida-
    tions." (V. Ullrich *et al.*, eds.), pp. 709-720. Perga-
    mon, Oxford.
Kahl, G. F., Zimmer, B., Galinsky, T., Jonen, H. G., and Kahl,
    R. (1977). Induction of cytochrome P-448 by 3-methylchol-
    anthrene in the rat during inhibition of protein synthesis
    in vivo. In "Microsome and Drug Oxidations." (V. Ullrich,
    et al., eds.), pp. 551-558. Pergamon, Oxford.
Kessel, D., Sykes, E., and Henderson, M. (1977). Glycosyl-
    transferase levels in tumours metastatic to liver and in
    uninvolved liver tissue, *J. Natl. Cancer Inst. 59,* 29-32.
Lake, B. G., Hopkins, R., Chakraborty, J., Bridges, J. W.,
    and Parke, D. V. (1973). The influence of some hepatic
    inducers and inhibitors on extrahepatic drug metabolism.
    Drug Metab. Disp. 1, 342-349.
Levin, W., Ryan, D., West, S., and Lu, A. Y. H. (1973). N-
    Octylamine difference spectra of cytochrome P-450 and
    P-448 from rat and mouse liver: A species difference.
    Drug Metab. Disp. 1, 602-605.
McPherson, F. J., Bridges, J. W., and Parke, D. V. (1974).
    In vitro enhancement of hepatic microsomal biphenyl 2-
    hydroxylation by carcinogens, *Nature (London) 252,*
    448-489.
McPherson, F. J., Bridges, J. W., and Parke, D. V. (1976).
    Studies on the nature of the *in vitro* enhancement of bi-
    phenyl 2-hydroxylation provoked by some chemical carcino-
    gens. *Biochem. Pharmacol. 25,* 1345-1350.

Magee, P. N., and Barnes, J. M. (1967). Carcinogenic nitroso
    compounds. *Adv. Cancer Res., 10,* 163-246.
Parke, D. V. (1975). Induction of the drug-metabolizing
    enzymes. *In* "Enzyme Induction." (D. V. Parke, ed.),
    pp. 207-271. Plenum, New York.
Parke, D. V. (1977a). The activation and induction of bi-
    phenyl hydroxylation and chemical carcinogenesis. *In*
    "Microsomes and Drug Oxidations." (V. Ullrich *et al.,*
    eds.), pp. 721-729. Pergamon, Oxford.
Parke, D. V. (1977b). Biochemical aspects. *In* "Principles
    of Surgical Oncology." (R. W. Raven, ed.), pp. 113-156.
    Plenum, London.
Parke, D. V. (1977c). Regulation of the drug-metabolizing
    enzymes. *In* "Drug Metabolism--From Microbe to Man."
    (D. V. Parke and R. L. Smith, eds.), pp. 55-70. Taylor
    and Francis, London.
Parke, D. V., and Gray, T. J. B. (1978). A comparative study
    of the enzymic and morphological changes of livers of
    rats fed butylated hydroxytoluene, safrole, Ponceau MX or
    2-acetamidofluorene. *In* "Primary Liver Tumors." MTP
    Press, Lancaster (in press).
Parke, D. V., and Symons, A. M. (1977). The biochemical
    pharmacology of mucus. *In* "Mucus in Health and Disease."
    (M. Elstein and D. V. Parke, eds.), pp. 423-441, Plenum,
    New York.
Parks, W. P., Scolnick, E. M., and Kozikowski, E. H. (1974).
    Dexamethasone stimulation of murine mammary tumor virus
    expression: A tissue source of virus. *Science 184,*
    158-160.
Price, P. J., Suk, W. A., and Freeman, A. E. (1972). Type c
    RNA tumor viruses as determinants of chemical carcino-
    genesis. Effects of sequence of treatment. *Science 177,*
    1003-1004.
Radomski, J. L., and Brill, E. (1970). Bladder cancer induc-
    tion by aromatic amines: Role of N-hydroxy metabolites.
    Science 167, 992-993.
Roe, F. J. C., and Rowson, K. E. K. (1968). The induction of
    cancer by combinations of viruses and other agents.
    Intern. Rev. Exp. Pathol. 6, 181-227.
Schachter, H. (1974). The subcellular sites of glycosylation.
    Biochem. Soc. Symp. 40, 57-71.
Sims, P., Grover, P. L. Swaisland, A. J., Pal, K., and Hewer,
    A. (1974). Metabolic activation of benzo(a)pyrene pro-
    ceeds via a diol epoxide. *Nature (London) 252,* 326-328.
Suarez, K. A., Carlson, G. P., Fuller, G. C., and Fausto, N.
    (1972). Differential acute effects of phenobarbital and
    3-methylcholanthrene pretreatment on carbon tetrachloride-
    induced hepatotoxicity in rats. *Toxicol. Appl. Pharmacol.
    23,* 171-181.

Swaisland, A. J., Hewer, A., Pal, K., Keysell, G. R., Booth, J., Grover, P. L., and Sims, P. (1974). Polycyclic hydrocarbon epoxides: The involvement of 8,9-dihydro-8,9-dihydroxybenz(a)anthrancene 10,11-oxide in reactions with the DNA of benz(a)anthracene treated embryo hamster cells. *FEBS Lett.* *47*, 34-38.

Takagi, M. (1977). Binding of polysomes *in vitro* with endoplasmic reticulum prepared from rat liver. II. Possible involvement of cytochrome P-450 in the binding. *J. Biochem.* *82*, 1077-1084.

Takahashi, S., Shigematsu, M. D., and Furukawa, T. (1974). Interaction of volatile anaesthetics with rat hepatic microsomal cytochrome P-450. *Anaesthesiology 41*, 375-379.

Tardiff, R. G., Carlson, G. P., and Simon, V. (1975). *In* "The Environmental Impact of Water Chlorination." (R. L. Jolley, ed.), pp. 213-227. National Technical Information Service, U.S. Dept. of Commerce, Springfield, Virginia.

Uehleke, H., Werner, T., Greim, H., and Kramer, M. (1977). Metabolic activation of haloalkanes and tests *in vitro* for mutagenicity. *Xenobiotica 7*, 393-400.

Ullrich, V. (1977). The mechanism of cytochrome P-450 action. *In* "Microsomes and Drug Oxidations." (V. Ullrich *et al.*, eds.), pp. 192-201, Pergamon, Oxford.

Van Beek, W. P., Smets, L. A., and Emmelot, P. (1973). Increased sialic acid density in surface glycoprotein of transformed and malignant cells - a general phenomenon? *Cancer Res.* *33*, 2913-2922.

Warren, L., Fuhrer, J. P., Tuszynski, G. P., and Buck, C. A. (1974). Cell-surface glycoproteins in normal and transformed cells, *Biochem. Soc. Symp.* *40*, 147-157.

Weisburger, J. H., Yamamoto, R. S., Williams, G. M., Grantham, P. H., Matsushima, T., and Weisburger, E. K. (1972). On the sulphate ester of N-hydroxy-N-2-fluorenylacetamide as a key ultimate hepatocarcinogen in the rat. *Cancer Res.* *32*, 491-500.

Williams, D. J., and Rabin, B. R. (1971). Disruption by carcinogens of the hormone dependent association of membranes with polysomes, *Nature (London) 232*, 102-105.

Wolf, C. R., Mansuy, D., Nastainczyk, W., Deutschmann, G., and Ullrich, V. (1977). The reduction of polyhalogenated methanes by liver microsomal cytochrome P-450. *Mol. Pharmacol.* *13*, 698-705.

CHEMICALLY INDUCED TUMORS--
THEIR RELATIONSHIP TO HUMAN CANCER

W. H. Butler

So far, many aspects of the Delaney Clause have been
discussed with much critical comment. This has been directed
at both the regulatory and fundamental scientific basis of
the amendment. One can agree with much of what Hutt and
Kolbye said. Both emphasized that existing public health
regulations were adequate to cover the risks of chemicals in
the food, and indeed, Kolbye emphasized that the Clause,
in practice, had seldom been used. I would consider that a
major objection to the Delaney Clause is it elevates the
public health aspects of cancer risk above other equally,
or possibly more important, medical problems. This seems to
be totally unjustified. The Clause itself refers exclusively
to food additives but it is inescapable that, at present,
it is invoked to cover wide aspects of both the food and the
industrial scene. If one just considers nonnutritive food
additives, which do not contribute to the nutritional value
or safe preservation of food, no adverse toxicological effect
should be acceptable. It is unfortunate that the view that
cancer presents the preeminent public health risk in technological society pervades these societies. These diseases
are not new, as there is evidence of malignant neoplasia
going back 120 million years to the Jurassic period.
As has already been mentioned in this symposium, preventive medicine and, in particular, those aspects related
to public health have been preeminent in increasing the
expectation of life of the population. One could reasonably
expect that in the field of cancer, preventive medicine will
play as an important role as it has in the field of infectious diseases. In planning the policies that should be
followed by society as a whole, it is possibly unreasonable to
expect that society will necessarily be rational in its
judgments. It is obviously preferable that they be rational
but what is most important, in retaining the credibility of

ISBN 0-12-192750-4

science, is that the advice given to society be based on a rational assessment of data. This, however, immediately raises the issue, which I will not discuss, that what appears rational to me is not necessarily rational to a colleague. In order to give this advice, it is necessary to acquire the raw data as a starting point. The acquisition of these data should in the first instance be independent of the uses to which it is put in society. The collection of raw data, and the first assessment of these data should not be prejudiced by the uses to which they are put.

I was asked to talk about chemically-induced tumors and their relationship to human cancer. This immediately raises two issues that had been touched upon by both Kolbye and Clayson in discussion. These are (1) What is a carcinogen? (2) How are these recognized?

It has been said that approximately 25 industrial carcinogens, or groups of carcinogens, have been recognized for man. To this must be added certain pharmaceutical products and dietary contaminants. All this information is based on epidemiology. It is also said that there are approximately 2000 animal carcinogens based upon the evidence of experimentation in laboratory animals. Does the difference between these two figures lie in the failure of epidemiology, species variation, or overrecognition of carcinogens in laboratory species? It is probable that all three play a part in the wide discrepancy between the number of human carcinogens and the apparent number of animal carcinogens. I wish here to consider the view that there has been overrecognition of carcinogens in laboratory animals. This immediately raises the question as to what is a carcinogen.

Many people have attempted to define a carcinogen with greater or less success, although so far at this meeting little attempt has been made to do so. Broadly based on Foulds (1969), I would consider a carcinogen as follows:

> a chemical carcinogen is a substance that causes a cell or group of normal cells, which would not otherwise have shown this property, to change its biological behavior and demonstrate autonomous growth of a malignant character.

This is easy to say, but difficult to recognize. In terms of this definition, a chemical carcinogen can only be defined within the test species and the protocol in which it has been assessed. Many carcinogens satisfying this criteria have been shown to affect many species but this is not to say that one should consider carcinogens in a global sense affecting all species in any situation.

Many biologically active chemicals, including those which I would recognize as carcinogenic for a certain species, produce a range of effects. All of these are not of equal relevance to the process of induced neoplasia. Until the mechanism and biology are understood, it is a matter of judgment whether one is satisfied that my definition is fulfilled and it can be said that malignant neoplasias have been induced within the test species.

With a wide range of substances such as the aflatoxins (Butler and Barnes, 1963) and nitrosamines (Magee and Barnes, 1967), there is little doubt that they cause normal liver parenchymal cells of the rat to become neoplastic and demonstrate the biological behavior of malignant neoplasia. One can be reasonably confident that aflatoxin modifies, *de novo*, such cells to have altered biological properties. The same can be said for many nitrosamines and other chemical agents, but can the same statement be made with confidence when considering the wide range of data that are available from various bioassays that have been undertaken? Two problems arise, which give concern to fulfilling the criteria for malignant neoplasia. The first is the pathological assessment of individual lesions, and the second is the distribution of such lesions within the experimental groups. Considering the pathological assessment of lesions, the problems facing a pathologist are clearly illustrated in interpreting lesions of the mouse liver.

While not wishing to rehearse the detailed arguments concerning the differential diagnosis of nodular lesions of the mouse liver, it illustrates the problems that a pathologist has in interpreting animal data. The mouse liver is only illustrative as such problems arise in many different organs in many species. Broadly, the controversy is centered around whether nodular lesions should be considered to be hyperplasia, a reactive phenomenon, benign neoplasia, or malignant neoplasia. Until comparatively recently, little attempt has been made to characterize this lesion and correlate the histological appearances with biological behavior. There is now increasing evidence that this can be done (Walker *et al.*, 1972), and that the biological behavior associated with malignant neoplasia, that is, invasion and or metastasis, is only associated with one histological pattern. This has been discussed fully (Jones *et al.*, 1975), and the current information shows that such behavior, substantiated by Vesselinovitch (1978), is only associated with a lesion showing an abnormal trabecular pattern (Fig. 1). More recently, the National Academy of Sciences, PIREC, in reviewing the experimental data on feeding trials of heptachlor and chlordane to rats and mice, concluded that the

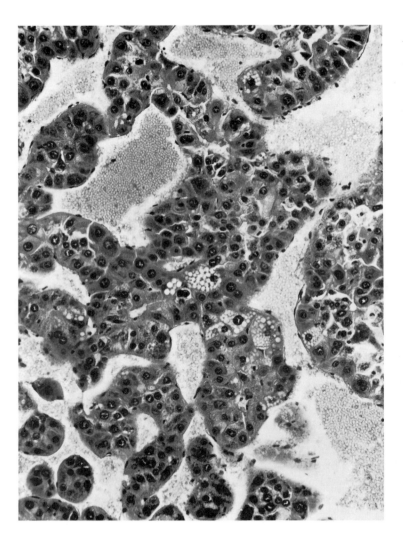

Fig. 1. Low-power photomicrograph of a nodule induced by phenobarbitone showing an abnormal trabecular pattern. H and E. 150X.

lesions showing this abnormal trabecular pattern were those
with the histological features of malignant neoplasia, which
would be recognized by most pathologists and should be considered to be hepatocellular carcinoma. The simpler lesions
(Fig. 2) also described by Walker *et al.* (1972) and Jones *et
al.* (1975), should be considered to be hyperplasia. Unless
this type of assessment is done on every study, one will not
know the true incidence of malignant neoplasia within the
test groups. Without making this critical distinction and
pooling as a broad group all nodular lesions, it has been
demonstrated that the overall incidence of nodular lesions
may be modified by environmental factors. This has been
discussed in this symposium by Kraybill. However, most of
the earlier experiments are difficult to interpret as it is
now impossible to make the distinction between nodular hyperplasia and malignant neoplasia. In one study, Gellatly
(1975) has shown that dietary modification alters the incidence of the simple hyperplastic lesions but does not modify
the malignant neoplasia. There is less problem with the rat,
where there is a good correlation between the final neoplasm
and biological behavior. However, the observation that many
chemicals are carcinogens is based upon mouse data. Unless
the correct assessment is made of the pathology, the incidence
of the carcinogens will be overassessed.

The other factor that modifies the opinion as to whether
a chemical is a carcinogen is the distribution of the lesions
within groups related to the treatment. Two examples of this
can be given. In the evidence of the carcinogenicity of
DDT, the incidence of nodular lesions within the liver has
increased from 20 to 80% in one dose group, but there is no
evidence within the other gruops of a dose response (Tomatis
and Turosov, 1975). There is no published evidence as the
distribution of the types of lesion in this experiment. The
second example of this phenomenon is of the drug metronidazole, in which the same phenomenon of increasing an already
high incidence of both pulmonary adenomas and lymphomas in
mice has been demonstrated (Rustia and Shubik, 1972). This
type of evidence does not give one confidence that the chemical under investigation has *de novo* caused a group of normal
cells that would not otherwise have demonstrated this biological property to have done so. Therefore, these results
would not satisfy the criteria that the chemical is a carcinogen, that is, induced neoplasia. The degree of uncertainty is too great to allow such an opinion. Arising from
the observation that the age-specific incidence of a neoplasm may be modified by treatment, has been the view that
the shortening of the latent period is evidence of a carcinogenic effect. This concept suffers from certain disadvan-

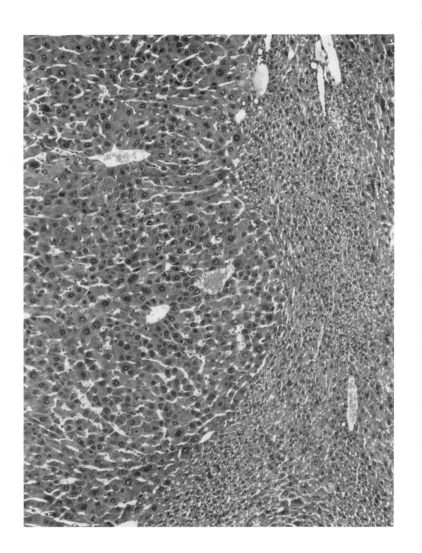

Fig. 2. Low-power photomicrograph of a nodule from a control animal showing a simple trabe-cular pattern. H and E. 100X.

tages. First of all, the concept itself is ill-defined, in that where carefully looked for, there is little evidence that malignancy arises after a prolonged period of no change. The time to death may be prolonged but not necessarily time of induction. Indeed, in at least one system related to the kidney, it is possible to demonstrate neoplastic cells very early in the course of development (Hard et al., 1971), but the macroscopic recognition of the final tumor requires many weeks (Hard and Butler, 1970). The other problem in using such loosely defined concepts as a definitive statement of carcinogenicity is the manner in which much of such data is obtained. Unless the experiment is designed to deliberately kill equally large groups of animals at stages throughout the experiment, such data cannot be derived, with any confidence, from animals that die from other causes. Such data just reflect the occurrence of incidental disease such as pneumonia, and the fact that a neoplasm was observed at that time. Possibly the best definition of a latent period is that period which has not been looked at.

The further problem related to interpretation of such tests is the weight and emphasis to be placed upon associated events. If such observations are to be used in determining whether a compound is a carcinogen, it must be clearly understood that this is based upon a hypothesis of mechanism of action. As variable hypotheses are common observations in all branches of science, they should be discounted for the purposes of recognition of an endpoint of a titration. Carcinogenic testing should be considered to be a titration of a chemical against a known effect with a known and recognizable endpoint. It is important that the endpoint be defined prior to the initiation of the test and not after the data have been reviewed. A clear example of this can be taken from the liver, and in my view has been used incorrectly to assess the carcinogenicity of a compound. As previously mentioned, the National Academy of Sciences PIREC review of the heptachlor/chlordane data concluded that there was no statistically significant increase on paired analysis of the incidence of malignant neoplasia. However, they concluded that there was an increase of nodular hyperplasia and that when this was added to the incidence of carcinoma, the pooled results became statistically significant. This process would appear to me to be in error as it is based upon hypothesis. The first hypothesis, for which there is basis in the experiments of Farber (1973) and Teebor and Becker (1971), is that there is progression of hyperplasia to carcinoma. But hyperplasia does not inevitably progress to carcinoma (Teebor and Becker, 1971). It has also been shown that hepatocellular carcinoma may develop in a liver that does not show nodular hyperplasia (Butler, 1976; Bannasch, 1968). It has, however, been a most useful hypothesis and led to much interesting work. It does not distinguish, however, between the problem of recognizing sequential from parallel phenomena. However, added to this first hypothesis is a second, that is, the development of neoplasia in a mouse is similar to that of a rat. While both are rodents, there is remarkably little published evidence on the sequential development of mouse hepatic neoplasia. It may indeed develop through the same stages as the rat, and it is possible that other mammals do the same, but at present there is no good published evidence to support this.

The recognition of these problems inevitably means that in order to say a chemical is a carcinogen, within a given test protocol, requires judgment, and in that opinion, one may be wrong. It is necessary to accept this fallibility, and in agreement with many speakers at this meeting one cannot have absolute safety. Many factors are involved in coming to a final decision for regulation and some of these have been described by Congressman Martin in the clauses of the proposed act. It follows from this that an assessment of a chemical and its likely effect upon public health requires an individual judgment and cannot be achieved in a general way.

So far, I have concentrated entirely on the experimental basis for concluding whether a compound is a carcinogen for a test animal. One has to consider the relationship of these findings to man. As I indicated in my preliminary remarks, this association is at the best tenuous. One has to accept that human carcinogens, with possibly one exception, are also animal carcinogens, I do not think that this should be at all surprising, but one cannot conclude the corollary of this. The evidence in man is largely based upon industrial experience of rare tumors occurring in selected populations. There is little evidence of the dose-response characteristics that can be so clearly demonstrated in animal experiments, but there are two examples of this. First is the evidence from smoking, although the definitive carcinogen has not been identified, and the other is that of aflatoxin. There is a good correlation between the incidence of hepatocellular carcinoma and the estimated daily intake of aflatoxin (Peers and Linnsell, 1977). The dose-response characteristics at low levels of exposure are so poorly understood, and the biological mechanisms involved have hardly been studied. Inevitably one wonders whether the various mathematical models that are at present used to make these dose-response extrapolations have any validity.

There is undoubted evidence that chemicals are carcinogenic for man. The industrial chemicals are good examples of this, but they can hardly be considered to play a major role in the present incidence of malignant neoplasia. The evidence for smoking and aflatoxin is much more compelling. One can argue from the aspect of preventive medicine and public health that if people did not ingest aflatoxin and did not smoke, the incidence of the major hazards of bronchogenic carcinoma and liver cell carcinoma would be drastically reduced. It is necessary to be cautious in making such statements, advocating restrictive control measures, as any assessment becomes a balance of risks. This is particularly so in the case of aflatoxin. The Protein Advisory Group of WHO recommends levels of aflatoxin in supplemented food for children that if given to rats would induce liver cell carcinoma. This advice is given, recognizing animal data, but also realizing that the threat of malnutrition represents a much greater risk to the children, and this risk must be taken. In the case of smoking, it is harder to define the benefits. The examples given so far have been related to known or likely human carcinogens. However, other substances have been considered to be animal carcinogens, notably in the mouse, but epidemiological evidence indicates no human risk. An example of this is phenobarbitone, which in certain experiments, may be considered to be a hepatocarcinogen for the mouse (Walker *et al.*, 1972), and has been shown by Clemmesen *et al.* (1974) not to induce neoplasms in man. Similarly, dieldrin, which again some consider to be a carcinogen for the mouse (Walker *et al.*, 1972), has not been shown to induce carcinoma in man (Versteeg and Jager, 1973). There is comparable evidence available for DDT (Lawes *et al.*, 1967). The nitrosamines, which are such an important group of experimental carcinogens, have been considered to be carcinogens in search of a neoplasm in the sense that there is no definitive evidence they are related to human disease.

Therefore, we can now recognize industrial chemicals, natural contaminants, food in the form of alcohol, drugs, and smoking all as carcinogens for man. Of the drugs, there is now increasing evidence that the contraceptive steroids cause an increase of liver tumors in young women (Mayes, 1976). Also, if one considers the theoretical implications of steroids for fertility control, one could envisage a situation where with a declining birth rate there would be an increase of mammary cancer and also with a decline of other forms of contraception, an increase of cervical neoplasia. Returning to matters of policy, it would seem reasonable to expect that industrial exposure to noxious chemicals should be limited as far as practically possible. However, society is not rational in its approach to the risk of chemicals to

health. They will willingly accept the risks of smoking and
many other dangerous activities. They will willingly accept
the ingestion of alcohol, and yet they demand the reduction
of industrial exposure to chemicals, such as vinyl chloride,
which on a worldwide basis over 15 years has induced something like 60 cases of angiosarcoma. It is possible that a
distinction can be made between intentional and unintentional
exposure. However, if one is choosing to regulate within the
field of public health and to achieve the aims of preventive
medicine, it would seem somewhat illogical to accept the distinction between these two forms of exposure to noxious substances. It is argued that in certain cases, the benefits
outweigh the risks. It is obviously important in every
instance to make some sort of risk/benefit analysis, but it
should be realized that this is inevitably done from a given
standpoint, and in giving this opinion it may differ from a
second opinion from a different standpoint. This is clearly
illustrated in the problems over fertility control with the
contraceptive steriods.

Wynder, in a rhetorical statement, said "We do not wish
to have one carcinoma induced by a chemical." This may be
the ideal state, but there is now abundant evidence that this
is not the wish of society. Society is prepared to accept a
varying rate of neoplasm, induced by recognized agents, if
they consider the benefits, either economic or social, to
outweigh the risks. What is most important to recognize is
that these issues cannot be resolved by legislation or small
groups, no matter how well-meaning, but must be debated at
large.

REFERENCES

Bannasch, P. (1968). The cytoplasm of hepatocytes during
    carcinogenesis. *Recent Results Cancer Res.* *19*. Springer-
    Verlag, New York.
Butler, W. H. (1976). Early cell changes in the course of
    chemical carcinogenesis. *In* "Fundamentals of cancer pre-
    vention," pp. 89-102. Univ. of Tokyo Press, Tokyo.
Butler, W. H., and Barnes, J. M. (1963). Toxic effects of
    groundnut meal containing   to rats and guinea
    pigs. *Br. J. Cancer 17,* 699-710.
Clemmesen, J., Fuglsang-Frederiksen, V., and Plum, C. M.
    (1974). Are anticonvulsants oncogenic? *Lancet 7860,*
    705-707.
Farber, E. (1973). Hyperplastic liver nodules. *Methods
    Cancer Res. 7,* 345-375.

Foulds, L. (1969). "Neoplastic Development," Vol. 1.
Academic Press, New York.

Gellatly, J. B. M. (1975). The natural history of hepatic
parenchymal nodule formation in a colony of CJ7 BL mice
with reference to the effect of diet. *In* "Mouse Hepatic
Neoplasia" (W. H. Butler and P. M. Newberne, eds.), pp.
77-109. Elsevier, Amsterdam.

Hard, G. C., and Butler, W. H. (1970). Cellular analysis of
renal neoplasia. Light microscope study of the development of interstitial lesions induced in the rat kidney by
a single carcinogenic dose of dimethylintrosamine.
Cancer Res. 30, 2806-2815.

Hard, G. C., Borland, R., and Butler, W. H. (1971). Altered
morphology and behavior of kidney fibroblasts *in vitro*
following *in vivo* treatment of rats with a carcinogenic
dose of dimethylnitrosamine. *Experentia 27*, 1208-1209.

Jones, G., and Butler, W. H. (1975). Morphology of spontan-
eous and induced neoplasia. *In* "Mouse Hepatic Neoplasia"
(W. H. Butler and P. M. Newberne, eds.), pp. 21-59.
Elsevier, Amsterdam.

Lawes, E. R., Curley, A., and Biros, F. J. (1967). Men with
intensive occupational exposure to DDT. A clinical and
chemical study. *Arch. Environ. Health 15*, 766-775.

Magee, P. N., and Barnes, J. M. (1967). Carcinogenic nitroso compounds. *Adv. Cancer Res. 10*, 163-246.

Mayes, E. (1976). Standard nomenclature for primary hepatic
tumors. *JAMA 236*, 1469-1470.

Peers, F. G., and Linnsell, C. A. (1977). Dietary aflatoxins
and human primary liver cancer. *Proc. IUPAC Symp. Myco-
tosius Foodstuffs*.

Rustia, M., and Shubik, P. (1972). Induction of lung tumors
and malignant lymphomas in mice by metronidazole. *J.
Nat. Cancer Inst. 48*, 721-729.

Teebor, G. W., and Becker, F. F. (1971). Regression and persistence of hyperplastic nodules induced by N-2-fluoren-
ylacetamide and their relationship to hepatocarcinogene-
sis. *Cancer Res. 31*, 1-3.

Tomatis, L., and Turosov, V. (1975). Studies on the carcinogenicity of DDT. *Gann Monogr. Cancer Res. 17*, 219-241.

Versteeg, J. P. J., and Jager, K. W. (1973). Long term
occupational exposure to the insecticides aldrin, dieldrin, endrin and telodrin. *Br. J. Ind. Med. 30*, 201-202.

Vesselinovitch, S. D. (1978). Morphology and metastatic
nature of induced hepatic nodular lesions in C57 BL ×
C3H F_1 mice. *Cancer Res.* (in press).

Walker, A. I. T., Thorpe, E., and Stevenson, D. E. (1972).
The toxicology of dieldrin (HEOD). I. Long term oral
toxicity studies in mice. *Food Cosmet. Toxicol 11*, 415-
432.

GENERAL DISCUSSION

DR. CLAYSON: I want to direct my remarks to Dr. Lijinsky, and before I do so I would like to make it absolutely clear that I respect his liberty to have any opinions he wishes on this subject. However, he is present today as a scientist, I hope, and I think it is absolutely essential (a) that he presents the correct facts, and (b) when he states these facts, he draws balanced conclusions from them.

I became alarmed during his paper when he told us, a trivial point maybe, that azo dyes are relatively inactive in mice. I would remind him that nearly one of the first carcinogens to be discovered, aminoazotoluene, in fact is far more carcinogenic in mice than in rats, as was exquisitely shown by work in the late 1940s and early 1950s. I think the reason why we know very little about azo compounds in mice is really that practically everybody has done the testing in rats.

Dr. Lijinsky further told us that he thought the correlation between the carcinogens known in man and shown to be carcinogenic in animals was satisfactory. To my knowledge, there are only two chemicals accepted as being carcinogens in humans, namely arsenic and benzene, and no animal data have been published yet. There are, in fact, about 26 established human carcinogens. I find then two exceptions in this number 26. It does not show a very good correlation in my view; 90% is worrisome, 99% would be far more adequate.

Dr. Lijinsky repeated the well-known adage that the fetus is particularly susceptible to carcinogens. This is based on original observations with methyl and ethyl nitrosourea, where undoubtedly the fetus is far more susceptible. My colleague Dr. Yalter has just tested a series of 10 nitrosamines--he cannot really say from this whether the fetus is more or less sensitive. The mothers were also kept for the rest of their lives to see if they developed tumors.

In another experiment that we are just completing in the Eppley Institute, we have studied the potent carcinogen niridazole. We have tried it as a drug; we have studied it transplacentally. We have had a zero response during pregnancy. This is an area where I think we would be very unwise to assume necessarily that the fetus is all that more sensitive to carcinogens, which require metabolic activation, than is the adult animal. We need far more information in this area.

I speculate whether DES given to these unfortunate young women, and since repeated in animals, is a carcinogen per se or is inducing its carcinogenic effect by a teratogenic mechanism.

ISBN 0-12-192750-4

My colleague has examined the lesions he found in hamsters given this compound transplacentally and came to the conclusion that the major factor appeared to be a continuance of secretion of estrogen by the female offspring of these hamsters. So, if it is a teratogenic effect, we would look on I think the dose-response relationship very, very differently than if it were a true carcinogenic effect.

Dr. Lijinsky again talked down about any correlations between chemical structure and carcinogenic activity. I find this to be extreme! I would ask Dr. Lijinsky if, in a situation we both find ourselves in from time to time, he was presented with the alternative of testing an aromatic amine or a straight-chain alcohol, which one would he choose in the hope of getting a positive result, which could be used in the regulatory sense?

Dr. Lijinsky suggested, and I believed this at one time, that the latency of cancer is related to the lifespan of the species concerned. Even in man, now, we know of some instances where the latency of cancer is exceptionally short. For example, in immunosuppressed renal transplants, we find reticular cell sarcoma of the brain as soon as 5 months after the operation. When we use the drug chlormethazine, which was used in Denmark to control polycytocemia, we find bladder tumors turning up after large doses as soon as 2.5 years. I would suggest to you, Dr. Lijinsky, that the reason we find tumors with different latencies or with latencies that appear to be related to the lifespan of the animal is merely that the longer-lived animals take longer to develop tumors more easily.

Finally, we have given a lot of discussion as to what is the contribution of chemicals to cancer. I think it can be accepted that environmental factors play a large part in the induction of human cancer, and it is environmental factors, rather than genetic factors, with which we should be concerned. But to equate this with synthetic chemicals I think may very well be wrong. I think we have to maintain a cautious stance when we talk about synthetic chemicals. I would ask you to contrast the fact that in 1975 more than 81,000 males in this country died from lung cancer and that the majority of these are probably directly associated with cigarette smoking, with 10, possibly 15 cases of vinyl chloride-induced angiosarcoma of the liver, which we have made so much fuss and so much publicity about.

I think the occupational carcinogens are terribly important because they affect small communities to a large extent.
Goldwater in a report from his country, for example, found
that in a population of 48 men exposed to 2-methylamine
alone, 12 developed bladder cancer. It is an immense tragedy
in that area, but when we look at the problem in the community, maybe we have got to cast our net a little bit further.

Now Dr. Lijinsky, I dislike making these remarks particularly as I'm going to give my paper subsequently. I only
hope that I have not fallen into some of the traps as you
have.

DR. LIJINSKY: Yes. But not quite in the same vein. I think
you might have misunderstood me. I deliberately said that I
did not believe the time of occurrence of tumors was related
to lifespan. In fact, quite the reverse, that is part of my
argument. I confess to misleading you on the azo dyes. I
really should have talked about butter yellow, which is not
carcinogenic in mice. I think you will agree on that.

I was using the term chemical in the very broad sense. I
am sure that most of lung cancer from cigarette smoking is due
to chemicals. I was not specifically talking about industrial
chemicals. I was talking about the chemical induction of
cancer and using that as a reason for looking at carcinogenic
chemicals, as habits therefore to be eliminated. I think most
cancer is caused by chemicals. That is all I said: whether
these be endogenous chemicals or chemicals from cigarette
smoking or industrial chemicals. In fact, I agree that the
number of industrial cancers that we know is a handful compared with the overall incidence of cancer. But my concern
is that the exposure to the industrial chemicals even at
levels that do not produce more than a small number of cancers,
enhances the response, perhaps, to cigarette smoking. I think
the example of asbestos and cigarette smoking is a very good
one. Asbestos in itself is possibly a weak carcinogen, but
combined with cigarette smoking it is a very lethal combination. That is why we have to eliminate exposure to any chemical carcinogen to the extent that it is possible, reduce the
exposure, I mean. And I am sure that you will believe me
when I say I do not believe vinyl chloride manufacturers
should be stopped. But, I do see a very good reason for reducing the maximum limit of exposure from 100 ppm to 1 ppm.

DR. COULSTON: Or protect the workers.

DR. LIJINSKY: And to protect the workers. Yes. I said I do
not believe the Delaney approach should be applied to anything
but food additives. That is my feeling. The other things I
think are not terribly important.

DR. PARKE: I would like to address these comments to Dr.
Lijinsky.   A little bit of epidemiology seems to be a requirement for this and further, some remarks that you made yourself
this morning about DES.   There is not a single documented case
of DES-related cancer in the United Kingdom.   They looked into
this very, very carefully and the explanation that has been
given is that the dose administered to the women in the
United Kingdom was considerably lower than that which was prevalent in the United States.   But there was one case that I
knew of personally, never documented, because the husband of
the woman concerned was a very distinguished medical man and
wished to keep it confidential.   But Sir Charles Dodds, who
was the physician in charge of this woman's mother, recalled
the case when I discussed it with him, and said that this
woman had three to five times that which became the common
dose used in subsequent cases in the United Kingdom.   I would
like to ask whether Dr. Lijinsky considers that this might be
some degree of evidence of a dose-response in humans?

    Another thing that he briefly mentioned: he said that we
cannot yet determine the people in a population most susceptible to carcinogens.   I think we are not far short of doing
this.   Some very interesting work has been done in mice and a
number of people have done similar work in other species.   I
think it is only a matter of time before we are going to be
able to type people into those who are highly susceptible and
those who are not.

    What I would like to ask Dr. Lijinsky is, if we are able
to do this in say 10 years' time what use would you make of
this information?

DR. LIJINSKY: I have absolutely no idea.   I think you might
caution certain susceptible people not to use saccharin, even
if it was sold over the counter.   That is the sort of thing
I would do.   Or not to expose themselves to certain drugs
that have a low-level carcinogenic effect or a weak carcinogenic effect, but that nevertheless are used because they are
very useful.   There are many drugs that are used that are
possible carcinogens and nevertheless are useful.   I do not
mean only the life-saving anticancer drugs, but other drugs.
If such a drug is used and not prohibited by the FDA, people
who might be susceptible should be cautioned.   I am very
pleased to hear that it is so close at hand.

DR. COULSTON: This is a very important issue. Dr. Parke has pointed out that in Great Britain, contrary to the United States, they have not found a single case, except perhaps one, of a woman who gave birth to a child that later developed vaginal carcinoma. He attributes this to the fact that it was common in England to use half or less of the clinical DES dose that was used in the United States. In fact, he cites one case where it did occur and that was a woman who took three to four times the dose that the doctors generally prescribed.

Now that alone is not too exciting, but let's couple that with the smoking problem. You know that the inhalation of cigarette smoke is the one, firm example in humans of a dose-response curve to chemical carcinogen. It is well established that if you only smoke half a pack a day you get so much cancer, if you smoke a pack you get more, three packs a day increases your chances. The epidemiologists have firm data on this now. It is a dose-response curve with a no-effect level even though we do not know what chemical causes the cancer in humans from the smoking. Although one has to accept that inhalation of smoke from cigarettes has a causal relationship to lung cancer, you cannot reproduce this experimentally in animals. There are no data in animals but you have a dose-response in man.

I don't know why there is reluctance to accept dose-response curves for carcinogenic studies. At the regulatory agencies, they seem not to understand what we are talking about. We have human evidence of a dose-response repeated time and time again. Even the vinyl chloride story is a clear example of a dose-response in man.

DR. KRAYBILL: Getting back to these percentage figures, they derived I believe from two sources. One is Boyland and the other is Higginson. The derivation and all the calculations surrounding this appear in a proceedings that we had from one of our inner agency groups. It was called "Cancer Is a Social Disease." At that particular meeting, there was a lot of refutations of these figures, particularly by Dr. Joe Fraumni to state a figure of 9%. I think this is what Boyland arrived at, deducting radiation variances, but there are other factors that come into play here.

So I for one and many of us have refrained from using this figure of 80-90%. Mike Shimkin, for example, said I will give you any figure ranging from 40 on up to 100%, if you want. But I think it would be better to say that many of our cancers may have their origin from etiologic agents in the environment, and many of them are preventable. But I think we should get away from using this figure of 90%, because it does not have a sound basis, really.

DR. KOLBYE: I should like to make two brief comments, one with respect to ethanol. It is my impression that with pure ethanol, so to speak, if it is acting primarily in the capacity of a promoter rather than an initiator, and if we are talking about "booze," there may be low-level substances present that may have in some instances initiating capabilities, but I have no way of knowing.

Shifting slightly, I should like to compliment Dr. Parke for an excellent presentation. I should like to suggest that a very concerted effort be made on an international basis to attempt to identify those substances that can influence the expression of cancer in humans or in animals, even if those substances are not "carcinogenic" in the sense of initiators.

I wonder if we could meet again 100 years from today, how much of our current thinking about preventing cancer in humans would be somewhat analogous to the thinking of 150 years ago on the advantages of bleeding patients. What disturbs me is that with all the pressures to identify carcinogens, test bioassay this, bioassay that, we still are not learning very much in any sort of a systematic way. I suggest to you that the presentation of Dr. Parke gives us a direction and an interrelated hypotheses actually that deserves very, very serious consideration, and by very serious consideration, I mean some very well-directed and well-executed research efforts.

DR. CLEGG: I would like to pick up something that Dr. Clayson said. He was talking about the susceptibility of the fetus and the effect on the fetus with regard to carcinogenesis. I would like just to insert a warning into the discussion. At the present moment, we are using rodent species for most of our tests. The rodent placenta does not even remotely follow the pattern of the human placenta. We know in the yolk-sac placenta, for instance, that trypan blue will cause teratogenic effects. There is no evidence that trypan blue will cause this type of effect in non-yolk-sac placental animals. I would suggest, therefore, when we are looking at the susceptibility of the fetus to carcinogens we do need to be very careful in choice of our species in diagnosis.

DR. KENSLER: I'd like to agree with Dr. Kolbye that Dr. Parke did a beautiful job of synthesizing some rather complicated approaches to the problem. He did mention in asking Lijinsky a question that we were very close to identifying the genetically or otherwise highly susceptible high-risk part of the population. I wonder, are you talking about the aryl hydroxylases?

DR. PARKE: Yes, I am talking about the aryl hydroxylases but
I have discussed this with Dan Nabors and various scientists
and I think they have been using the wrong substrate. In
using benzopyrene hydroxylase, they are measuring about a
mixture of half a dozen or more products. They have a very
complex substrate relationship, which was not understood, but
which Jim Fouts explained in a publication.

I think this biphenyl hydroxylase of ours is a much more
specific enzyme and much more specific substrate. We are
also working with human lymphocytes and the human monocytes.
We have evidence that there is a genetic variation in the
stability of the endoplasmic reticulum in humans. We are
also using this as a model to screen drugs, which might
stabilize in the endoplasmic reticulum and possibly even act
as, what shall I say, an inhibitor of carcinogenesis, and
maybe even reverse the malignant process.

DR. KENSLER: The reason I asked the question was because
while that group had this very clean, genetic control mechanism, which they demonstrated with possibly the wrong substrate, in the work that was done at Buffalo on man, there
did not seem to be any correlation using the same substrate,
as I recall, between the susceptible individuals among the
smokers and the nonsusceptible.

And we in Cambridge when we were doing a lot of mouse
skin painting. Over the years, we had a large control group
about 40% of which developed tumors while the other 60% did
not. We took these two groups and looked for inducibility of
aryl hydroxylase, etc., thinking we did not care which way it
went, but hopefully that there be a difference. Well, there
wasn't any difference, which said to me that as far as tobacco
is concerned, as far as this substrate was concerned, that
that was not a critical event.

DR. PARKE: Well I think it possibly was the wrong substrate,
but cancer must basically come down to enzymes; one or maybe
100 enzymes are involved. We know that no enzyme is precisely
identical in structure or activities throughout the population.
There are genetic variations in every enzyme that have been
observed. There must be genetic variations in the enzymes
that are involved in cancer and therefore, there has got to
be a genetic distribution. I know this very well. Half of
my siblings are very prone to cancer; I belong to the half
that are pretty resistant to it. I have seen this in my own
family.

DR. KENSLER: I have one other question, if I may. In the
early 1940s we were measuring the great difference of susceptibility as a function of diet to the azo dyes. We found
that with our carcinogenic diet, that where the animals
developed liver tumors, their riboflavin, DPN, and so on, were
markedly reduced. If we supplemented the diet with riboflavin, which increased the FAD levels and is part of your inte-
grated series, the animal livers looked great. They did not
develop tumors at maybe three times the dose, because they
ate an awful lot more than the animals that were going to get
cancer. I am just wondering how you see the flavin-added nu-
cleotide playing a role in conjunction with the P-450, P-448
in that sort of situation.

DR. PARKE: I think there is an optimum level for riboflavin
in the system. Too much and you would have the reductase
acting on its own in an uncoupled fashion, which would tend
toward this P-448 system. So that up to a certain level,
riboflavin is essential and beneficial; over that level it may
not be. We see exactly the same with vitamin C, which again
is involved in this same hydroxylase system. A minimum level
is necessary, an excess level is deleterious. We are beginning to see the same for vitamin A, and I think possibly a
number of other components. There is an optimum level for the
input of these chemicals.

DR. KENSLER: Actually the riboflavin was the other way
around. With the low level you got the tumors, with the high
level you did not.

DR. COULSTON: This subject should have been discovered and
worked on over 10, maybe 15 years ago. When I presented the
first data that the rough endoplasmic reticulum was involved
with so-called drug metabolizing enzymes, I was literally
laughed out of a committee of the National Academy of Sciences.
The issue is that the pharmacologist and biochemist looked
only at the smooth endoplasmic reticulum, which is often
formed in excess many days later, after the chemical has been
absorbed in the body, and after the rough endoplasmic reticulum has been stimulated. This is a case where scientists did
not look early enough.
    Dennis Parke and his group should be given a great deal
of credit. Those researchers who got fixed into the smooth
endoplasmic reticulum concept without considering the rest
of the story should be and I hope are embarrassed.

DR. ZAWEL: It is such a hard platform for someone like me to
take. I would like to explain that I am not a scientist.

I have to apologize because it is a terrible problem in trying to establish a dialog. I believe I was invited here because the issue under consideration was primarily the Delaney Clause, which is an expression of public policy regarding physical illness, in this case cancer.

What I would like to point out and hope that it has some ability to be related to the scientific dialog that is going on, is that the Delaney Clause appears to me, more today than ever before, a social expression of concern about scientific uncertainty and social progress and the uncertainty as to where that progress is taking us.

Where people have clearly defined risks, they make personal decisions. The risks we are talking about now, particularly in regard to the Delaney Clause, are in relationship to a food supply that is changed by the nature of the way we live. A lifestyle and the social dynamics are deeply tied to our economic system and who makes money and who does not.

I believe that (even though the Delaney Clause has been in existence for a long time) the increased public dialog and extreme concern regarding these issues among the public is an expression not only about health and cancer, but an expression about where our society is going and how we are going to make decisions related to that society. Dr. Kolbye, very early yesterday I thought, gave terms that are understandable to the public. There is no way, and I consider myself fairly articulate, that I could describe what I heard at this conference to any public person who was interested in cancer or its prevention or individual choice of policy. Absolutely no way could I bring it forward. And, I would like to go back to Dr. Kolbye's original statement that we need some kind of definition of safe and dangerous, or not shown to be safe, clearly not safe, unsafe, or distinctly potentially harmful.

We are living right now, particularly in regard to food, which has clear sociological psychological relationships to people's lives. Food isn't just something you eat, it is a way you raise your family, and we have seen families break down. It is an expression of religious ideas, and we are out of touch with our religious and sociocultural bases. We have had reliance on progress to save us from endemic disease caused by microbes or by an environment untamed. We are now facing a time where it is our man-made environment that now presents to us the greatest doubt and the greatest fear about the future. I feel this is a symbolic issue of social and public concern as well as a scientific one. And, within the context of the discussions that are going on here, I would like to hear more discussions as to how we define in social policy goal terms, not only the scientific uncertainty, but also the scientific certainties that we do have.

You know, newspapers hand it to consumers. If talking about cigarette smoking is not news, then the newspaper people are not only telling you that they can only communicate one thing or are interested in only communicating one thing, they are also telling you that people are only interested in being communicated to about certain kinds of things.

And these are the kinds of dialog that are so very difficult to come by. I would urge, never a group this size, a smaller body of people to pull together the kinds of considerations about social policy, the Delaney Clause, and those kinds of issues surrounding science that provide a broader representation, so dialog can exist. I feel totally inadequate to representing any position. I understand why Dr. Sidney Wolfe did not come and he certainly knows a great deal more about the pure science than I do, but I really appreciate it.

DR. COULSTON: We appreciate very much what you are saying. However, the title of the meeting is clearly, the scientific basis. In other words, we have heard Peter Hutt say the Delaney Clause has nothing to do with science, as you say perhaps rightly; it is a social problem, a political thing, but we, as scientists are trying to say, there is a scientific basis too that must be considered, and you ought to listen. Perhaps I do not understand your social philosophy, and maybe you do not understand our scientific considerations. I am agreeing with you that we should have a meeting where you can sit around the table and talk about social welfare, but that would be a different kind of meeting than this is. There is no question what you say is absolutely correct and we all honor this and respect it, but this meeting is clearly a scientific meeting. We are trying to establish whether scientists have any role in the future, ever again, to make decisions. This is what we are asking really. We are trying to see what we know about the scientific data and basic concepts of safety evaluation to enable us to become involved with decision-making and not leave it to nonscientists. We have been told by Peter Hutt to get out of this decision process; he said we have nothing to say anymore. You just said the same thing, it is a social decision. It cannot be a social decision alone. There has to be some kind of scientific fact to make a basis for balanced and reasonable decisions.

DR. UPHOLT: Dr. Lijinsky made a very strong case for the lack of organ specificity of cancer, which I doubt that many people will disagree with, very wholeheartedly, anyway. This seems to imply that from a preventive standpoint, we must consider all cancers as a single disease. This bothers me very much. I wonder if there is anyway in which we can, from a

statistical standpoint, make use of the fact that deaths from
cancer are only increasing in a very few situations. This
seems to me to be a very significant fact. As far as I know,
this has really not been mentioned here in this meeting. Yet,
it seems to me that it should have some major significance to
us as a regulatory group from a preventive standpoint. Yet I
have seen no evidence that it is useful so far.

DR. COULSTON: I did mention it in passing. I referred to the
fact that cancer is not increasing except for the smoking problem.

DR. LIJINSKY: What I was particularly saying, Dr. Upholt, was
you cannot predict from an animal experiment what the target
organ of the carcinogen in man would be. Therefore, we should
be very cautious in our epidemiological searches. But to say
a word or two about the differences in organ response, the
differences in pattern of cancer, we see in experimental animals in experiments that have been done that as you change the
dose of the same compound or the regimen of the same carcinogen, you can change the tumor response. You can get more of
one tumor and fewer of another tumor, in the same species.
There obviously is an upsetting of some delicate balance.
This is perhaps what cancer is about. And it is the interplay
of these various insults, if you will, that produces the
particular patterns of cancer we see, I believe. That is why
we have to be very cautious, again I have to repeat it, about
exposure and unnecessary exposure to any carcinogen. Carcinogenesis is a cumulative toxic response.

DR. COULSTON: Well you do not know this really, it is speculation.

DR. YODAIKEN: I think it is very difficult, despite the
erudite presentations we have had here this morning, to
separate science from social issues in a clear-cut way. I
am, as I said, from NIOSH. I am formerly a professor at a
distinguished university in the South. I am now in government,
and I find that I am going to make some mundane comments by
scientific standards, but the descent from the esoteric to
the mundane is a pleasant, soporific experience.
    I would like to congratulate Dr. Lijinsky on many of the
things that he said. I think he made very good points several
times in dealing with not only a population that has a choice,
but a population that does not have a choice. Despite the
warning remarks from Canada and the forceful arguments of Dr.
Clayson, we very well know that many of the carcinogens we
introduce nowadays do in fact cross the placental barrier and

put the fetus at risk. PCB is a classical example. We know that PCB is picked up in the fetus of primates and there are many others and I know that this audience is fully aware of them.

Concerning the fetal population, you also drew attention to the fact that we are talking about young people who do not have a choice and I would like to enlarge it. You are talking about people who are handicapped, talking about the elderly, talking about the disadvantaged, and an international community, not represented here. This country exports an enormous amount of foods to other countries. Powdered milk, infant formulas, and attached to those foods in one way or another are our deliberations on the Delaney Clause.

I would like to come back to the very interesting story of EDB, which I spoke about yesterday. Ethylene dibromide, as I pointed out, is an additive to gasoline and as such we are all in contact with it more or less, certainly in this country. I would like to tell you the very interesting story of EDB. I don't know where it began, but somewhere along the line, the National Cancer Institute ran an experiment, a classical experiment, which Dr. Cranmer said was the sort of thing we should not do and which other people would agree should not be done. They literally poured EDB down the throats of rats and it produced cancer of the stomach in 100% of cases. I might point out that in animals on larger doses, there was no development of cancer. It was on the smaller of the two very high doses that the cancer was produced and the reason was that in the larger doses the tissue was simply destroyed. Nevertheless it produced 100% of cancers, and a public interest group got hold of this result and they said, "Hey, what are you doing about it?" And EPA turned to Dow Chemical among others, and said, "Look, you make EDB, get a move on and do an epidemiological study on this," and Dow Chemical obliged. In the meantime these things take time; the public interest group does not always take cognizance of this; they said if you don't do something about it we are going to sue you. So the EPA, Dr. Anderson is unfortunately not here, the EPA decided to do something about it and they did two things. They took Dow Chemical's epidemiological data and they took the cancer of the stomach experiment and they got the epidemiologists--epidemiologists come in many more shapes and sizes than Dr. Wynder tended to indicate. They made a model and they extrapolated from the animal cancer experiments to Dow's results. And Dow's results showed that there was no significant increase of cancer among workers exposed to inhalation. The EPA group said well they came up with a result that shows that EDB should only be tolerated in doses of something like one part per million. One part per

billion, I don't know, trillion, whatever. An amount that,
of course, Dow Chemical and other manufacturers of EDB would
find great difficulty in meeting. They also did something
else at the EPA. They came out with a flyer that they sent
off to people who use EDB as a fumigant. I didn't read the
flyer, I just read the first sentence before my hair stood on
end: It said 100% of animals exposed to EDB died of cancer.
Something like that.

Now the reaction was a social concern. Their response
was to a social concern. They did right. They did what they
thought they should do. Meanwhile back in Cincinnati, NIOSH
was doing an experiment with EDB in the levels that one would
expect workers to be exposed to, because our focus is on
worker populations. We were doing the experiment and we were
exposing animals to EDB at TLV or threshold level values, 20
parts per million, and we got no cancers. But when we added
disulfuran, which I mentioned yesterday and which is an anti-
booze, to the diet of these animals we got an enormous outbreak
of various types of cancers. The significance of this is not
in the two compounds that have been used. The significance is
that they are possibly classes of compounds that are being
used.

EDB comes also as many analogs perhaps that are important;
disulfuran is an inhibitor of hepatic enzymes. By the use
of these two substances together, we got a significant outbreak of cancer, so much so, as I told you yesterday, that we
have alerted the scientific community and the government
agencies around the country.

To come back to the social issues, the question is, is
EDB necessary? I don't know. Dow Chemical, Houston Chemical,
Great Lakes Chemical, and ICI in England are going to have to
decide.

DR. COULSTON: What is it used for?

DR. YODAIKEN: It is used as a scavenger in gasoline. It is
used to scavenge the lead in gasoline.

DR. COULSTON: You mean when you take the lead out.

DR. YODAIKEN: No, when the lead is in, it combines with the
lead.

DR. COULSTON: They're taking the lead out so they will not
use it soon.

DR. YODAIKEN: Well that may be--I'm not in a position to
judge when they take the lead out.

DR. COULSTON: I'm not being facetious. I'm curious for the audience to know what it's there for. Dr. Blair, what is it there for, since we are making a big issue here?

DR. BLAIR: Ethylene dibromide is added to gasoline that contains tetraethyllead. It is used as a scavenger to remove the lead within the combustion chamber of the engine and it goes out as a volatile material. So as the tetraethyllead content goes down, the EDB content goes down, and the benzene content goes up.

DR. YODAIKEN: I am going to conclude. The point is that it is being used. It is used as a fumigant, as indicated, and used in other ways, but that is not important. The number of people involved with EDB, working with EDB, is small. But the importance lies in the types of experiment that were conducted. We had two compounds.

DR. COULSTON: I think this is a very important point and many people have heard me speak on chemical-chemical interactions, chemical-drug interactions for the past ten years, so I appreciate the point you are making. If we think we know what one chemical does, we have not got the faintest idea what two, three, four together do. Beyond two, I do not even think we can judge.
    You are really asking for a kind of research to be done that must be done. It is in the public interest to do it and I can understand the concern.

DR. YODAIKEN: Well I was also going to say and I will say this in conclusion: in this country, and I have been a citizen of the United States long enough to say it and paid taxes for long enough to say it, we are different in many ways. We may not be as good as other societies, but we are different in that we make demands on our industrial society to produce a variety of choices. We do not always give them a chance to use that variety of choices, but we do produce a variety of choices. Now we are turning around and saying, "What the hell are you doing? Why are you producing so many carcinogens?" It is the cost/benefit, which Dr. Darby so eruditely spoke about, that is the crux of the problem.

DR. COULSTON: I find no objection to your point. I only object when we try to sell our philosophy and products overseas, based on our regulations. This is where I begin to object, because the overseas people have different regulations and social needs, which we must honor. If India needs DDT, let us give it to them and not deny it on the decisions, whatever, of a U.S. regulatory agency.

DR. FURMAN: Dr. Coulston, I just wanted to comment on Dr.
Parke's observation of Sir Charles Dodd's patient with vaginal
carcinoma, which he suggested might be related to the fact
that her mother took a relatively large dose of DES, in comparison with other women in England.

If you believe that DES is etiologically related to the
lesion, then examining the data in the registry in the United
States indicates that, where the information is available, the
total dose that these women were exposed to or given, then
pregnant women, ranged from 3 to 1800 mg. So, there is an
enormous range of dosage. And, I doubt, assuming ever that
DES plays a role, that one can ascribe a dose-effect relationship to this at all.

DR. ABRAHAM: I just want to make a couple of comments about
Dr. Parke's presentation, and ask a question. You mentioned
that in the fetus there is an increase in SER, and I was just
wondering what species, because I thought it was just to the
contrary that there was more RER and less SER.

Second, regarding starvation, it is very common that when
you starve the animal you lose a lot of glycogen and this ex-
poses a lot of SER. So I'm just wondering, was the method
you used quantitative to distinguish between SER and RER?

And one further point if I may. We were talking about
glycoprotein synthesis. Are you relating it to DNA synthesis,
for instance? Are you taking into account the role of the
Golgi body and the Golgi vesicles, where I think most of the
terminal sugars are added onto the proteins? And what role
does this have later on in your scheme?

DR. PARKE: The work on the neonates and fetuses was not done
in my laboratory. Quite an amount of work has been published
on this, mostly in the rat and the mouse. I think the hamster has also been looked at. In all cases, as far as I recall, there is a marked increase in the smooth endoplasmic
reticulum that one finds in the liver. That is the organ
that has been looked at. The explanation of this is that
this gives rise to a much greater synthesis of intracellular
proteins than normally occurs and, therefore, to an increased
rate of growth of the cells and an increased rate of mitosis.

Coming to the other point about the glycoproteins, this
work was done in Toronto. They showed that in order for glycoprotein synthesis to proceed, you had to get the polypeptide synthesis beginning in the rough endoplasmic reticulum.
The first sugars are then put on in the smooth endoplasmic
reticulum and the final and the most important ones of course
are the terminal--fucose, sialic acid, etc.--they go on in the
Golgi complex. The Golgi complex then moves to the periphery

of the cell, where some of these glycoproteins are incorporated in the plasma membrane. Others are excluded from the
cell like serum albumin from the liver or immunoglobulins
from lymphocytes or mucous from epithelial tissue. It is the
nature of the terminal sugars, whether sialic acid or fucose
predominates in the glycoproteins of the glycocalyx, that
determines whether a cell divides or not.

It is a question here really of what came first, the
chicken or the egg, and it is not decided whether the nucleus
controls the glycoproteins and the terminal sugars in the
glycocalyx or vice versa.

This is a very large area of biochemistry, which is as
complex a problem as the genetic code itself, and it has not
been resolved. But one thing that has been ignored by most
cancer specialists, when they talk about alkylation of DNA,
is that some years ago it was shown that DNA does not only
exist in the nucleus, it is also found in the mitochondria
and indeed in the endoplasmic reticulum. What happens when
the DNA of the endoplasmic reticulum gets alkylated is anyone's guess.

DR. SCOTT: I would like to interject another thought for
your consideration here during the next day of your deliberation. I might background this by saying I am an industrial
researcher whose primary purpose in life is to discover new
synthetic chemicals that are safe and may be beneficial to
society.

I think we must address this aspect of science as well as
all the past ones that we have had around. I listened to Dr.
Kraybill talk about using proper procedures; I heard Dr.
Kolbye talk about all the unanswered questions. I have heard
all the discussions this morning among many scientists about
how to determine whether something is a carcinogen, or talk
about benefit:risk, and in the meantime our society is being
stymied in the area of research in making progress.

Our social scientists should pay attention to this as
well, because it is almost impossible today, if you have got
a new chemical entity, to say it is safe. This is a real
problem and I hope that the group will address itself to this
aspect. What are we going to do while we are making up our
minds as to how to determine and interpret the Delaney Clause
and interpret safety? Are we just going to stand still in
progress? Because until we can come up with some interim
system accepting certain minimal data as evidence of safety
and continue work, we are stymied.

The advancement of new science will be stopped and we are
not going to need so many organic, synthetic chemists in the
future, if we do not utilize these people.

DR. COULSTON: You're making a good point, of course. It is just as valid as the adversary position the other way around, to get rid of all these chemicals. We are trying to develop right here in the next days, a body of knowledge to point the way to the future, so that we can answer both questions.

Widely diverse views, such as those of Congressman Martin, Dr. Lijinsky, and myself are important as they are discussed not in an adversary way but because from this discussion will come a reasonable approach. What is wrong in my opinion right now is that we are always placed in adversary positions by regulatory decision-makers, between government, industry, academia, and the consumer. This is nonsense. Consumer groups do not have to be advocates any more than scientists have to be advocates. Let us find the best way so that all the people can benefit. That is what we are talking about.

DR. ZAWEL: I do not want to disagree at all. I think that part of the problem is that we are in a social crisis as to whether we are going to go backwards or whether we are going to go forward. The relationship of our basic economy to continued research in new product development in a broad spectrum of areas is a significant related question to this area. Let me just bring to your attention, and I do not know how many of you had followed it, something for which there is increasing public sentiment and support, at least in the State of New Jersey. There is somebody running for Congress on the premise that our primary social goal should be to identify carcinogens and deal with preventing cancer. The plan is to take money out of direct cancer research, looking for cures and other kinds of things, to put an emphasis on looking at our socioeconomic environment, changing it in order to gain a particular kind of health for society.

These are very serious public questions, which have only scientific bases for real answering, but unfortunately our social efforts move much more quickly than science does, as is so beautifully exemplified here. What I would like to see more discussion of, which I did not hear today, was things like, if I understood you correctly Dr. Parke, greater understanding as to the processes and the development of cancers and why they form rather than the mere interplay of a chemical to the formation of carcinoma. I have heard virtually no discussion. It may simply be inappropriate of how we treat cancer and how we live with it.

I assure you I am very confident that there would be much greater public support for long-range study if there were also public confidence that there was some hope of curing the disease in a substantial way rather than adding greater stress to the human organism.

INTRODUCTORY REMARKS
(SESSION II, AFTERNOON)

DR. EGAN: I would like to discuss public issues and the carcinogenic factors. Going back a little to what Dr. Harrison said, we are in danger of at least misleading the public when we talk of environmental chemicals and synthetic chemicals and just chemicals and do not distinguish between them.

Chemical reactions, in fact, influence the course of enzyme activity and involve molecular and electron transfer processes. These apply whether we are talking about synthetic chemicals or natural contaminants or chemical species derived from irradiation. For example, in a sense, we are talking about causation from chemical reactions and, in a sense, one can say that all cancers are chemically induced.

I do not really want to say more than that, but whether that goes any way to helping us clarify the situation is perhaps questionable. At least chemical reactions are probably always involved. These at the molecular and submolecular levels involve enzymes that themselves are chemicals and electron transfer processes. I think that we are getting very near again now to the presentation that Dr. Parke gave us this morning. In a way as you said, Dr. Coulston, we might easily have another meeting beginning at that point. Perhaps we can keep that in mind for a future occasion.

Dr. Lijinsky raised a question of chemical structure in carcinogenesis and that has often been looked at. I was reminded by his words of the Swedish filter papers on which maybe still today, but certainly at one time, the box depicted a picture of an alchemist saying that these filter papers are used by the most clever chemists. These chemists have given a lot of consideration with other scientists to the simple relation between chemical structure and carcinogenesis, and no one can pretend that carcinogenesis can be predicted from a structure. The same is largely true of structure and biological activity in general, I would think. It is good to know that drug companies and pesticide companies, for example, have prepared and tested many hundreds of thousands of substances in the quest for the predetermination of biological properties.

I might, however, point out that some relatively noncarcinogenic substances can themselves be very conveniently used, for want of better knowledge, as convenient indicators of carcinogenic activity. One of the most familiar examples, I think, is the range of polynuclear aromatic hydrocarbons that are normally studied by the retinue of trace chemists when more than benzopyrene is concerned. There are among the 10 or 13

219

ISBN 0-12-192750-4

indicator polynuclear aromatic hydrocarbons some that certainly are much less carcinogenic or noncarcinogenic than others.
Even noncarcinogens can be useful indicators in the trace
analysis field.

Finally, I think once again, to go back to the stance of
the analytical chemist who has to defend the pressure that is
often or has been often on him in developing more and more
sensitive methods, Dr. Kolbye was quite right, of course, in
speaking of the limitation a generation ago of trace methods
to the detection of the order of one part in 1000, although
even some 200 years ago it was possible in special cases then
to detect, perhaps not to measure, but to detect traces of
lead, for example, in mortar, of the order of a few parts per
million. I was pleased to hear that the line is not going to
be drawn at a single molecule. The analyst has really found
out more and more about less and less, until now he knows
everything about nothing. I do not think there is any more
to find out. Today the levels of practical interest are, as
I think is widely recognized now, in the order of either one
part per million or in the range of one to ten parts per
million for solid foods, or mostly for the carcinogens and
potential carcinogens in a range somewhat below that from one
hundred parts per thousand million. Those are the normal
practical limits for regulatory action, although somewhat
lower levels may be appropriate to liquid foods and to water,
and parts per trillion level, parts per 10^{12}, are largely
found in the literature, but they are mainly in the area of
research and investigation.

As Dr. Darby said, safety cannot be the subject of positive experimental proof. Here we have a set of obviously
sophisticated trace methods of analysis focusing on these two
ranges, say between one and one thousand parts per billion.
We have already reviewed some of the limitations in our ability to interpret results obtained from those and it is too
easy, and in any case not particularly profitable, to ask the
negative question, how do we know that subclinical and massive
levels of say one-hundredth of a part per thousand million,
that's 10 parts per trillion, are not harmful?

I have sympathy with the view that has already been expressed that since we have established additives that contam-
inate species, at least we should be making more attempts to
make much better use of the epidemiological evidence that can
be made available and used by us.

I did want to touch on the numbers game and show how
difficult and misleading it can be, in a way that I have done
on other occasions. We take a relatively pure food substance
like sugar and suppose a total level impurity of only 1/100%,
that is say, 100 parts per million, made up of impurity perhaps of other sugars and carbohydrates. Now, if the impurity

is present as a single substance in this relatively pure
chemical food it is fed to the extent of 0.10% of 100 parts
per million. If we make a different assumption that there
are ten different impurities and then further assume that all
those are there at the same level, then there are ten impurities present at 0.001% level or 10 parts per million. When
you are considering impurities at the one part per million
level in this same model on this basis there could be 100
different chemical substances or 100,000 different chemical
substances if they are all present at one part per billion,
one part in 10^9. And where we talk of one part per trillion
there could conceivably be in the same model, 1000 million
different substances. Well, what is the significance of a
single impurity of the part per trillion level? If we know
this what is then the difference of significance of 27 such
compounds in water? What about our limitations in the evaluation of synergism of potentiation to which you referred to
yesterday? This is merely to illustrate the way I think you
can be misled! I think I would be misleading you if I tried
to play that numbers game!

DR. KRAYBILL: I would just like to make a few remarks based
on what I have heard thus far because there is a level of
ignorance relevant to a comprehensive understanding of the
mechanisms of carcinogenesis. I believe there is a situation
developed where there is obviously much disagreement in the
scientific community. We certainly have an area of dialog
where the arguments are not unlike those, at least in my view,
in politics and religion, reflected in the vehemence of the
arguments that are advanced. Now this would be considered a
polarization and, indeed, I felt that during the last four or
five years among my colleagues at NCI, you sort of divided
into two camps. I think this polarization that is developed
is most unfortunate. Some people may think it is fortunate
to have this dialog, but I think the paper that Dr. Parke
presented goes a long way to getting us beyond this speculative area and give us some explanation of what is going on.
And maybe we can get out of this so-called area of religion
and politics and get down to facts.
        Actually the presentation he made was in line with a con-
cept, I think it was three years ago, I made to one of my
colleagues. I said this is the sort of thing we need. I was
very much impressed with Perry Gehring's work on the glutathione and the sulfhydryl groupings and explained the sopping
up of these groups at different doses. I believe this is what
we need and, when I said this is what we need rather than
having to resort to mathematics, he said, "Oh that's ten years

away." But, I think maybe we are closer than ten years. I would say we make a plea to our adversaries to listen carefully to our admonitions on many of the design features of our carcinogenic testing; this is what bothers me.

We are all biochemists or toxicologists. We have taken training at various universities and, certainly, if the kind of training and the concepts we have developed over the years are wrong, then we have wasted all this background of experience. So how can we disagree so widely? I guess I am like Dr. Wynder, who makes a plea for the epidemiologists to be a part of this team.

I would like to make a plea to my colleagues who are biochemists (we are all biochemists working in this general area in the field of carcinogenesis) to heed the warnings of their biochemist colleagues in the field of nutrition. Now that is my bias, I came out of that field. I spent most of my career in the field of nutrition. I feel, as nutritionists, we have made some mistakes, many mistakes--certainly, when we fed 35% solids of irradiated spleens, as we did at Fitzsimmons Army Hospital because that is what the guideline said we had to do. And we did it, and a chap over in the laboratory was pumping thiamine and riboflavin into these animals, because he thought they had a vitamin deficiency. When the people at Chicago learned about this, they said "Those dummies, don't they know any better?" All that oxalic acid in the spinach! So we had animals with calcium tetany. There again, the effect of the massive doses, what this can do! As nutritionists, I think we have made some mistakes years ago in the design of our chronic toxicity experiments. We probably learned more about vitamin K metabolism from the irradiated food program than we learned about the toxicity of irradiated foods. We found that the dosing here, the radiation dose, whether you are at 6 or 3 Mrad, makes a lot of difference, in terms of the hydrcxyhydroperoxides that are formed and the effect of those hydroxyhydroperoxides on vitamin K and vitamin E.

Therefore, I would like to plead with our colleagues who insist on the principles of massive dosing to please profit by our mistakes. Listen to us, we have something to tell you! I think this is long overdue. I noticed the other day that some mention was made about getting as a part of this team some nutritionists. I think Dr. Darby and people like that can make a tremendous contribution to this area and into elucidation of some of the concepts with regard to carcinogenesis and the testing of chemical carcinogens.

METABOLISM AND PHOTODEGRADATION OF CHLORINATED HYDROCARBONS
AND THEIR POTENTIAL IN CARCINOGENESIS

F. Korte

It is my understanding from the many discussions on this
topic, where I have been involved, that there exist numerous
hypotheses as to how chemicals produce carcinoma. It seems
that the discussion will continue on how to extrapolate from
experimental to environmental levels, on the interpretation of
species differences, and especially on the relevance of laboratory data for man. It seems to be generally accepted that
the development of carcinoma in man follows long-term exposure
to small doses of carcinogens. This is not contradictory to
the finding that single high doses are more potent in this
respect than repetitive small doses. I do not feel competent
to discuss these questions in detail. There is one question,
frequently claimed to be involved in chemical carcinogenesis;
namely, metabolic activation.

I would not like to discuss the chemistry that seems to
be involved in the action of the different classes of carcinogens, like alkylating agents--a term defined by different
disciplines--polynuclear aromatics and arene epoxides. The
real knowledge we have in this area has been critically discussed at the conference on "Human Epidemiology and Animal
Laboratory Correlations in Chemical Carcinogenesis" in Mescalero.

Since we have much information on our topic for today,
i.e., chlorinated hydrocarbons, I would like to mention here
the *in vitro* data of Henschler and others on chlorinated
solvents. Chlorinated ethylenes are metabolized by a major
route to the respective alcohols and acids via their oxiranes.
With increasing number of chlorines the rate of oxirane formation decreases. This fact is understood, since chlorine substitution increases the persistence of a molecule against
metabolic attack. The biological activity of chlorinated
ethylenes has been correlated with their alkylating action,
whereas the other metabolic conversions, like reductions,
hydrolyses, conjugations, and rearrangements, are detoxications.

223

ISBN 0-12-192750-4

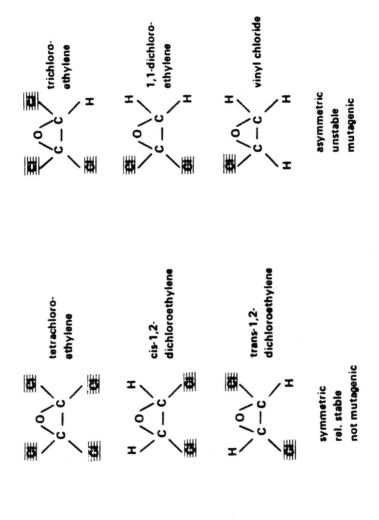

Fig. 1. Oxiranes of chlorinated ethylenes and their mutagenic potential in vitro.

Three chlorinated ethylenes; namely, vinyl chloride, vinylidene chloride, and trichloroethylene, are mutagenic *in vitro*. This fact is frequently interpreted as an indication of carcinogenicity. As can be seen from Fig. 1 , the mutagenic isomers form asymmetric oxiranes that are unstable. The oxiranes of the nonmutagenic compounds are relatively stable. If the reaction of the oxiranes with constituents of the animal cell is the basis for genetoxic effects, this correlation between symmetry/asymmetry and stability/instability may be a criterion to predict biological activity. Furthermore, this finding would indicate--if it could be generalized--that stability against metabolic attack would be a favorable characteristic of a chemical.

Using as an example my own experiments with the model chemical dieldrin, I would like to demonstrate that species differences in metabolism might be the cause of different species responses even if the metabolic pattern is qualitatively identical in the different species. It is well established that dieldrin, like many other chlorinated and non-chlorinated chemicals, induces the formation of liver hepatoma in sensitive strains of mice. These hepatomas have been found neither in rats nor in other laboratory animals and primates. Qualitatively, mammalian metabolism of dieldrin is identical in all species apart from "exotic" minor products.

In Fig. 2, I have summarized all known metabolites of aldrin and dieldrin formed by animals, plants, and microbes. The animal metabolites are shown in the second line. The major products in any animal species by far are the 12-hydroxydieldrin and the 6,7'-dihydroaldrin-*trans*-diol. The pentachloroketone and the diacid have no significance as mammalian metabolites. In order to elucidate the quantitative metabolic pattern of the two major animal metabolites, we had to do an experimental investigation, despite the fact that numerous studies on dieldrin metabolism in many animal species had been done, but each investigator used different experimental conditions, so that literature data are not comparable. A single oral dose of 0.5 mg/kg bodyweight of ^{14}C dieldrin was administered to male and female mice, rats, and rabbits, to two male rhesus monkeys, and to a female chimpanzee (Table I). The radioactivity, excreted within 10 days after administration, was analyzed for unchanged parent compound and the two major metabolites. Conjugates of the diol were calculated as diol. Regarding the ratio of the two metabolites, the rat is very well comparable to the primates, whereas in the mouse and the rabbit diol formation is the predominant pathway. This could be an indication that the diol is involved in hepatoma formation in mice, which does not occur in rats. But this would not explain the fact that hepatomas are not formed in rabbits, since in rabbits the diol to 12-dihydroxydieldrin

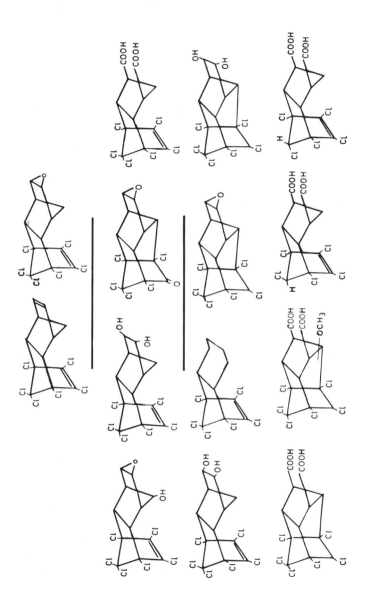

Fig. 2. Survey of dieldrin metabolites.

Table I.
Comparative metabolism of dieldrin in mammals [a]

animal		dieldrin % in excreta	12-OH-dieldrin % in excreta	aldrin-trans-diol % in excreta	excreted within 10 days % of applied
mice	male	14.3	33.8	51.9	38.5
	female	8.7	20.3	70.5	36.9
rats	male	6.7	73.9	19.3	11.9
	female	28.6	46.9	24.5	9.8
rabbits	male	16.7	n.d.	83.3	1.8
	female	18.5	7.4	74.1	2.7
rhesus	male	44.1	46.1	9.8	20.4
chimpanzee	female	50.8	31.7	17.5	6.3

[a] Single oral dose of 0.5 mg/kg bodyweight.

ratio is even higher than in mice. When looking at the total
excretion of metabolites, this discrepancy is resolved. Total
excretion of 37 to 38% and biotransformation of 33 to 34% of
the applied dose is by far the highest in the mouse. This
must necessarily result in relatively high concentrations of
the diol in the mouse liver. Since excretion-metabolism in
the rabbit is lowest within the investigated species, diol
concentrations in the rabbit liver are only minute--about the
same level as rats and rhesus monkeys--although the diol is a
major rabbit metabolite. Consequently, the data shown in
Table I would be a chemical explanation for the data of biological experiments on hepatoma formation by dieldrin in animal species. In order to verify whether this explanation is
really valid, feeding experiments with a pure 12-hydroxy-
dieldrin and the *trans*-diol are in progress. This example
was given to show how complicated the interpretation of biological data may be.

The difficulties even increase with interdisciplinary discussion of the evaluation of experimental data. This increase
is caused by the fact that the disciplines involved in the
evaluation use identical terms with different definitions.
With the involvement of socioeconomic, political, and even
philosophical criteria in the assessment of environmental
chemicals, this situation becomes even more complex. It must
be accepted that these criteria have to be considered in
addition to the scientific and technological facts. In order
to achieve the necessary coordination, all parties involved
should speak the same language. This is not yet the case even
in the natural science and technology disciplines. In order
to enhance this urgent communication, the Federal Republic of
Germany is taking the initiative to organize an international
meeting on UNO basis in 1979. This meeting should come to a
generally accepted glossary of terms relevant for the assessment of environmental chemicals.

I would like to explain the confusion in terms, which make
scientific discussions so difficult, by a few examples. Let
us start with the term "toxicity." Frequently, it is exclusively used for target organisms only; but what about non-
targets, what about effects "in use"? Has acute toxicity any
significance for environmental chemicals? What about the
types of harm? What does the term "toxic compounds" mean?
Is it exclusively negative?

Another example is organochlorines. For a chemist this
term means all organic chemicals that have chlorine in their
molecules. For many other disciplines the term includes only
a few insecticides, PCB, and HCB. In this context, I would
like to mention that 20% of the pesticidal chemicals contain

halogen in their molecules. Figure 3 only shows three examples of pesticides that are not properly classified according to chemical definitions.

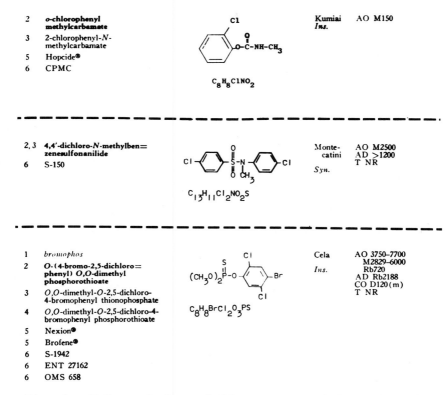

2	*o*-chlorophenyl methylcarbamate	Kumiai *Ins.*	AO M150
3	2-chlorophenyl-*N*-methylcarbamate		
5	Hopcide®		
6	CPMC		

2, 3	4,4′-dichloro-*N*-methylben= zenesulfonanilide	Monte- catini	AO M2500
6	S-150	*Syn.*	AD >1200 T NR

1	*bromophos*	Cela	AO 3750–7700 M2829–6000
2	O-(4-bromo-2,5-dichloro= phenyl) O,O-dimethyl phosphorothioate	*Ins.*	Rb720 AD Rb2188 CO D120(m) T NR
3	O,O-dimethyl-O-2,5-dichloro-4-bromophenyl thionophosphate		
4	O,O-dimethyl-O-2,5-dichloro-4-bromophenyl phosphorothioate		
5	Nexion®		
5	Brofene®		
6	S-1942		
6	ENT 27162		
6	OMS 658		

*Fig. 3.   Halogenated pesticides not classified as organo-
halogens.*

The term "residue" has been defined for pesticides by the
joint FAO/WHO committee but, nevertheless, it is sometimes
used for the parent compound only, and sometimes one or the
other metabolites are included. What about all environmental
conversion products?
Regarding the topic of this meeting, let's have a look at
the term "carcinogen". Does it mean carcinogen to men or to
rodents? Does it include the level of exposure? I had already mentioned alkylating agents. For a chemist this is a

fast number of reactants that can be used in syntheses to
yield alkyl derivatives of starting materials. Even ethanol
is an alkylating agent in this sense. For those involved in
the mechanism of chemical carcinogenicity, alkylating agents
are just a few chemicals that have the potential to alkylate
DNA or RNA molecules.

Other terms that would need definitions are "metabolism,"
"degradation," "accumulation," and last but not least "persistence." I will not continue this list, but will discuss
one further important point on the example of persistence.
Although undue persistence has been defined by IUPAC, this
term is mostly used in a quite vague sense to express the
characteristics of a chemical as to stability against enzymatic degradation. This is frequently connected with a bio-
accumulation potential.

One fact has been disregarded over the years; namely, the
possible correlation between persistence and abiotic processes
in the real environment. It is known that organic chemicals
exist in the atmosphere, on the surface of plants, and in soil,
and that in these situations they are exposed to UV irradiation, which may lead to changes. The aim of this research was
to study how hydrocarbons like aldrin, dieldrin, and photodieldrin, chosen as model substances, chemically change under
simulated atmospheric conditions and what kind of degradation
products result.

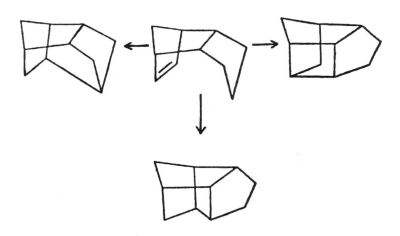

Fig. 4. Photoisomerization of cyclodienes.

Various photoreaction types of cyclodiene insecticides are known:

(a) Photoisomerization reactions (Fig. 4): Cyclodiene insecticides contain a chlorinated double bond, which can be excited by long-wave UV light (~300 nm). By interaction with the methylene bridge in the nonchlorinated part of the molecule, the corresponding photoisomeric product can be formed by a $\pi\sigma-2\sigma$ reaction. The excited double bond abstracts the opposite H atom, whereby a new σ-bond is formed (bridging).

When sensitized, these reactions proceed almost quantitatively. However, they can also be observed unsensitized in the solid and gas phase.

(b) Dechlorination reaction: In protonated solvents as well as in the solid phase, aldrin and dieldrin are photochemically dechlorinated at the chlorinated double bond. It may be assumed that the dechlorination reactions proceed from the singlet state of the molecule. In contrast to the above-described photoisomerization reactions, in this case an inter-molecular reaction also occurs whereby, after the photolysis of the C-Cl compound, the abstraction of a nearby H· (from solvent) is effected.

(c) Reactions with reactive oxidizing agents: It is known that mainly aldrin and dieldrin in the atmosphere are found after their application (gaseous or adsorbed to aerosols). These substances change their light absorption characteristics under adsorbed conditions; for instance, the absorption maximum of photodieldrin (195 nm in hexane) under adsorbed conditions on SiO_2 moves to longer wavelengths (260 nm). This absorption shift reveals that even photodieldrin in an adsorbed form can be excited by wavelengths higher than 260 nm, so that a possible reaction under the photochemical conditions of the lower atmosphere (with wavelength >290 nm) cannot be excluded. Photoreactions in the atmosphere further depend on the height (because of the increasing radiation intensity of the short-wave UV irradiation in higher altitude as well as on the condition of the local atmosphere presence of other reactive substances). The vast number of possible reactions in the atmosphere and the lack of experimental information at the present time make it impossible to estimate the rates of the individual steps (Table II).

Irradiation of aldrin in solution revealed that the photoisomerization product photoaldrin is formed primarily together with a smaller amount of the monodechlorination product. Contrary to dieldrin, aldrin is dechlorinated by wavelengths above 300 nm, which is possible in the lower atmos-

Table II.
Photochemical reactions of aldrin

Phase	Time of irradiation	Type of lamp	Wavelength (nm)	Applied amount (mg)	Conversion (%)	Photoproduct (yield, %)
Film on glass	1 month	Sunlight	290	–	97	I (24), II (4), III (9.5), polymer (not identified) (60)
Adsorbed on Al_2O_3 + benzophenone	10 hours	TK 150	290	50	100	I (90)
Adsorbed on SiO_2	16 hours	HPK 125	290	200	84	I (4), II (55), III (2), IV (4)
Film on glass	2 hours	HPK 125	290	200	74	I (41), II (5), III (2), polymer (not identified) (25)
Ethyl-acetate	7 days	–	–	–	–	I, V

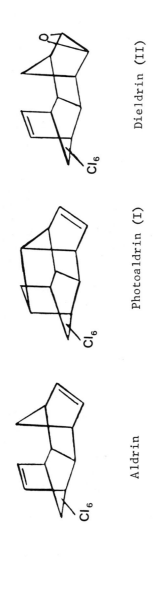

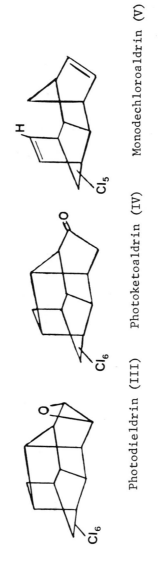

Aldrin

Photoaldrin (I)

Dieldrin (II)

Photodieldrin (III)

Photoketoaldrin (IV)

Monodechloroaldrin (V)

Table II: Photochemical Reactions of Aldrin (continued)

Table III.
Total balance of UV irradiation of aldrin

	Applied amount (mg)	Conversion (%)
Aldrin	266.8	100
Dieldrin	110.1	63.9
Photoaldrin	8.5	5.1
Photodieldrin	4.6	2.7
Photoketoaldrin	8.1	4.7
Unidentified products	-	23.6

phere. The irradiation of aldrin in the solid phase (on glass) by wavelengths above 300 nm revealed that aldrin under such conditions produces preferably photoisomers. Photodieldrin could be isolated and identified in lower yields. Irradiation of aldrin adsorbed on SiO_2 using wavelengths above 300 nm, however, revealed that the bridging product (photoaldrin) exists only in small yield. The main product is dieldrin (yield 64%). Table III indicates the total balance of this irradiation.

Nitrogen dioxide is of primary importance among the inorganic components of air pollution because of its generation during the combustion of fossil fuels. A connection between the daily variation of ozone concentration and sunlight intensity with the daily fluctuation of nitrogen dioxide and hydrocarbon concentrations was first proposed during the 1950's based on measurements in the Los Angeles area. Experimental evidence confirmed that photolysis of nitrogen dioxide is a primary step for the increase in ozone concentration. Both nitrogen dioxide and ozone absorb in the ultraviolet and visible regions of the spectrum. Photodissociations at wavelengths greater than 300 nm are known to occur with both compounds. The compounds in Fig. 5 were isolated from solutions of dieldrin and nitrogen tetroxide in fluorocarbon after irradiation with UV light having long wavelengths (λ >300 nm). Irradiation of solutions of dieldrin and nitrogen tetroxide in carbon tetrachloride under similar conditions yielded the same compounds. The OH compound was detected after irradiation of a solution of dieldrin and ozone in fluorocarbon. In the presence of nitrogen dioxide, dieldrin distributed in nitrogen in gaseous state was converted mainly into photodieldrin by UV light.

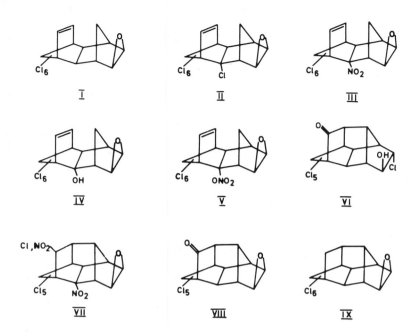

Fig. 5. Photochemical reactions of dieldrin in the presence of NO_2.

Irradiation of dieldrin gaseously distributed in ozonized air yielded, as in ozonic solution, several products of higher molecular weight, which, however, were not isolated (Fig. 6). It is well established that such reactions also take place with any nonchlorinated organics. This is just one example of a chemical occurring from man-made and natural sources in the atmosphere, which forms an epoxide and ketones during its pathway to mineralization.

The importance of abiotic degradation under atmospheric conditions has to be emphasized. In the past, research has been done on the photochemical changes and degradation of organic chemicals, including the investigation of the reaction mechanisms and the kinetics of such reactions. The results of these studies derive from laboratory experiments that did not attempt to simulate actual atmospheric conditions. It has been known for a long time that organic molecules are subject to isomerization, conversion, and incomplete degradation reactions through UV irradiation. It has not generally been recognized that mineralization may take place even in diffuse daylight. So far, only some organic chemicals such as methane,

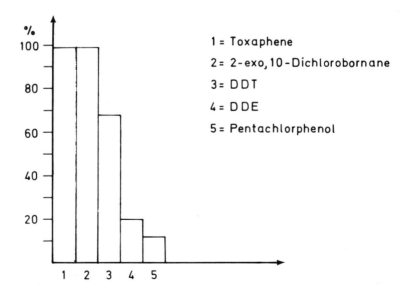

$$\begin{array}{c} H_3C \\ \\ H \end{array} C=C \begin{array}{c} C_2H_5 \\ \\ H \end{array} + O(^3P) \longrightarrow \begin{array}{c} \text{excited} \\ \text{intermediates,} \\ \text{radicals} \end{array} \longrightarrow \text{degradation}$$

$$CH_3\overset{\|}{\underset{O}{C}}CH_2C_2H_5 \qquad \begin{array}{c} H_3C \\ \\ H \end{array}\overset{\triangle}{\underset{O}{}}\begin{array}{c} C_2H_5 \\ \\ H \end{array} \qquad H\overset{CH_3}{\underset{}{C}}\overset{}{\underset{\|}{C}}HC_2H_5$$

cis+trans

Fig. 6.   2-Pentene photochemistry.

propane, and unsubstituted arenes that participate in the formation of the so-called photochemical smog have been studied in great detail. Our studies demonstrate that even persistent chemical substances (e.g., photodieldrin) can be mineralized within relatively short periods if irradiated with light of wavelengths between 230 and 300 nm, as well as above 300 nm, the latter being a wavelength also found in diffuse daylight (Fig. 7).

%
100

80

60

40

20

1   2   3   4   5

1 = Toxaphene

2 = 2-exo,10-Dichlorobornane

3 = DDT

4 = DDE

5 = Pentachlorphenol

Fig. 7.   Photomineralization of organochlorine chemicals adsorbed on silica gel with wavelengths $\lambda > 290$ nm (7 days).

Table IV.
A survey on photomineralization[a]

Wavelength:	>230 nm		>290 nm	
Duration of irradiation:	2-3 days		6 days	
Mineralization products:	CO_2	HCℓ	CO_2	HCℓ
Aldrin	+	+	+	+
Dieldrin	+	+	+	+
Photodieldrin	+	+	+	+
DDT	+	+	+	+
DDE	+	+	+	+
Hexachlorobenzene	+	+	−	−
Pentachlorophenol	+	+	+	+
2,4,5,2',4',5'-Hexachlorobiphenyl	+	+	−	−
2,5,2',5'-Tetrachlorobiphenyl	+	+	−	−
Toxaphene	+	+	−	−
2,10-Dichlorobornane	+	+	−	−
2,6-Dichlorobornane	+	+	−	−
Hexachlorobutadiene	+	+	+	+
1,1-Dichloropropene	+	+	+	+
1,2-Dichloropropene	+	+	+	+
Tetrachloroethylene	+	+	+	+
1,1-Dichloroethylene	+	+	+	+
Trichlorfluoromethane	+	+	+	+
Dichlorfluoromethane	+	+	+	+

[a] +, Detected; −, not detected (<1%).

Considering the fact that an instant availability of energy exists in the atmosphere, the abiotic (photochemical) degradation is probably a more important process than bio-degradation, which in most cases only leads to conversion products easily excreted by the living organism. Therefore, the atmosphere can be regarded as a large sink for organic persistent chemicals. Comparative studies of today's global concentrations of some persistent chemicals such as DDT and dieldrin, and the total amounts released suggest that the bulk may be mineralized.

Thus, it may be possible to estimate permissible emission levels by determining the rates of photochemical mineralization reactions in the atmosphere. Table IV shows a comparative diagram of the mineralization rate of several organochlorine chemicals. As you see, these are exactly the chemicals that are persistent against metabolic attack. From Table IV you see that most of the persistent chemicals we have investigated can be mineralized by tropospheric sunlight. Thus, sunlight not only helps in the welfare of the biosphere by supplying abundant energy to induce life, but the energy of the sun is also used to lead man-made chemicals back into natural mineral cycles.

It is especially the potential of the atmosphere to mineralize chemicals that calls for a final and conclusive statement; namely, that we should not have one fixed-forever system of evaluating environmental chemicals. Following new scientific results that change the light in which a criterion has to be seen, we should allow for new or adopted evaluation systems. Photochemical mineralization results in a reduced importance of the criteria of persistence and accumulation. Accumulation potential is reduced by photochemical mineralization of biologically persistent compounds and, consequently, the risk for a long-term hazard is reduced, too.

EXPERIMENTAL PROCEDURES IN THE EVALUATION OF CHEMICAL CARCINOGENS

Charles J. Kensler

In this chapter, attention will be directed toward approaches to the evaluation of chemical carcinogens rather than a description of the care and feeding of animals exposed to potential carcinogens and the pathologic examination of the tissues at the end of the experiment. As is consistent with the central issue of this volume, the emphasis will be on the importance of dosage, particularly low doses, in evaluating potential carcinogens.

Among the approaches to the selection of compounds for study in animal experiments, considerable effort is underway to develop short-term tests to enable one to choose those compounds most likely to be carcinogenic for animals and man. The National Cancer Institute has been supporting a program in this area, which includes study of systems based on (1) mutagenesis, (2) DNA damage/repair, and (3) cell transformation. These systems are listed in Fig. 1. Although claims have been made for the predictive utility of several of the tests, it is clear that the predictive validity of these screening tests is still in the process of evaluation. The *Salmonella* system (1) has been most widely studied, and the Ames group claims that it is predictive for 90% of the positive animal carcinogens while picking up only 20% false positives. On the other hand, Heddle *et al.* (2), working with another group of compounds detected only 65% of the animal positives and 19% false positives. The hope is that further exploration of combinations of the short-term tests listed in Fig. 1 will lead to a useful predictive set of procedures. This is an important approach but will require years of work before any combination can be ensured to provide valid predictive information.

Once a compound is selected for tests in animals for whatever reason (short-term test results, large-scale human exposure, suggestive epidemiologic inputs, structural similarity

ISBN 0-12-192750-4

MUTAGENESIS-BASED TESTS

Salmonella - histidine locus ("Ames")

E. coli - WP2, uvr a$^-$

Mouse lymphoma L5I78Y, Thymidine kinase$^{+/-}$

DNA DAMAGE/REPAIR TESTS

E. coli - Pol A$^{+/-}$

Rat hepatocyte - excision repair

CELL TRANSFORMATION-BASED TESTS

Syrian hamster embryo - direct clonal assay

Syrian hamster embryo - in vivo/in vitro assay

BALB/c-3T3 mouse - focus assay

MLV-Fischer rat - focus assay

Epithelial cells - developmental work

Fig. 1. Short-term approaches, National Cancer Institute bioassay program.

to established animal carcinogens), it is important to select the doses to be tested to be those most relevant to the objectives of preventing human disease. If, as is the usual case, massive doses are the only ones tested or are the only ones used in attempts to assess human risk, it appears likely that compounds with little or no potential for cancer induction at low doses will be banned because of observed activity at high doses. Admittedly, the high-dose tests tend to compensate for the small animal population at risk, but if the metabolism of the compound is different at high doses than low doses, we may be testing metabolite activity that is not present at low doses.

If the pharmacokinetics of the compound are linear (first order), then the biologic half-life, the composition of the excretory products, and the routes of excretion are independent of the dose, and the area under the blood level curve is proportional to the dose. If, however, the kinetics are nonlinear (dose-dependent) then the decline of body levels is not exponential. The time to eliminate 50% of the dose increases with increasing doses, the composition of excretory products changes with the dose and the routes of excretion. In the case of compounds handled in the nonlinear fashion, one will inevitably be testing combinations of metabolites that do not exist at low doses and hence are not relevant to the risk assessment task.

Regardless of whether the compound is handled by linear or nonlinear kinetics, the cells at risk have defense mechanisms as illustrated schematically in Fig. 2. First, detoxification mechanisms exist that can handle low-level exposures, and second, macromolecular repair mechanisms (including DNA) can repair low levels of damage.

An example showing the relationship between the levels of linear nonprotein sulfhydryl groups and the binding of {C14} vinyl chloride (VC) to liver protein is shown in Fig. 3. At low doses, the reactive metabolite of VC is detoxified by re-action with the SH group of glutathione (GSH) and protein binding prevented or minimized. As the dose of VC is raised, GSH in the liver is lowered and interaction with liver protein is increased. Studies of the closely related vinylidine chloride (C14-VDC) have shown a similar relationship between GSH depletion and C14-VDC liver protein binding. Data showing the relationship between covalent binding of reactive metabo-lites of VC with protein and the incidence of hepatic angio-sarcomas (Fig. 4) have been reported by Watanabe et al. (3). The protein binding data are theirs and the tumor incidence data are those of Maltoni. My interpretation of these data is that low doses that do not result in labeled protein do not represent a carcinogenic hazard even though VC is clearly a carcinogen at high doses.

1, 4-Dioxane has been reported to produce hepatomas in rats but only in doses that produce severe renal and hepatic pathology (4). A metabolic study (5) has shown that the metabolism of dioxane follows nonlinear kinetics and that de-toxification step of conversion to β-hydroxyethoxyacetic acid is a saturable step. Thus high-dose extrapolation of carcino-genic effect to low doses that neither saturate the detoxifi-cation mechanism nor produce hepatic or renal pathology is not appropriate to the assessment of the potential hazard associ-ated with low doses.

Studies have been underway in our laboratories (6) on the metabolism of methylene chloride, a widely used solvent, to ascertain its metabolic behavior in order to establish an appropriate dose range for long-term carcinogen tests. Meth-ylene chloride is known to be metabolized to CO and CO_2 pre-sumably by different pathways (7,8). Whether or not an active alkylating intermediate is formed is not known, but work on this is in progress. Preliminary data on $^{14}CO_2$ and ^{14}CO in expired air (Fig. 5) (6) indicate that, with both corn oil and water vehicles as solvents, the amount of expired CO_2 and CO

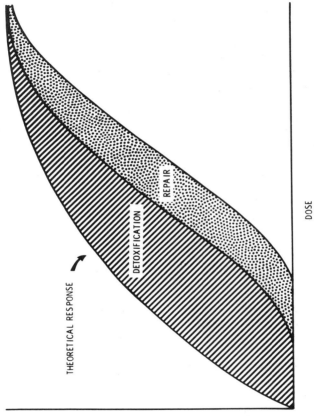

Fig. 2. *Defense mechanisms of cells at risk.*

242

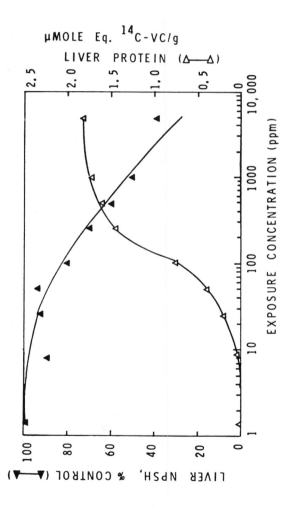

Fig. 3. Relationship between detoxification and covalent binding of reactive metabolite of vinyl chloride.

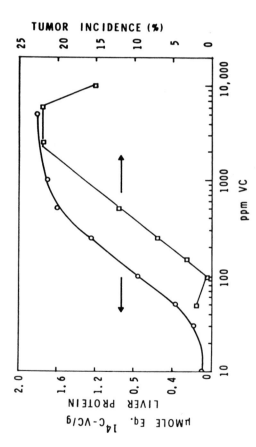

Fig. 4. Relationship between covalent binding of reactive metabolite of VC and incidence of hepatic angiosarcomas.

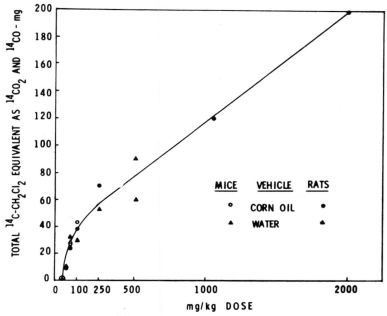

Fig. 5. $^{14}C\text{-}CH_2Cl_2$ equivalents ($^{14}CO_2$ and ^{14}CO) in expired air.

as a fraction of the dose shows a break in the curve around 100 mg/kg. Data (Fig. 6) (6) for the amount of unchanged methylene chloride expired as a fraction of the dose also indicate an inflection in the curve at about 100 mg/kg. This suggests that overloading or saturation of metabolic pathways occurs at doses over 100 mg/kg/day. An experiment was performed administering ^{14}C-labeled methylene chloride to rats that had been exposed to methylene chloride for 30 days. The chronically exposed animals appeared to handle the compound the same as the previously unexposed animals (Table I). In the absence of information (which we are in the process of developing) on labeled macromolecules as a function of dose, it would appear desirable to test methylene chloride at least four dose levels--two on the curve above 100 mg/kg and two below this level, as indicated in Fig. 7. Two doses would be above the apparent metabolic overload dose and two below. If current National Cancer Institute bioassay guidelines were followed, only the two higher dose levels would be included.

If one looks for low-dose-response data for human exposure to carcinogenic stimuli, one is hard put to find any. Even the cigarette data for low doses are essentially nonexistent, although the practical equivalent of a safe dose or the (my words) practical equivalent of a threshold is indicated (9).

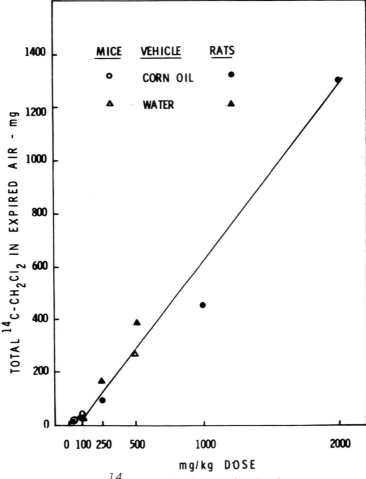

Fig. 6. ^{14}C-CH_2Cl_2 in expired air.

Studies of aflatoxin intake represent far and away the best
data we have for low-dose human exposure to an established
animal carcinogen. The doses reported for man based on total
diet analysis of daily food intake are lower for the highest
population intake identified (Mozambique) than the lowest
active level in the sensitive rodent species, the rat, on a
semisynthetic diet. Meselson (10) has used currently popular
(or at least widely used) statistical extrapolations to estimate probable maximal human incidence and found that the predicted incidence for liver cancer from aflatoxin is ten times
that observed in man from all causes. A plot of the available
human dose--male incidence data by I. Miller of our laborator-
ies--is shown in Fig. 8. The data are based on African and
Asian population studies and suggest the existence of a thres-

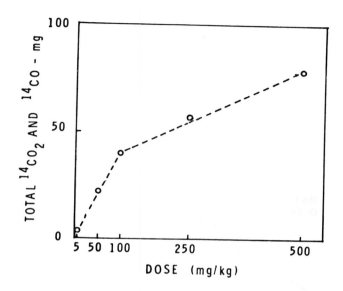

*Fig. 7. Choosing doses for toxicology studies
based upon break in CO_2 and CO production.*

hold. The exposure of the U.S. population is probably equivalent to -0.2 or less on the plot. A recent FDA document (11)
supporting a 15-ppb tolerance for aflatoxin in peanuts, as
opposed to a 5-ppb level, concludes that the reduction from 15
to 5 ppb would show relatively no significant gain in the protection of public health. They further note that the population at highest risk from aflatoxin in the southeastern U.S.
shows a lower liver cancer incidence than the U.S. average.
Alabama and Georgia have incidence rates of 100 and 103 per
100,000, whereas the U.S. average is 161 per 100,000.
    In summary, it appears (1) that the statistical extrapola-
tion of the results of high dose testing of compounds is inadequate to predict the effects of low dose exposure to these
agents; (2) that the acquisition of pharmacokinetic and metabolic data as a function of dose in several species would be a
great aid in interpreting the significance of low dose exposures; (3) that the development and refinement of more sophisticated statistical extrapolation models such as that recently
published by Cornfield (12) should be encouraged. This model
will not help, however, without the acquisition of low- and

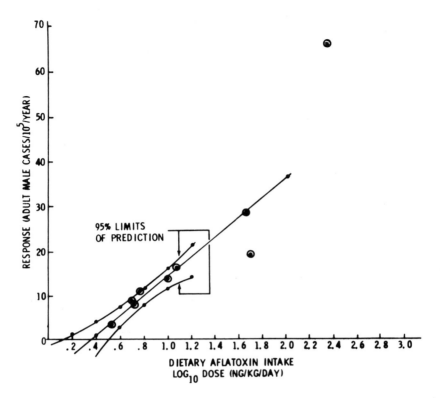

Fig. 8. Liver cancer in man.

TABLE I

^{14}C-CH$_2$C$_2$ Equivalents in Expired Air in Previously Unexposed or Chronically Dosed Rats

Dose of: ^{14}C-CH$_2$Cl$_2$	^{14}C-CH$_2$Cl$_2$	CO$_2$	CO
(mg/kg)	(mg)	(mg)	(mg)
Previously exposed			
1	0.07	0.4	0.5
50	15	15	13
1000	545	87	73
30-day chronically dosed rats			
1	0.2	0.4	0.4
50	16	12	15
1000	625	105	95

high-dose metabolic and kinetic data; (4) that the Delaney
Clause philosophy, which simplistically applies high dose
results to low dose exposures, is (a) encouraging regulators
and administrators to devote scarce resources to high dose
testing only; (b) discouraging in-depth application of our
most advanced technologies and sciences to low as well as
high dose studies; and (c) causing economic hardship without
any assurance that the cancer incidence in man will be lowered
by removal of low dose exposures. High dose exposures should,
of course, be reduced if animal experiments are positive for
carcinogenicity but society does not need the Delaney Clause
to proceed to deal constructively with this issue.

REFERENCES

1. McCann, J., and B. N. Ames, *Cold Spring Harbor Conf.
    Cell Proliferation C,* 1431-1450 (1977).
2. Heddle, J. A., and W. R. Bruce, *Cold Spring Harbor Conf.
    Cell Proliferation C,* 1549-1560 (1977).
3. Watanabe, P. G., M. J. McKenna, and P. J. Gehring,
    "Chemobiokinetic Perspectives on Mechanisms of Chemical
    Carcinogenesis, Industrial Toxicology." Elsevier, New
    York (1978).
4. Kociba, R. J., S. B. McCollister, C. Park, T. R.
    Torkelson, and P. J. Gehring, *Toxicol. Appl. Pharmacol.
    30:* 275-286 (1974).
5. Young, J. D., W. H. Braun, P. J. Gehring, B. S. Horvath,
    and R. L. Daniel, *Toxicol. Appl. Pharmacol. 38:* 643-646
    (1976).
6. Yesair, D. W., D. Jaques, P. Schepis, and R. H. Liss,
    Fed. Proc. 36: 998 (1977).
7. Kubic, V. L., M. W. Anders, R. R. Engel, C. H. Barlow,
    and W. S. Caughey, *Drug Metab. Disp. 2:* 53-57 (1974).
8. Ahmed, A. E., and M. W. Anders, *Drug Metab. Disp. 4:*
    357-361 (1976).
9. Gori, G. B., *Science 194:* 1243-1246 (1973).
10. Meselson, M., "Contemporary Pest Control Practices and
    Prospects: Report of the Executive Committee," National
    Research Council, Vol. 1, p. 82 (1976).
11. Bureau of Foods, F.D.A., "Assessment of Estimated Risk
    Resulting from Aflatoxins in Consumer Peanut Products
    and Other Food Commodities," January 19, 1978.
12. Cornfield, J., *Science 198:* 693-699 (1977).

REMARKS

DR. KRAYBILL: Thank you, Dr. Kensler. I don't think I will make any comment publicly on what NCI should be doing except to say, unofficially, I rather share your views. But I don't know whether that's accepted policy, so I will say nothing more. Henschler had a system of calculating the threshold and indeed he did. I gave an example in one of my papers of a molecular weight of a chemical at 500; he came up with a threshold of 5 ppb. About 2 years ago, the late Dr. Leo Friedman went through this process, this procedure, and he calculated it for aflatoxin, which I thought was interesting. One of my colleagues at NCI was shocked at the very idea of a threshold for aflatoxin.

DR. WAND: Thank you, Dr. Kraybill. Just one point that I would like to follow up on Dr. Kensler's excellent presentation. Everything we have heard so far indicates that we are focusing on translating these animal data to humans, but I have yet to hear anything mentioned, including Kensler's paper, on comparative pharmacology, comparative biochemistry, or comparative pharmacokinetics between the animals that we test and the humans who are the target for protection.

I would like to urge that as we proceed in our discussions we do consider this because this is the kind of necessary bridge between animals and humans that our epidemiologists have been pushing for, the kind of thing all of us, like Lijinsky and Coulston, are concerned about, the human end target. We need not only human epidemiologic data that will compare with the animal studies but we also need to know the comparative biochemistry, metabolism, pharmacology, and pharmacokinetics.

DR. KRAYBILL: Back in the days when Arnold Lehman was at FDA, and I was there with him, I remember him always saying man does not squeak like a mouse or bark like a dog. There was a symposium held right in this city about 10 years ago dealing with comparative pharmacology and I will never forget it. They went up the phylogenetic series, testing various drugs, and it was amazing to see the different responses to the drugs on a biochemical, pharmacological basis from the rodent up to the new and the old world monkey; there was a difference here. And finally, a determined fellow in the back jumped up and said, "Yeah, what about man himself?" I think it was Coulston. We even have a divergence in man. I think we have to keep that in mind and I think your point is well taken.

DR. SCHMIDT-BLEEK: I have been listening with great interest these two days to all the testing that has been done with animals. We are talking about extrapolation of all available data from animals to man, but a point on which we should like to get more information, if possible, is the validity of all the data.

Let me say it another way. We have undertaken a program in Europe in the nine member states of the EEC with more or less 60 laboratories that will be discussed in detail later. We did a comparative study between laboratories on the quality of the studies performed in each laboratory. We sent those laboratories a number of chemical compounds and asked every laboratory to do the testing. I am talking now particularly with respect to single administration toxicity, LD_{50} testing. In fact, we were very surprised by the great difference in results obtained in those laboratories. Let us say that for one compound we found a difference of a factor of 12.

The point I want to make is that we are talking in a very sophisticated way today about extrapolation from animals to humans. But what is the quality of the exercise performed in the different laboratories and how do we know the results of one laboratory are comparable with those in another laboratory? We are accumulating these days all types of information, but is there really comparability between all these data? What is being done here in the United States to assure that studies performed are of good quality?

DR. KRAYBILL: I think that is a good point. I think in the National Cancer Institute bioassay studies, we have attempted to do some of this, in that we do have some control over the animals; that is, the rats and the mice all come from a common source. There is a Fisher strain rat and the B63F1 mouse, and diet is from one central source. The animal husbandry, of course, could vary: temperature, humidity. We have about ten subcontract laboratories and we could make a comparative study here. We talked about this recently. We would have thousands of animals that would fall in the control population, on which we can observe the spontaneous tumor incidence, for example. We think it might be a useful exercise because the data are readily available. All we need to do is bring them together. Someone talked to us about looking at the water supply. Now this is predicated on the idea that one could keep the air pollutants constant. I do not know if we can. We can fairly well prescribe the diet to our specifications, so that we are all studying the same diet. Therefore, this would lead us down to the variation in the water that is given to these animals to see whether that has any impact on the spontaneous tumor incidence.

I guess the way we go about this is to run a GLC and mass spectra on various samples of water from these different labs and see first whether it is worth it. These are the variants of many of the laboratories. There were three studies done on bistrichlorethyl ether (I remember going down to the hearings on this chemical at EPA) and as you know from the literature, two studies came up negative and one positive. I wondered how they were going to deal with this. Well they dealt with it and they did indeed set up a tolerance level for bistrichlorethyl ether in the Philadelphia water supply. I'm glad they did not set it at zero, because they would have had to shut down the Philadelphia water plant. They did set a tolerance of 1/10 ppm.

DR. SCHMIDT-BLEEK: You informed me that you have a number of subcontractors. Are you telling me that every subcontractor is working with exactly the same protocol? There are so many parameters that should be considered in the comparability of studies, but do they all work with the same ones?

DR. KRAYBILL: Absolutely. They use what we call a bioassay screen guideline. That is our technical report number 1, which is made available to all the laboratories. It is only a guideline; we do not want to use this as a bible. Some people said we shouldn't send this everywhere and leave that impression, but we all use the same protocol.

NEGATIVE HEALTH IMPLICATIONS OF
THE DELANEY PHILOSOPHY

John P. Frawley

Many of you have asked me why I agreed to give a paper at
this conference. I've asked myself that same question many
times since I agreed to participate. For several years I
have steadfastly refrained from participating in any debate
on the Delaney Clause/cancer issue. First of all because I
had nothing to contribute, nothing to say that others haven't
said dozens of times, in a more convincing manner than I ever
could. Second, these symposia or debates are almost like TV
reruns. The cast of actors is usually the same, their role
and lines are always the same, and the outcome is always the
same--no one convinces anybody to alter his point of view
on the Delaney Clause from the point of view he possessed
when he came. The issue many years ago degenerated into an
adversary proceeding and no amount of science has won a con-
vert from one side to the other.

It is this protracted debate that I must categorize as
the first negative health implication of the Delaney philosophy.

Let me briefly interrupt myself and define what I mean by
the Delaney philosophy as contrasted with the Delaney Clause.
As we know, the "Clause" is specific legal language contained
in the Food, Drug and Cosmetic Act, more specifically in the
Food Additives Amendment and the Color Additives Amendment.
The absolute prohibition against use of chemicals that cause
cancer in animals is not contained in legislation that is
enforced by the other health regulatory agencies--EPA, OSHA,
CPSC, Department of Agriculture. Nevertheless, these agencies
have adopted essentially the same concept in administering
their laws. In essence, we have a national cancer policy,
which I call the Delaney philosophy, which prescribes that
the level of exposure to any chemical that causes cancer in

253

ISBN 0-12-192750-4

man or animals, regardless of dose, shall be reduced to the
lowest technically feasible level, and to zero if there is a
suitable alternative. It is this broader government policy
about which I am speaking today, not just the Food, Drug and
Cosmetic Act.

Let me return to my suggestion that the protracted debate,
which has gone on for over 20 years now, has had negative
health implications. Just about every month of the year,
there is a 2- or 3-day meeting on the subject, someplace not
unlike this one. It has become big business. These meetings
occupy the time and attention of many of our most talented
scientists, not just during the meetings, but in preparation
and travel for the meetings. And because nobody ever convinces anybody else, I contend that their time could be spent
more productively in the more mundane practice of toxicology,
that is, in the investigation of toxic effects of chemicals.
On balance this has been unproductive, but now I think it has
become counterproductive to protection of public health,
especially when we know our words are not influencing anyone.
I think we are actors in the longest running play on Broadway
or in D.C.

I'd like to say more about the inflexible positions that
we all have on this issue later, but I'd first like to continue to describe some of the less obvious negative health
implications of our current position. For the moment, place
yourself in the shoes of a businessman. Government advises
you that if you test your products for carcinogenicity and if
they are positive, you must take them off the market. If you
are a small businessman with only a few major products, you
can't afford to take that chance of testing your products and
possibly going out of business. Even if you are a large
businessman, you will be very reluctant voluntarily to test
those which are vital to your corporate profitability and
future. It is a form of product Russian roulette! If a law
said that everyone who weighed more than 150 pounds would be
executed, I don't think many of us would be weighing ourselves
in public. Some of you may consider this attitude as irresponsible, but only if you deny the existence of economic
responsibilities. If a positive result in a carcinogenic
bioassay triggered a scientific discussion and a legitimate
evaluation of the hazard rather than a ban, this reluctance
to test would be largely removed. Even more unfortunately,
because some protagonists are recommending that regulatory
agencies include mutagenic results within the Delaney philosophy, industry scientists are reluctant to conduct an Ames
test on their company's existing products, but are not at all
reticent to conduct one on their competitors' products. Is
this the kind of scientific jungle we want? I don't think so!
I think it is counterproductive.

That is "negative" number two. The next is the one I personally consider the most disruptive of all to an orderly society and to responsible government. To me it is comparable to political anarchism.

To paraphrase the Delaney philosophy, we cannot determine a safe level for a carcinogen for all humans. A corollary to the Delaney philosophy is that because we cannot determine a safe level for a carcinogen for all humans, we cannot assume that any level of an untested chemical doesn't pose some risk of cancer to man. This is the denial of the principle of "toxicological insignificance," which can be debated as emotionally as the Delaney philosophy as is obvious from Peter Hutt's comment that it is knee-jerk toxicology. It's not toxicology at all.

In reality it is the same debate, or Delaney philosophy, but when expressed in the negative it results in even more serious consequences. In effect, denial of the principle of toxicological insignificance eliminates a major component in the equation of safety evaluation that is necessary to intelligent setting of health hazard priorities.

I don't intend to redebate the technical issues of the concept of toxicological insignificance. However, without it regulatory agencies are reduced to blindfolded children playing pin the tail on the donkey when it comes to pinning health hazards on the correct chemicals. Most tragically, EPA has been rendered helpless in the administration of the Toxic Substances Control Act. Because it will not accept a level of insignificance, it cannot set priorities and consequently is forced into trying to regulate 30,000 chemicals at one time as if they all constituted an equal threat to health. I think we can invest our professional time and financial resources more productively if we learn to say "so what" to a part per trillion while we direct our efforts to the parts per million chemicals. Someday we may get around to the parts per trillion, after the parts per million and parts per billion levels of exposure are evaluated, but to try to regulate them all concurrently is a chaotic approach to public health, and directly the result of the Delaney philosophy.

My fourth negative health implication is the potential substitution of otherwise more toxic chemicals for animal carcinogens. It is well known to research toxicologists that chemicals that are relatively nontoxic on acute and chronic exposure, are more likely to be carcinogenic in a maximum tolerated dose bioassay than relatively toxic chemicals. In the case of the latter, you simply can't administer enough to be carcinogenic without killing the animal by some other toxic response. For example, I believe this is a real consequence of the banning of DDT. It is a known fact that substitute pesticides are much more toxic acutely and thus present a far greater risk to the pesticide applicator. This seems a high price to pay for a theoretical increase in the safety of food to the consumer. In fact, EPA seems to have adopted a policy that the risk of cancer to the consumer is more important than death to a pesticide applicator or agricultural worker. I maintain that all risks should be subject to rational safety evaluation. If you dictate that one mechanism of death is more important than all others, I think you have introduced a negative and irrational health implication.

Now let's look at a fifth negative implication of the Delaney philosophy, which was exemplified by the ban on cyclamate and proposed ban on saccharin. This is the denial to the public of the beneficial aspects of a chemical without a reasoned weighing of that benefit vs. the risk. The automatic nature of a Delaney ban is more awesome than the laws of nature. Even in the case of the law of gravity, intervention is possible. You can place your hand under a falling object and prevent it from hitting the ground, which is attracting it. But in the case of Delaney, only Congress can stop its application, and you know that has happened only in the case of saccharin.

DDT is another example of denial of benefit, which in the judgment of many and I believe the majority of safety evaluators, constituted sacrifice of greater benefit to eliminate lesser risk. I realize that some regulatory officials will deny that the Delaney Clause was the reason for the ban of DDT, because the pesticide legislation does not contain the Delaney Clause. However, those of us who were close to the situation know that Delaney philosophy was paramount in the decision. In addition to the need to replace DDT with more hazardous substitutes, as I mentioned earlier, we have no good substitutes for DDT for some insect problems--Dutch Elm disease being only one. But far more serious is the incredible policy that developed proposing that we cannot sell DDT to other countries who need it for malaria control. Such denial of health benefits of DDT to the peoples of other parts of the world without allowing them to make their own benefit vs. risk assessment would be contrary to health protection viewed on a global basis. But even if we have no concern for the health of the native population of these countries, which have endemic malaria problems, there are American tourists and businessmen in those countries who have an increased risk to malaria. I have had malaria and I want no part of giving it to others. Personally, I find this proposed extension of the Delaney philosophy immoral and embarrassing to me as an American. Something seems wrong when we will sell a country guns but not DDT.

My sixth selection of a negative implication is along a little lighter vein and may seem almost facetious to some. If so, consider it comic relief in the middle of a rather serious discussion. In some circumstances the public exposure to a chemical is extremely low, for example, a nonvolatile resin used as a stabilizer in floor tile. By demanding that a maximum tolerated dose carcinogenic bioassay be conducted before making a judgment of safety, in effect we create a small group of citizens with much greater exposure than would ever occur from commercial use, namely the toxicology technicians who mix the diets and feed the laboratory animals. Even if you have no reason to suspect a chemical of being carcinogenic, this is a group at potential high risk on every bioassay. Under TSCA, the public will not be exposed to any new chemical until many toxicologists and technicians have been exposed. As incredible as it may seem, some agencies are requesting concurrent known carcinogen control groups to ensure the risk is high. It seems to me that any person who has ever worked in a toxicology laboratory that handled a carcinogen and who develops cancer, would have little difficulty collecting a workman's compensation claim against the laboratory or university and on a negligence claim against the sponsor, either industry or government. My point is that the Delaney philosophy in some cases forces unnecessary testing and greater exposure to risk for some individuals than would ever happen in commercial use. This is a negative health implication, even if you don't consider that toxicologists are worth protecting. I have my own doubts.

Number seven. Let us take the following situation: You have a group of plant workers who have been exposed to traces of a chemical for many years. Your industrial hygiene practices have been good and you have confirmed by air sampling that worker exposure is well below all known toxic levels. An obscure journal carries a report on a poorly controlled animal study concluding that this chemical is a weak carcinogen in rats. As a toxicologist you are convinced beyond any reasonable doubt that these workers are not subject to any significant health risk. But under a Delaney philosophy you cannot conclude this, because these employees have a right to know that they have been exposed to an animal carcinogen that might induce cancer in them. It is just possible that the anxiety syndrome that you create in these employees and their families is a greater health hazard than the exposure to the chemical. I have a vivid mental picture of an employee telling his wife and children at the dinner table that he may have cancer from exposure to a chemical on the job, when in fact his risk of cancer may be no greater than from eating a charcoal steak or a handful of peanuts, or smoking one cigarette, or playing tennis in the sun last summer.

I also think we are doing a disservice to the American
public by our carcinogen of the week policy, sometimes two a
week. The chronic anxiety created by the thought of living
in a sea of carcinogens about which we can do nothing is
counterproductive to public health. What is accomplished by
telling the public that their drinking water supply contains
traces of several cancer-causing chemicals, when we have no
evidence whatsoever that these levels will cause cancer in
man? It is almost sadistic! And we should not forget that
anxiety itself has been shown to be carcinogenic in mice.

In reality, we seem to have created two extremes of public
opinion, both of which are just that "extreme"--and in my
opinion wrong. The first is the neurotic who is afraid to
eat, drink, or breathe. The second is the cynic who has heard
government cry wolf so often, who has lived through the cran-
berry to saccharin era and is convinced that if you stuff
enough into a mouse, you will get tumors with everything.
This attitude is seriously counterproductive, not principally
because it challenges the integrity and credibility of the
profession of toxicology and of the regulatory agencies,
which it has, but because that individual is no longer motivated to avoid known health hazards. Why should he give up
smoking when he'll get cancer from maraschino cherries? I
have high school- and college-age children and this is what
they and their friends are telling me. Smoking is no worse
than saccharin.

Shall we try number eight? I realize that scientists in
many walks of life don't believe they should consider the
economic consequences of their recommendations. I can accept
this for laboratory toxicologists and oncologists, but not
for those of us who have responsibility for making safety
judgments on specific issues. The economic and social consequences must be a vital part of that decision. Whenever we
invest money in unnecessary toxicological testing or place
unnecessary restrictions, as monitoring on a chemical, or ban
a chemical, we are not only wasting resources but we are con-
tributing unnecessarily to the inflationary spiral. And many
of these actions in recent years have contributed directly to
the inflationary movement of food prices. I believe to contribute unnecessarily to the cost of food via the Delaney
philosophy is a negative health implication, especially when
examined in relation to the poor and those on fixed low in-
come.

I'm going to stop counting. I'm sure I've overlooked some negative health implications that occur to you. If you have more, send them to me or mention them later in our discussion period so that if I am ever motivated again to give such a talk, I can have additional citations to discuss. If you would like, I'll serve as the international respository for NHIDPs.

As promised, I want to return to my first concern, that of the big business that has developed because of the debate and the vested interests that are deeply entrenched on both sides of the issue. I am less concerned about the hundreds of millions of dollars per year spent as a result of the Delaney philosophy than the diversionary effect on our best talent. There are other important health hazards from chemicals, and our techniques for measuring these are far from infallible and need our attention.

I believe we have reached the stage in development in the profession of toxicology that we need statesmen rather than political toxicologists or political oncologists. To continue to debate the issue of whether there is a safe level of an animal carcinogen is reminiscent of the philosophers of old debating how many angels can stand on the head of a pin. I recommend a moratorium on Delaney debates. We all know that we shall never be able to establish a finite level of any chemical with 100% guarantee of safety. Society accepts risks associated with every other endeavor in life on this planet. If we, as professionals, are not willing to decide what is an acceptable degree of risk in health hazard from chemicals, others less qualified to evaluate our data will do it. I believe we should help them rather than continue to tell them that it can't be done. We must accept some degree of risk, even though it will by necessity be based on some arbitrary assumptions. By this I mean an extrapolation or a margin of safety. Once a policy has been adopted, we would have something worthy of significant research and debate, namely, those assumptions. And as more knowledge becomes available on these assumptions, safety conclusions can be modified accordingly.

We can adopt this approach as an interagency national policy for chemical carcinogens in all regulatory circumstances except the food and color additives sections of the Food, Drug and Cosmetic Act. Thus, we could reduce the Delaney philosophy to the Delaney Clause. If we did that, I think the Clause would be repealed by Congress Commissioner of FDA would find it within his discretionary power to consider certain animal tests inappropriate to the evaluation of safety.

In summary, I've tried to identify a few of the counter-productive aspects of the Delaney philosophy and debate. I've also suggested that we stop debating the issue and devote our talents to a better balanced job of health protection from all toxic properties of chemicals.

I sincerely believe we have made an obsession of our theory of the chemical etiology of cancer. I believe it is time to rethink the issue. As toxicologists we should play an important role. But we won't unless we are realists and regain our credibility.

GENERAL DISCUSSION

DR. KRAYBILL: That is a very impassioned speech. Dr. Coulston, I think it tops your presentations really. It is almost a state of the union type of speech. Having said that does that catalog me as being proindustry?

You know I mentioned that the other day. If you take a certain posture you are either considered proindustry or anti-industry. I think that Dr. Harrison, Dr. Coulston, and many of the rest of us when we retire, should get out of the clutches of the government or industry, should get together and write a book on what makes scientists take certain stands.

This has bothered me and it is perplexing. Maybe I am naive, because I always thought that science was a true maiden and we had to be true. "This above all else to thine own self be true." I think I try to follow that philosophy and I must be a damn fool from where I sit. But I think we must have truth in science and I wonder why scientists take certain postures. I have facetiously figured it this way. If they are in academia or perhaps even the government, they do it because they honestly believe some of these views. Or do they take this position because they know it will get them more money for their program? For the government, this could happen. It gets them more staff, gets them more visibility, and last but not least, we always see the chap like I saw years ago who had a meeting at the New York Academy of Science. I went downstairs and there he was leafing through the New York Times to see whether his name was in the paper.

There are a lot of fellows who like to make headlines. And, we have to sort all these things out. But one observation I did notice, Dr. Frawley, was of all the papers presented here thus far, there was not a single soul in this room that was nodding when you were talking and I saw smiles come on the faces of many of these people that were serious up until now.

So you have introduced two degrees of meaning. Some levity, but I also think you introduced some serious thoughts for us to consider and we thank you very much for your presentation.

DR. COULSTON: I do not think anybody could disagree with what you said, Dr. Frawley. You reaffirmed the intention of this meeting. We talked about the Delaney philosophy in our invitation and I appreciate your reaffirming that. We said we were concerned not so much with the Delaney Clause as it affects FDA but with the philosophy that affects OSHA and NIOSH, all the other agencies. That is the real purpose of this meeting, to get it all laid out.

261

ISBN 0-12-192750-4

DR. KRAYBILL: Second point was my naivete. I thought during these sessions that Dr. Lijinsky swayed me, and I believe you did, you helped to influence my thinking, Dr. Clayson's, Dr. Darby's, Dr. Coulston's, and everybody else. I go to many meetings and, if this is true Dr. Frawley, we are all locked in, we have got our plugs in our ears and we have not loosened up on our hearing aid or something of that sort, we are not listening. I get that frustration sometimes, I feel that if they are not listening, then we are doomed. But if that is the case, then I do not think any future meetings are going to help us. But I hope that we have a flexibility in our think-ing and that we do listen and we do change our thinking. Is that not the role we are supposed to play as scientists? To have an open mind! I hope so.

DR. HARRISON: Could I suggest that there is a difference between listening and hearing?

DR. KRAYBILL: Very good point.
    I would like to bring to your attention an article I read just the other day in *Newsweek*. I want to read one paragraph from it because I think it is relevant to our discussions here, particularly what Dr. Darby had to say the other day. This was written by a physician, Michael Kraemer. One section of this he calls the "War on Common Sense."

    The danger of the war on cancer is that it is threat-ening to become a war on common sense. The proposed ban on saccharin will almost certainly adversely affect the health and perhaps even shorten the lives of diabetics, to say nothing of the potential increase in obesity in the rest of the population. The fact that all that sugar and all those calories may represent a substantially greater risk to health in terms of diabetes, cardiovascular di-sease, and tooth decay does not even appear to have been contemplated by our valiant cancer vigilantes. This may all sound a bit strange coming from a physician. To some it may even smack of therapeutic nihilism. But this is not my intention. It does not seem inconsistent with the role of a physician to advocate proof of any measures in vigorous research in question where proof is lacking. A plea for common sense, yes. For skepticism, perhaps. But for nihilism, definitely not. We cannot prevent the common cold by taking vitamin C or dressing warmly and we cannot prevent mental illness by avoiding food additives. We can probably do better with heart disease, emphysema, and lung cancer, and there are of course known poisons, infectious agents, and carcinogens that can and should be

avoided. And this is what I underline. The agreement
here is that we acknowledge our ignorance and stop acting
as if we know more than we do. I think that is the key
point. We are all victimized by unknown causes, by fate,
and even by ordinary bad luck.

DR. COULSTON: I don't have any questions but I would like to
attempt to interpret what Dr. Korte said so everyone, including some of the people who are not scientists in the room,
will understand what his statement was all about.

He has demonstrated now that certain chemicals that have
been considered by the environmentalists and other groups as
imminent carcinogenic hazards to man, and regulatory agencies
have considered them dangerous in that sense, that these compounds in nature break down. He has shown for the first time,
he and his group, that these chemicals break down relatively
quickly, even these so-called nondegradable, nonbiodegradable
substances like DDT and so on.

This research explains why there is no buildup of DDT
worldwide, but rather a decrease worldwide of the amount of
DDT found in the environment. I think this is incredible at
this point in time, after we have already restricted or banned
important chemicals. How does the Delaney Clause handle new
information like this? Shall we reinstate the chemicals for
their intended use? It applies, you see, to all of the fat-
soluble chemicals that have been put into limbo primarily
because of the fact that they may build up and remain in our
environment.

It is important to recognize now that these data show that
this concept is not true. These chemicals will break down, as
Dr. Korte said, even quicker than some so-called biodegradable
substances.

His data also show very clearly the kind of thing that
some people have been talking about such as the dieldrin example, where the data clearly indicate that there is a different major metabolite in the mouse vs. let us say monkey or man.

Now he did not present the human data, but the data are
there. The major metabolites in man are very similar to those
in the chimpanzee or the rhesus monkey, but quite distinctly
different than those in the mouse or the rat.

The fact that there are differences in how the compounds
are handled between mice and rats alone and then comparing
that to man and nonhuman primates is a remarkable bit of evidence to indicate that the question of Delaney interpretation,
or chemical carcinogenesis, should be based on a comparative
consideration.

This should be built into the new Delaney Clause, if it ever comes. The comparative nature of toxicology, comparative toxicology, should be considered and decisions should not be made based on one species or strain of one species.

DR. SCHMIDT-BLEEK: As you know, I am from the West German Environmental Protection Agency. I would like to follow up on what Dr. Coulston just said in telling you that it is the work of Dr. Korte and his group that has led to the new points that are in the directives of the EEC. We have in fact now taken out the word biodegradability and replaced it with degradability, because we are indeed pursuing this idea very strongly. In fact in our regulations, at least I can say this specifically for West Germany, we shall take account of this new development.

DR. CLESSERI: As I understand it Dr. Korte, the work you were doing was looking at photodegradation, but I think a lot of the DDT is in the sediments of lakes and streams. I wondered what work you have that would show us the rate of degradation under those conditions.

DR. KORTE: We have investigated specifically the photodegradation and we do not investigate actually degradation in sediments, although there is such a program going on in the laboratory. However, it was found recently, or let us say three years ago, by many others and by us, too, that DDT evaporates very easily from the water surface and then you have it in the air. It is decomposed. Does that answer your question? DDT is only one example. Or even PCB; this is the most interesting example because it does not have any double bond. It cannot be excited by light itself, but it will be decomposed completely in a short time to CO_2. PCB is the same as all these compounds. The more halogen you have in the molecule, the faster its degradation to CO_2 and water. You can correlate the speeds or rate very likely with the polarization of the molecule. Just the other way around as in metabolism activation in animals.

DR. CLESSERI: This discussion relates to a very practical situation that exists in New York State today. There have been a considerable amount of PCB's put into the Hudson River and there is a lot of concern about the ultimate impact of this on man, as well as the ecosystem.

Recently a study was made over an eight-month period and it recommended that the sludge in the river should be dredged at a cost of $70 or 80 million. The question then is, a lot of people would have the concern that this may do more harm than good. I am wondering about whether we just have assumed that these things are so refractory, they persist for such a long time, and this may be just a myth, so to speak. Yet, I would like to say that this procedure is something that may not be very practical, but might very well occur in the next few years on a grand scale and could spoil the condition for many other rivers to be similarly treated.

DR. KORTE: Sure. I would recommend that you study the evaporation rate of the chemicals in the river and then wait a couple of years and you will find that everything is over. We did a similar study in Italy and could show the degradation, with this enormous sunlight in north Italy, was so fast that very likely in five years no PCB would any longer be detected on the ground.

Jokingly, if you could install ultraviolet lights all over this would very likely take one year to accomplish. Seriously, I would think even for technical purposes this could become a cleaning method for the future, even in houses.

DR. CLESSERI: Is that also true of the polybrominated biphenyls?

DR. KORTE: Sure.

DR. CLESSERI: But of course one of the problems is that people are very carefully burying these compounds and of course they are extremely insoluble in water and they are sinking down to the bottom of the river and are not really subject to water degradation.

DR. KORTE: We are not so certain that they would not evaporate out of the water surface. You know everything is soluble; this concentration of insoluble PCB does not exist. Everything is soluble. It is not so much a question of the vapor pressure, but the codistillation effect. This is not very well understood, actually, but that is a major point. The elimination also from the soil is more based on codistillation with water, if the sun shines on the soil.

DR. CLESSERI: For example, with the PCBs bury these animals and they're sitting down there and the chemical is leaching out. In other cases, water is washed down and it is now sediment, with the sand and so forth. I think it is a long time coming to the surface. I am not in favor of dredging.

DR. KORTE: If you have the chemical in a kind of heavy mud
and sediment, like stones on the bottom, then you can't do
anything, that's clear. But that is not the normal situation.
The normal situation is that these chemicals are in solution
and then they move and by codistillation come into the air.

DR. DICARLO: I think another aspect of Dr. Korte's presentation is that you do not really need to have more than one
substance administered to have a drug interaction; various
metabolites and chemicals in an animal may interact as well.
There is another point I think going back to the conceptual
presentation of Dr. Parke, which interrelates here. Dr.
Korte is really talking about the biotransformation, often
necessary to produce the carcinogen by a system that is
otherwise induced. So what he is saying, and this comes to
what Dr. Upholt answered yesterday, he is producing a situation where he has induced enzymes. Now you can have all kinds
of effects relative to synergism and so on, which was the
original question yesterday, because the range of substrate
specificity is just fantastic. So you can have synergism in
these situations. You have all kinds of effects.

DR. KORTE: There are two different things: In one situation
there is metabolic activation. Then everything happens. When
you give one compound to an animal you may find 20 or 30 chemicals; it depends on the method. Then you can have small
changes, never complete decomposition, with the persistent
chemicals at least. But, if you put these chemicals in the air
and they are in the air, the main part is in the air. I gave
the figure for aldrin and dieldrin under practical conditions,
50% goes in the air, then they decompose like DDT.

DR. DICARLO: That is my point. You are in the atmosphere
and you are inhaling even synthesized chemicals. That is
what I am talking about. You are also inhaling the photodegradation products. So now you have a whole gemisch and
obviously you are going to have all kinds of interactions,
inductions and so on.

DR. KORTE: As long as you have the chemical in the air, fine.
If you have it in the body then it is more complicated.

DR. DICARLO: Nothing happens until you absorb it obviously.

DR. KORTE: I think there are quite distinct differences. In
the body you only can have reaction with so much chemical for
each kilogram. You cannot expect more; there is no more avail-

able. In the air you easily have photodegradation and this
composes the chemical very easily, even the CH bond completely
breaks down to CO_2 then.

DR. DICARLO: What I was going to ask you originally was can
you cite something on photodegradation of polybrominated biphenyls?

DR. KORTE: Oh yes. Even bromine bonds and the freons are
decomposed to CO_2. This might influence the thinking on the
ozone layer. It is foolish to think that the freons will
affect the ozone layer, since they are photodegraded long before they reach the ozone layer.
     We have just published in *Nature* a summary of this re-
search.

DR. KOLBYE: There are also publications based on research by
Michigan State University on the photodegradation of polybrominated biphenyls.

DR. KORTE: This is much easier to demonstrate and I agree.

DR. KRAYBILL: Dr. Kensler, I was interested in some of your
statements, particularly the work that is going on at Dow.
As I recall, below 100 parts per million, down around 50 parts
per million, you get this considerable macromolecular binding
with sulfhydryl groups, glutathione. Now, you may have heard
a statement about the low dose. You sort of see a disappearance of tumors, I believe, below 50 parts per million. You
would no longer observe angiosarcoma of the liver, at least.
But weren't there other tumors at other sites below that
level?

DR. KENSLER: Well I'm not totally familiar with the tumor incidence in those.

DR. KRAYBILL: Well 50 parts per million did produce angio-
carcinoma.

DR. KENSLER: Well on that slide, I think, it went down to
10 ppm, where there were tumors. Those were Maltone's data.
The protein binding data were the Dow data.

DR. COULSTON: Dr. Kensler implied this, but in fact I think
it is worth underscoring the need to accumulate basic information concerning the metabolic changes, at different dosage
levels. I think this may well be one of the more important

considerations in assessing whether or not large doses of
test materials are giving us information that is applicable
to the ordinary low intakes that we would observe in practice.

DR. KENSLER: Well I certainly tried to make that point. I
think it is important to develop a body of data on some of
these materials, which are in, shall I say, the range to which
man is actually exposed, in significant numbers and significant quantities. We ought to try to prioritize what we are
going after and develop background information to see whether
or not we can use methods such as this to help arrive at
better decisions on new materials, later.

DR. KRAYBILL: I think that is a very significant point. I
gave one example yesterday, perhaps two, methylparathion was
another, that I did not include in the slide, but I think
we need this. But Dr. Suara, who was formerly with us, said
we can also follow this trend down as you go on a dose-response
curve with histopathology and histochemical measurements.
This would be interesting, too, to show at different dosages
as you go down the scale what different lesions you are seeing from a pathological viewpoint and to chemically see if
you can detect differences. We need all this kind of data to
reassure us in our decision-making process.

DR. ZIMBELMAN: I know there is not a consensus, but it seems
to me that we are causing tumors in a lot of instances by at
least three indirect mechanisms that have been alluded to,
one of which is the liver damage caused by, for example,
selenium, disturbing the metabolism, or altering the hormonal
status.
    It seems to me that we do not really need to do lifetime
studies that are going to repeat and exaggerate one of these
mechanisms. For example, if your compound in short-term test
increases prolactin in mice, you might as well admit *a priori*
that if you do a study in the C3H mouse, you are going to have
an increased incidence of tumors. It seems that we could
save a lot of natural resources by just avoiding doing those
kinds of studies, admitting that that kind of result would
happen, leaving us to interpret the results. If that kills
a compound, well then, you might as well do it on that basis.
If it is going to be considered for use you know that can be
tolerated, well then, you can determine this on short-term
doses. So, it seems to me that we are at the point where we
ought to do studies in normal animals, that is, animals that
do not have toxicologic, metabolic, or hormonal alterations.
Then we would be measuring the effect of the drug directly on
cancer and that is really what we are talking about.

Now I know this might not always be in accord with Dr. Lijinsky's approach to these kind of problems, but it seems to me that we might as well admit that those indirect mechanisms, which have already been defined, would occur again, if our drug in the short-term test produces these changes.

DR. KOLBYE: Dr. Kensler, I think in terms of safrole it goes the other way. I remember serving on a committee about 20 years ago looking at safrole data. It was declared a carcinogen based upon the incidence of liver tumors, but at the doses where you observe liver tumors, there is a decrease in mammary tumors so that the net tumor per population did not change. And yet for the liver of rodents obviously you had a carcinogen.

DR. LIJINSKY: Dr. Frawley, you certainly make your point with considerable humor. However, I have a question to ask. How, if we do not have this Delaney-type philosophy that safety should be established by some recognized test, and I don't mind which test you suggest for it, how to ensure that the public is protected from exposure to chemicals that might induce cancer or might increase the cancer risk? Most large companies simply do not have the resources to test. So what do we do about that? How do we protect, how do we make decisions or how does the government or society make decisions which protect it from economic and political pressures? That is what I have to ask. You do not have a philosophy that removes a large measure of discretion from the regulatory agency.

DR. FRAWLEY: I'm sorry if I mislead you by suggesting that I thought we should not test our chemicals. I was merely saying that this rigid Delaney philosophy discourages people from voluntarily testing, because as I said you know jokingly I called it Russian roulette. I think there would be more voluntary testing, if indeed you had a positive result. Then you could sit down with the regulatory agencies and take another look at the metabolic route, at the dosage level at which you got your positive effect to see whether or not you are faced with an overloading situation. If you have a huge margin of safety, not just to jump in and ban it, but do some definitive research on the compound. But I am not for a moment suggesting that we not test chemicals by the best available toxicological procedures.

DR. LIJINSKY: It has disturbed me recently to find that many
over-the-counter drugs in particular have never been tested
for safety. Some of them have only been given a 30-day test
and yet millions of people are being exposed to these preparations. Does that show responsibility on the part of industry?
That is an example of the thing that I want to correct.

DR. FRAWLEY: I really do not know whether that statement is
true. There certainly are a lot of drug products that are on
the market that have not been tested for cancer. Under the
old philosophy, if there was not chronic exposure you would
not have gotten chronic effects. These have been pharmaceuticals that have been taken for a short duration at a time.
    I think we have changed our philosophy in that direction.
I can attest for my own pharmaceutical division, I know that
we are testing everything that we have done research by multi-
generation reproduction studies, carcinogenic bioassays,
mutagenic tests, teratogenic tests, and everything else.
There is nothing going on the market today that is in the
situation you described.

DR. LIJINSKY: I am talking about the past. I know of two
compounds that are disturbing me and one is methapyrilene and
the other is chlorpheniramine, which are widely sold products
and have never been tested, never been given a test longer
than 30 days in animals and I think that is a disgrace.

DR. FRAWLEY: Do you know that for a fact?

DR. LIJINSKY: Absolutely.

DR. KRAYBILL: We are going to test them, right? We are going
to test them.

DR. KOLBYE: I would like to just briefly mention to those of
you who did not have much to do or did not know much about
the polybrominated biphenyl effect situation in Michigan, the
kinds of spill-over effects of the Delaney philosophy that
occurred out there. Just to give you a quick flavor; over
approximately three years, there was considerable public
panic, there was a need to guard FDA officers and scientists
from physical violence by state troopers, we testified approximately six times. Even Dr. Coulston testified for the FDA
position. It is the one instance that I am acutely aware of
where a state legislatively went beyond the recommendations
of the FDA going down to almost a zero tolerance in terms of

the practicability of the analytical method and was risking the reliability of the analytical findings, when the State of Michigan moved to reduce the action level from 0.2 parts per million down to 20 parts per billion.

It also involves the potential destruction of 40,000-100,000 more dairy cattle in the State of Michigan to recover, if you will or to avoid what has been variously estimated at maybe 300 to 450 gm of polybrominated biphenyl widely dispersed through those cattle. It threatened potentially the future of dairy agriculture in the State of Michigan, because while some degradation of polybrominated biphenyls could be expected to take place, it would be relatively insignificant in terms of some of the farms that were most adversely affected. This briefly means when they bring in new dairy cattle they will pick up a sufficient amount of low-level residues to again become detectable and hence illegal by operation of Michigan law.

So far, we do not know whether or not polybrominated biphenyls are carcinogenic, but it is interesting that some of our testing is done on, for example, PCBs, the commercial preparation of the compounds that contain a variety of other non-PCB isomeric compounds. In terms of polybrominated biphenyls we had to reason somewhat by analogy to PCBs and decide whether we had direct experimental evidence as to toxicity of PCBs. We still do not know whether polybrominated biphenyls can influence the expression of cancer in humans and a long-term epidemiological series of studies is in progress.

But to me, it was an absolute revelation of the degree to which the public in this country can be panicked out of their bleeding minds and go absolutely crazy. There were night-riding episodes and barn-burning threats. There were threats on the witnesses, such as Dr. Coulston. Private witnesses, some farmers were intimidated into certain positions. It was just something that one would expect to read out of Alistair McLean or Helen McGinnis, only it was real life in Michigan.

DR. KRAYBILL: Dr. Kolbye, thank you very much because I think that is a very, very sobering statement to a degree where I would state we have some hysteria in this country on some of these issues.

DR. COULSTON: Just as a corollary to what Dr. Kolbye has said, I was involved, I was threatened. I was bodily pushed around because I defended the position of the FDA. I can tell you the rule of reason did not prevail, it came down to an economic basis. Those who were poor farmers from an agricultural viewpoint used this, in my opinion, as a way to get monetary recompense for their cattle, which were sick from all kinds of other problems. The good farmers never had a problem with PBB and yet because of the lowering of the maximum residue limits they had to abandon their good cattle and kill them off just the same as the poor practice farmers had to do.

Now the worst part of this story: there appeared on national educational television a one-hour movie of this PBB episode, which was so biased and so nonfactual, which told a story of the bad farmers, who were not good agriculturalists. The media compounded this tragic error if you will, it is and was an unbelievable experience, as Dr. Kolbye said.

I just want to add this, Mr. Chairman. I said in testimony that if they really believed that PBB was an imminent carcinogenic hazard to man, and Al heard me say it, they should immediately dig up all the soil and clean it, kill all cattle and I estimated it would have cost them something in the order of $7 billion and then I said I do not know what you would do with the soil when you dig it all up, for there is no technology on that scale to accomplish it.

But I said if you really believe that this is such an imminent carcinogenic risk, you must do it. The committee at that time went along with the FDA's proposal, which was to leave the PBB level alone, at 0.2 of a part per million. But as you heard from Dr. Kolbye, the pressure on the Governor and on the rest of the State to lower the level down to the 20 parts per billion level was so great that it was recently done. They cannot even measure 20 parts per billion analytically quantitatively, it can only be done qualitatively.

DR. KRAYBILL: If you really want to get aroused and it takes a lot to arouse me, look at the situation on the mother's milk. I facetiously was kidding about this when I was at a meeting recently, Dr. Plumly (EPA) said about DDT and DDE in the mother's milk that these women are literally delivering to the infant PCB if they are nursing, which we talked about here. The report of the IARC monograph on the carcinogenicity testing on PCB, PBB is under test. Want to hedge your bets on how that stacks up? PBB will probably come out like PCB. But above all, DDT and DDE, which have been around for years, will spill over also. Now these women are going across state lines and they are transporting a food product contaminated with chemicals across state lines!

DR. CLAYSON: I would just like to refer briefly, not as a member or having any connection with the drug industry, to Dr. Lijinsky's fairly overt criticism for not having carcinogenicity tests on all their drugs.

I think it should be realized very clearly that at the time of the greatest development of the pharmaceutical industry between, should we say 1935, 1940 and 1960, our testing requirements were a lot less stringent than we would like them to be today. In consequence there are a lot of drugs commonly used in human beings that have not been tested. I think if we went to the popular prescribing index, we would be horrified in fact at the number that have not been adequately tested. I would also say that I find Dr. Frawley's comments on the "carcinogen of the week" extremely worrying. He thinks they are, of course, neuroses. We just heard that they are neuroses, but I am reminded of the man who read in *Reader's Digest* that cigarette smoking caused cancer and immediately gave up reading the *Reader's Digest*.

I would like to ask you, therefore, if anybody has any ideas: the public enjoys being disturbed, but does not give up reading the newspaper.

THE USE OF NONHUMAN PRIMATES FOR CHEMICAL
CARCINOGENESIS STUDIES

Richard H. Adamson and Susan M. Sieber

INTRODUCTION

There are many reasons why species differ in response to
the therapeutic, toxic, and carcinogenic effects of foreign
compounds, and an understanding of these factors is necessary
to predict whether a response in one species will be the same
in another. These factors are summarized in Table I.

*Table I. Factors Underlying Species Differences in
Response to the Carcinogenicity of Foreign Compounds*

1.	*Differences in disposition (absorption, distribution, biotransformation, and excretion)*
2.	*Differences in binding to the receptor or critical nucleophilic sites in macromolecules*
3.	*Anatomical, physiological, and microbiological differences*
4.	*Differences in the diet*
5.	*Presence or absence of C type virus*
6.	*Differences in the immune system*
7.	*Differences in DNA repair mechanisms*

Biotransformation differences are particularly important
with regard to extrapolating carcinogenesis data from animals
to humans. Most chemical carcinogens must be activated
(biotransformed) to highly reactive electrophiles, which initiate the process of carcinogenesis by interacting with
critical nucleophilic sites in cellular macromolecules of
susceptible organs. Therefore, differences in the rate or
extent of binding to the nucleophilic sites will have an

275

ISBN 0-12-192750-4

effect on the process of carcinogenesis. Anatomical, physiological, and microbiological differences among species are also very important; one must be careful in extrapolating data about carcinogens that are activated by bacteria of the gastrointestinal (GI) tract, since the GI flora of rodents differs from that of humans. Differences in the diet may also be critical. Laboratory feeds from various companies differ with respect to their protein, carbohydrate, fat, vitamin, and mineral content, and the same source for various components of the diet is not always used by the same company. In addition, the laboratory chow may be contaminated with a number of environmental pesticides, mycotoxins, etc. That differences in diet do influence the process of carcinogenesis is illustrated by data from Miller and Miller (1) in which the incidence of liver tumors in male rats fed 0.006% 2-acetylaminofluorene (AAF) varied depending on which nucleophile was added to the diet (Table II). Those rats whose diet was supplemented with cystine, methionine, or casein developed fewer tumors than those fed AAF in a standard grain diet.

Table II

Incidence of Liver Tumors in Rats Fed Diets Containing 2-Acetylaminofluorene (AAF) plus Various Nucleophiles[a]

Nucleophile added to the diet	Rats (%) with tumor at	
	9 months	18 months
None	28	83
2% Tryptophan	28	72
2% Cystine	0	33
2% Methionine	0	27
40% Casein	6	11

[a]*Groups of 18 rats were fed a standard grain diet containing 0.006% AAF and the nucleophile indicated for 9 months; after 9 months, all rats were fed standard diet. Data from Miller and Miller (1).*

Thus, species differences make extrapolating rodent carcinogenesis data to the human exceedingly difficult. For this reason, we began about 16 years ago to evaluate the carcinogenic potential of various compounds in the Old World monkey, a species phylogenetically closer to man than is the rodent. The use of nonhuman primates in carcinogenesis studies has several advantages over the use of rodent species. For example, certain metabolic pathways, including metabolism by bacterial flora, resemble those of humans more closely than do those of rodents (2). The relatively longer lifespan of monkeys makes it possible to administer test compounds at doses equivalent to estimated human exposure levels; furthermore it provides a more reasonable estimate of the latent period for tumor development following exposure to potential carcinogens than is possible with rodents. Finally, the incidence of spontaneous tumors in our monkey colony is comparatively low (1-2%) as compared to some rodent colonies in which the tumor incidence in control rats is 46% (3).

The objectives of our program are as follows: to obtain comparative data on the response of nonhuman primates to substances known to be carcinogenic in rodents and to materials suspected of being human carcinogens; to evaluate the long-term effects of clinically useful antineoplastic and immunosuppressive agents; to develop model tumor systems in primates for evaluating the potential usefulness of new antitumor agents against rodent tumors before these agents are administered to cancer patients; to explore the possibility of preventing and/or reversing the process of carcinogenesis, i.e., "chemoprevention" instead of chemotherapy; to develop biological markers and other diagnostic tests for detecting preneoplastic changes as well as overt neoplasia; and to make available normal and tumor-bearing primates for pharmacological, biochemical, immunological, and therapeutic studies.

METHODS

Animals

The present colony, consisting of approximately 600 animals, is comprised of four species: *Macaca mulatta* (rhesus), *Macaca fasicularis* (cynomolgus), *Cercopithecus aethiops* (African green) and *Galago crassicaudatus* (bushbabies). Eighty of these monkeys are adult breeders that supply the newborns for experimental studies. The majority

of the animals are housed in an isolated facility that contains only animals committed to this study, and with the
exception of the breeding colony, most animals are housed in
individual cages. Newborns produced by the breeding colony
are taken within 12 hours of birth to a nursery that is
staffed on a 24-hour basis. The administration of test compounds is usually initiated within 24 hours of birth and
continues until a tumor is diagnosed or until a predetermined
exposure period has been completed. A minimum of 30 animals
is usually allotted to each treatment group, since in a
sample of this size it is possible to detect a tumor incidence of 10% within 95% confidence limits.

A variety of clinical, biochemical, and hematological
parameters are monitored weekly or monthly, not only to
evaluate the general health status of each animal, but also
for the early detection of tumors. Surgical procedures are
performed under phencyclidine hydrochloride, Ketamine, or
sodium pentobarbital anesthesia. All animals that die or are
sacrificed are carefully necropsied and the tissues subjected
to histopathologic examination.

Compounds under Evaluation

A wide variety of substances have been, or are being,
evaluated (Table III). These substances can be categorized
as follows: model rodent carcinogens (3-methylcholanthrene,
dibenz[a,h]anthracene, 3,4,9,10-dibenzopyrene, N-2-fluorenyl-
acetamide, N,N-2,7-fluorenylenebisacetamide, N,N'-dimethyl-p-
phenylazoaniline, N,N'-dimethyl-p-(m-tolylazo)aniline, ethyl
carbamate); food additives and environmental contaminants
(cyclamate, saccharin, aflatoxin B_1, methylazoxymethanol
acetate, sterigmatocystin, DDT, arsenic, cigarette smoke
condensate, low-density polyethylene plastic); therapeutic
agents (procarbazine, azathioprine, Adriamycin, 1-phenylala-
nine mustard); and nitroso compounds (1-methylnitrosourea,
N-nitrosodiethylamine, N-nitrosodipropylamine, 1-nitrosopip-
eridine, and N-methyl-N'-nitro-N-nitrosoguanidine).

The compounds are administered subcutaneously (sc),
intravenously (iv), intraperitoneally (ip), or orally (po).
For po administration to newborn monkeys, the compound is
added to the Similac formula at the time of feeding; when
the monkeys are 6 months old, carcinogens given po are incorporated into a vitamin mixture, which is given to monkeys as
a vitamin sandwich on a half slice of bread. The dose level
chosen is dependent on the chemical under evaluation. The
therapeutic agents under test are administered at doses
likely to be encountered in a clinical situation; other sub-

Table III. Substances Tested for Carcinogenic
            Activity in Nonhuman Primates
_____

Model Rodent Carcinogens
     3-Methylcholanthrene
     Dibenz [a,h]anthracene
     3,4,9,10-Dibenzopyrene
     N-2-Fluorenylacetamide
     N,N-2,7-Fluorenylenebisacetamide
     N,N'-Dimethyl-p-phenylazoaniline
     N,N'-Dimethyl-p-(m-tolylazo)aniline
     Ethyl carbamate

Food Additives and Environmental Contaminants
     Cyclamate
     Saccharin
     Aflatoxin B_1
     Methylazoxymethanol acetate
     Sterigmatocystin
     DDT
     Arsenic
     Cigarette smoke condensate
     Low-density polyethylene plastic

Therapeutic Agents
     Procarbazine
     Azathioprine
     Adriamycin
     l-Phenylalanine mustard

Nitroso Compounds
     l-Methylnitrosourea
     N-Nitrosodiethylamine
     N-Nitrosodipropylamine
     l-Nitrosopiperidine
     N-Methyl-N'-nitro-N-nitrosoguanidine
_____

stances, such as environmental contaminants and food additives,
are usually given at levels 10-40 times higher than the estimated human exposure level. The remainder of the chemicals
tested are administered at maximally tolerated doses, which,
on the basis of weight gain, blood chemistry and hematology
findings, and clinical observations, appear to be devoid of
acute toxicity.

RESULTS

Since the inception of this study 16 years ago, four
spontaneous tumors have been diagnosed in 211 nontreated
breeders and vehicle-treated controls, yielding a tumor incidence of about 2%. Studies on the carcinogenic effects of
26 chemicals have been initiated during this period, and
treatment with 19 of these has not yet been associated with
an increased incidence of tumors. The compounds that have
not demonstrated carcinogenic activity include 3-methylchol-
anthrene, dibenz a,h anthracene, 3,4,9,10-dibenzopyrene,
N-2-fluorenylacetamide, N,N-2,7-fluorenylenebisacetamide,
ethyl carbamate, N,N'-dimethyl-N-phenylazoaniline, N,N'-
dimethyl-p-(m-tolylazo)aniline, cyclamate, saccharin, DDT,
N-methyl-N'-nitro-N-nitrosoguanidine, low-density polyethylene plastic, cigarette smoke condensate, arsenic, sterigmatocystin, azathioprine, Adriamycin, and 1-phenylalanine
mustard. However, several of these compounds have been under
evaluation for less than 2 years.

Seven of the 26 substances are carcinogenic in nonhuman
primates, inducing tumors in 16-100% of the treated animals
(Table IV). The carcinogens include aflatoxin B_1, methylazoxymethanol acetate, procarbazine, methylnitrosourea, and three
nitrosamines, N-nitroso diethylamine, N-nitrosodipropylamine,
and 1-nitrosopiperidine.

Aflatoxin B_1 (AFB_1) is a contaminant of human foodstuffs
and a suspected hepatocarcinogen for humans. Nine of 45
monkeys (20%) treated with AFB_1 for longer than 2 years have
developed tumors (Table V). Two of the tumors were primary
liver carcinomas and three were adenocarcinomas of the bile
duct or gallbladder; in addition, single cases of hemangioendothelial sarcoma of the liver, hemangioendothelial sarcoma
of the pancreas, olfactory neuroepithelioma, and osteosarcoma
of the tibia were diagnosed. The latent period for tumor
induction ranged from 48 to 134 months, averaging 92 months;
the average total dose given to monkeys developing tumors was
551.7 mg (range 99-1354 mg).

MAM-acetate, the aglycone of the active carcinogenic compound in the cycad nut, cycasin, also induces carcinomas in
monkeys (Table VI). A total of 26 rhesus, cynomolgus, and
African green monkeys were given MAM-acetate by weekly ip
injections (10 mg/kg) or were given compound po (3 mg/kg, 5
days every week). To date seven animals (27%) have developed
primary liver cancer; in two of these monkeys liver hemangiosarcoma, bilateral renal carcinomas, and/or a squamous cell
sarcoma of the esophagus were also found. When administered

Table IV.

Chemical Compounds Inducing Tumors in Old World Monkeys[a]

Compounds	Route of administration	Total treated	No. alive	No. dead With tumor (%)	No. dead Without tumor
Controls[b]	--	211	145	4 (1.9)	64
Aflatoxin B_1	po, ip	45	17	9 (20)	19
MAM-acetate	po, ip	26	11	7 (27)	8
Procarbazine	sc, po, ip	50	13	11 (22)	26
MNU	po	43	25	7 (16)	11
N-Nitrosodiethylamine	ip	122	2	98 (80)	22
N-Nitrosodipropylamine	ip	6	0	6 (100)	0
1-Nitrosopiperidine	po	12	0	11 (92)	1

[a] In rhesus, cynomolgus, and African green monkeys surviving longer than 6 months following treatment.

[b] Includes both nontreated breeders and vehicle-treated controls.

Table V.

Tumors Induced in Monkeys Treated with Aflatoxin B_1 $(AFB_1)^a$

Monkey number	Species	Sex	Age at first dose	Route	Total dose (mg)	Latent period (months)[b]	Histological diagnosis
692I	Rh	F	birth	ip, po	99.19	48	Hepatic cell carcinoma
680H	Rh	M	3 days	ip, po	119.44	51	Hemangioendothelial sarcoma, liver
488F	Cyno	M	13 months	ip	130.80	50	Olfactory neuroepithelioma
590G	Afr Gr	M	birth	ip, po	291.83	107	Hemangioendothelial sarcoma, pancreas
582G	Cyno	M	9 days	ip, po	411.52	115	Osteosarcoma, tibia
500F	Rh	F	9.5 months	ip, po	463.55	118	Adenocarcinoma, intrahepatic bile duct with wide metastases
454F	Rh	F	8 days	po	842.36	74	Hepatic cell carcinoma
479F	Rh	M	birth	po	1252.28	130	Adenocarcinoma, gall bladder with invasion of liver and lung metastases
374E	Rh	F	birth	po	1354.24	134	Adenocarcinoma, bile duct

[a] AFB_1 was administered weekly by ip (0.125–0.250 mg/kg) or po (0.2–0.8 mg/kg) routes.

[b] Latent period is the time in months from the first dose of AFB_1 until the clinical diagnosis of tumor.

Table VI.

Carcinomas Induced with MAM-acetate

Monkey number	Species	Sex	Total dose ( )	Latent period (months)[b]	Histological diagnosis
671H	Rh	M	3.58	50	Primary liver carcinoma
664H	Afr Gr	M	3.88	63	Adenocarcinoma, liver
670H	Cyno	F	5.06	83	Hepatoma and liver hemangiosarcoma with lymph node metastases; bilateral renal carcinoma; squamous cell carcinoma, esophagus
672H	Rh	F	6.27	93	Hepatic cell carcinoma
665H	Rh	F	8.38	100	Hepatic cell carcinoma; bilateral renal carcinoma; adenocarcinoma, small intestine; squamous cell carcinoma, esophagus
669H	Rh	M	9.66	89	Hepatic cell carcinoma
309D	Cyno	M	48.53	95	Hepatic cell carcinoma

[a]Monkeys received the initial dose of MAM-acetate at birth; all received MAM-acetate by weekly intraperitoneal injections (10 mg/kg) except for 309D, which was given compound orally (3 mg/kg) on a vitamin sandwich five times every week.

[b]Latent period is the time in months from the first dose of MAM-acetate until the clinical diagnosis of tumor.

by the ip route the average latent period for tumor development was 81 months (ranging from 50 to 100 months), and the
average total dose received was 6.1 gm (ranging from 3.58 to
9.66 gm).

Of the chemicals classified as therapeutic agents, only
procarbazine has demonstrated carcinogenic activity in nonhuman primates (Table VII). Fifty monkeys have survived for
longer than 6 months after receiving procarbazine by sc, ip,
and/or oral routes at doses of 5-50 mg/kg. Thirty-seven
monkeys have been necropsied to date, of which 11 (22%) have
had malignant neoplasms. Six monkeys were diagnosed with
acute leukemia, all but one of the myelogenous type; the
other acute leukemia was undifferentiated. In addition to
the leukemias, one case of lymphocytic lymphoma, two cases of
hemangiosarcoma, and two cases of osteogenic sarcoma were
found. Figure 1 shows an X-ray of the osteosarcoma of the
humerus that developed in monkey 731I. When this animal was
necropsied, a pulmonary nodule was found that subsequent histopathological examination revealed to be composed of metastatic tumor cells with the same morphological features as
those in the primary osteosarcoma of the humerus.

The neoplasms induced by procarbazine developed after
treatment for an average of 75 months (ranging from 16 to 109
months); the total dose of procarbazine received by the monkeys developing tumors ranged from 2.6 to 101.6 gm, and
averaged 36.0 gm. Other adverse effects of long-term procarbazine treatment included vomiting, myelosuppression, and in
the males, testicular atrophy with complete aplasia of the
germinal epithelium.

The nitroso compounds as a class appear to be potent
carcinogens in nonhuman primates. Squamous cell carcinoma of
the mouth, pharynx, and esophagus developed (Table VIII) in
7 of 43 monkeys (16%) receiving oral doses (10-20 mg/kg) of
1-methyl-1-nitrosourea (MNU). Moreover, upper digestive tract
lesions such as atrophy or dyskeratosis of the esophageal
mucosa and esophagitis have been a consistent finding among
the 18 monkeys necropsied to date. All monkeys but one that
have received total doses of MNU exceeding 50 gm have developed carcinomas, whereas no malignant tumors have developed
in monkeys receiving a cumulative dose less than 50 gm. The
average latent period for tumor development was 84 months and
ranged between 63 and 124 months.

Several similarities between the esophageal tumors developing in our monkeys and human esophageal carcinoma were
apparent. For example, the monkeys developed many of the
common complications of esophageal carcinoma that are encountered in humans (e.g., regurgitation, aspiration, sepsis, and
hemorrhage). The order of appearance of the esophageal les-

Table VII.

Tumors Induced in Monkeys by Procarbazine[a]

Monkey number	Species	Sex	Total dose (gm)	Latent period (months)[b]	Histological diagnosis
267D	Rh	F	2.64	16	Acute myelogenous leukemia
733I	Cyno	M	7.29	57	Acute undifferentiated leukemia
726I	Cyno	M	16.24	68	Acute myelogenous leukemia
734I	Rh	M	17.17	71	Hemangioendothelial sarcoma, kidney
731I	Rh	F	24.22	68	Osteosarcoma, humerus, with lung metastases
314E	Cyno	F	32.88	97	Hemangiosarcoma, spleen, liver, intestine, kidney, and ovaries
313E	Rh	F	37.26	68	Acute myelogenous leukemia
567G	Rh	F	49.99	77	Acute myelogenous leukemia
315E	Cyno	M	50.19	98	Lymphocytic lymphoma
333E	Cyno	F	57.04	103	Osteosarcoma, jaw
13T	Rh	M	101.65	109	Acute myelogenous leukemia

[a] Monkeys were treated with procarbazine by sc, ip, or oral routes at 5–50 mg/kg; all monkeys received the initial dose of procarbazine within 48 hours of birth except for 13T, which received the first dose at age 5 months.

[b] Latent period is the time in months from the first dose of procarbazine until the clinical diagnosis of tumor.

Table VIII.

Tumors in Monkeys Receiving 1-Methylnitrosourea (MNU) by the Oral Route[a]

Monkey number	Species	Sex	Total dose (gm)	Latent period (months)[b]	Histological diagnosis
617H	Cyno	M	53.21	63	SCA[c], pharynx and esophagus with invasion of mediastinal lymph nodes; squamous metaplasia, trachea
622H	Rh	F	65.73	57	SCA, soft palate, tongue, and esophagus with invasion into stomach
539G	Rh	F	108.15	72	SCA, mouth; SCA in situ, pharynx; squamous papillomas, tongue, pharynx, and esophagus; dyskeratosis, esophageal mucosa
540G	Rh	M	129.28	83	SCA, mouth, and esophagus; squamous papilloma and hyperkeratosis, buccal mucosa
538G	Rh	M	133.81	72	SCA, mouth, pharynx, and esophagus; multiple squamous papillomas, pharynx, and esophagus
569G	Afr Gr	M	137.20	124	SCA, mouth, pharynx, and esophagus
579G	Rh	M	180.64	120	SCA, mouth, and esophagus

[a] MNU (10–20 mg/kg) was incorporated into a vitamin sandwich and given daily five times every week; dosing was initiated within 1 week of birth.

[b] Latent period is the time in months from the first dose of MNU until the clinical diagnosis of tumor.

[c] SCA, squamous cell carcinoma.

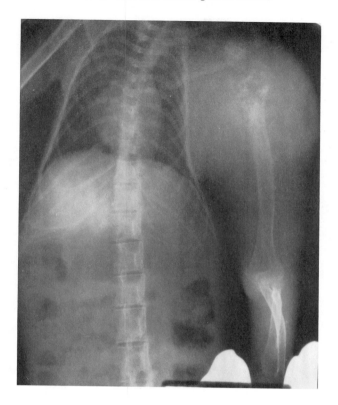

Fig. 1. X-Ray of osteosarcoma of the humerus in monkey 731I taken a few hours before the animal was sacrificed. The monkey had received a total procarbazine dose of 24.22 gm over a period of 68 months.

sions in the monkeys, as well as the clinical manifestation of the tumors, resembled that seen in humans and included difficulty in swallowing, frequent vomiting and subsequent weight loss, and sialorrhea. In addition, the radiographic appearance of these tumors was strikingly similar to human esophageal carcinoma (Fig. 2). Histological examination of these tumors revealed a morphology similar to that seen in human esophageal carcinomas, despite the highly variable nature of such tumors in both humans and monkeys, in which the morphology ranges from well-differentiated to highly anaplastic. In view of the many similarities between human esophageal carcinomas and those noted in the present study, we feel that MNU-induced esophageal carcinoma may prove to be a valuable model for the study of the human tumor.

Table IX.

Tumors Induced in Bushbabies (G. crassicaudatus) by Intraperitoneal Injections of DENA[a]

Monkey Number	Sex	Age at first dose (months)	Total dose (gm)	Latent period (months)[b]	Histological diagnosis
838K	M	1	0.295	13	Mucoepidermoid carcinoma, nasal cavity, with invasion of cranial bones and meninges, and lung metastases
935M	F	1	0.498	22	Well-differentiated mucoepidermoid carcinoma, nasal cavity
939M	M	3	0.525	19	Well-differentiated mucoepidermoid carcinoma, nasal cavity with invasion of bone and soft tissue of skull
907L	M	1	0.614	12	Poorly differentiated mucoepidermoid carcinoma, nasal mucosa, with metastases to brain
933M	F	1	0.636	22	Mucoepidermoid carcinoma, nasal mucosa, with invasion of brain
934M	M	1	0.730	27	Well-differentiated carcinoma, liver with lung metastases; mucoepidermoid carcinoma, nasal cavity

940M	M	0.769	16	3	Mucoepidermoid carcinoma, nasal cavity
936M	M	0.786	22	1	Anaplastic adenocarcinoma, nasal mucosa, with invasion of bone and soft tissue of skull and metastases to cervical lymph nodes
904L	M	1.130	22	1.5	Mucoepidermoid carcinoma, nasal mucosa
937M	M	1.485	25	5	Anaplastic carcinoma, liver with metastases to perigastric and -pancreatic lymph nodes; mucoepidermoid carcinoma, nasal mucosa, with invasion of bones and cartilage

[a]DENA was given bimonthly at doses of 10-30 mg/kg.

[b]Latent period is the time in months from initiation of treatment until clinical diagnosis of tumor.

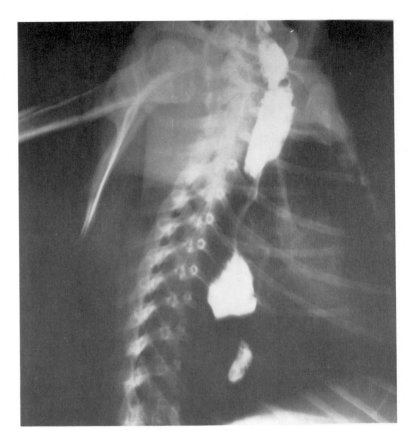

*Fig. 2. Radiographic findings in monkey 540G after bar-
ium swallow. The middle portion of the esophagus is con-
stricted by an invasive squamous cell carcinoma. The monkey
had received a total MNU dose of 129.28 gm over a period of
83 months.*

The remainder of the carcinogenic nitroso compounds are
hepatocarcinogens in Old World monkeys, and N-nitrosodiethylamine (DENA) also induces mucoepidermoid carcinomas of the
nasal cavity in the bushbaby (*G. crassicaudatus*).

DENA induced tumors in 32 out of 45 Old World monkeys
receiving po doses (40 mg/kg) five times a week. These
tumors, all of which were hepatocellular carcinomas, developed
in rhesus, cynomolgus, and African green monkeys that had
received the initial dose of DENA either at birth, at 1-8
months postpartum, or as adults. Figure 3 shows the typical

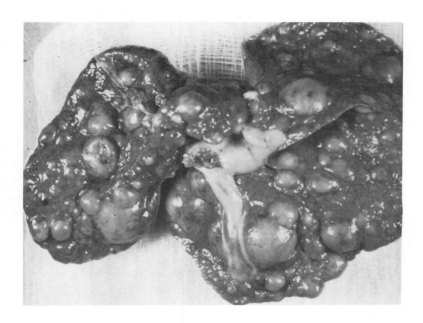

Fig. 3. Cut surface of liver from DENA-treated monkey showing numerous white tumor nodules measuring 0.3-2.0 cm in diameter.

appearance of the hepatocellular carcinomas found when these monkeys were necropsied. Numerous white tumor nodules of various sizes occupied all lobes of the liver, and metastatic tumor nodules in the lungs were almost invariably present (Fig. 4).

The average latent period for tumor development was 38 months, with a range of 14-148 months. The average total dose of DENA ingested by the monkeys developing tumor was 26.5 gm, and ranged from 6 to 58 gm. No clear species or sex difference was noted with regard to latent period or total dose of DENA required for tumor development. There is some indication that monkeys receiving the initial dose of DENA at birth required less total DENA for tumor development than those monkeys in which initial treatment was delayed until 1-8 months postpartum or until adulthood. However, the number of animals in each group is not sufficiently large to test the statistical significance of this apparent difference.

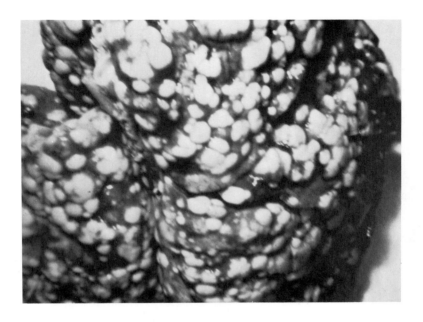

*Fig. 4. Numerous metastatic hepatocellular carcinoma
nodules on surface of lungs of DENA-treated monkey.*

DENA was also carcinogenic when given by the ip route.
A total of 98 monkeys given bimonthly ip doses (40 mg/kg) of
DENA developed hepatocellular carcinomas. The average latent
period was 17 months (range 6-32 months) and the average total
dose of DENA necessary for tumor development was 1.7 gm
(range 0.39-4.08 gm). The latent period and the total dose
required for tumor development appeared to be independent of
the age at which dosing was initiated as well as of the
species and sex of the animals.
      DENA also had carcinogenic activity in the more primative
primates *G. crassicaudatus* (Table IX). Ten out of a total of
12 animals have developed tumors after bimonthly ip injections
of DENA at doses of 10-30 mg/kg. In contrast to the DENA-
induced primary hepatocellular carcinomas in Old World monkeys,
all 10 of the bushbabies developed mucoepidermoid carcinomas
of the nasal cavity. In two of these 10 animals, carcinoma of
the liver was also present, and in both cases metastases to
the lungs or to intestinal lymph nodes was noted. The first
clinical symptom of the nasal cavity carcinomas was sneezing,
followed by audible nasal respiratory sounds, lacrimation,

Fig. 5. *Typical appearance of mucoepidermoid carcinomas developing in the nasal cavities of bushbabies receiving bimonthly ip injections of DENA. Upper photograph shows lesion in animal 907L, 15 months after initiation of DENA treatment and 15 months before death. Animal 940M received DENA for 20 months before developing nasal lesion shown in lower photograph.*

and a serous nasal discharge. Nodular development then became apparent on the nose just anterior to the medial canthus. The tumors continued to grow, often into the brain through the cribriform plate, or into the postorbital sinus. Photographs of the typical appearance of these tumors are presented in Fig. 5. The average total dose of DENA given the bushbabies was 0.747 gm, and ranged from 0.295 to 1.485 gm. The latent period averaged 20 months (range 12-27 months).

There is no obvious reason for the marked difference
noted between Old World monkeys and bushbabies with regard to
the site of DENA-induced tumors. It is possible that it is
related to differences between bushbabies and Old World monkeys in the metabolism and physiologic disposition of DENA,
and this possibility is currently being investigated in our
laboratory.

The carcinogenic effects of two other nitrosamines have
been evaluated in monkeys. N-nitrosodipropylamine (DPNA) has
induced primary liver cancer in rhesus and cynomolgus monkeys
when given bimonthly at ip doses of 40 mg/kg. All six macaques treated with DPNA developed hepatic cell carcinoma after
receiving an average total dose of 7.0 gm; the average latent
period for tumor development was 28.5 months. Hepatic cell
carcinomas were also induced in macaques by 1-nitrosopiperi-
dine (PIP), given by po or ip routes. Twelve monkeys received
po doses of PIP (400 mg/kg), and 11 (92%) have developed
hepatic cell carcinomas. In addition, 11 monkeys have been
treated with PIP (40 mg/kg) by the ip route, and four (36%)
have developed hepatocellular carcinomas. The average cumulative dose of PIP given orally was 1706.1 gm, and when given
ip it was 60.95 gm; the average latent period for tumor induction ranged between 22 and 101 months, depending on the
route of administration and the sex and the species of the
monkey. However, the number of animals in each group was too
small to draw any conclusions with regard to differences between species or sex in tumor susceptibility.

CONCLUSIONS

Nonhuman primates will probably never replace rodents as
the primary species for evaluating large numbers of chemicals
for carcinogenic activity, because such studies in monkeys are
expensive, time-consuming, and require larger amounts of test
chemicals. However, there are several advantages in using
nonhuman primates rather than (or in addition to) rodents in
carcinogenesis studies, and these advantages have been discussed in the introduction. Monkeys are phylogenetically
closer to man than are rodents, certain metabolic pathways in
nonhuman primates resemble those of humans more closely than
do those of rodents, the relatively longer lifespan of monkeys
makes it possible to administer test compounds at doses that
mimic human exposure levels, and a more reasonable estimate
of the latent period for a chemically induced tumor can be
obtained from nonhuman primates than from rodents.

Furthermore, the site at which some chemically induced tumors develop in primates appears to parallel the human situation more closely than is the case with rodents, as exemplified by studies with MNU. MNU or related nitroso compounds have been implicated as etiologic factors in human GI tract cancer (4,5). Nitrosamines may be synthesized in the upper GI tract of humans (6), and MNU is the final product of a reaction between nitrites in human saliva and methylguanidine, a constituent of several human foods (7). In nonhuman primates, MNU induces squamous cell carcinomas of the oropharynx and/or esophagus, whereas rats treated po with MNU develop kidney, skin, and jaw tumors as well as GI tract tumors (8). This similarity between nonhuman primates and humans with regard to site of chemically induced tumors also suggests that monkeys might represent a valuable model for the study of the corresponding human tumor. For example, many parallels were noted between the esophageal squamous cell carcinomas observed in the MNU-treated monkeys and human esophageal carcinoma; these parallels included the clinical manifestations of the tumor and its complications, radiographic appearance, and morphology.

Another advantage of primates over rodents in carcinogenesis testing is the comparatively low incidence of spontaneous neoplasms in the former species. In our colony the spontaneous tumor incidence during the past 16 years has been less than 2%. In contrast, results of rodent carcinogenesis testing are frequently obscured by the high incidence of tumors observed in control animals; in one rodent colony, this incidence was 26% for control mice and 46% for control rats (3).

Finally, results from studies with the hepatocarcinogens aflatoxin B_1, MAM-acetate, and the nitrosamines have made it possible to develop the α-fetoprotein assay as a tool for the early diagnosis of primary liver tumors and for monitoring the regrowth of hepatomas after surgical removal of the tumor or after chemotherapy (9,10). It is likely that other diagnostic tests, including additional oncofetal antigens, will be developed in nonhuman primates and will prove useful for the detection of preneoplastic lesions in man.

REFERENCES

1. Miller, E. C., and Miller, J. A., Approaches to the
    mechanisms and control of chemical carcinogenesis. *In*
    "Environment and Cancer" (The University of Texas M.D.
    Anderson Hospital and Tumor Institute at Houston),
    24th Annu. Symp. Fundamental Cancer Res., pp. 5-39.
    Williams & Wilkins Co., Baltimore (1972).
2. Smith, R. L., and Caldwell, J., Drug metabolism in non-
    human primates. *In* "Drug Metabolism--From Microbe to
    Man" (Parke, D. V., and Smith, R. L., eds.), pp. 331-356.
    Taylor and Francis, Ltd., London (1977).
3. Weisburger, J. H., Griswold, D. P., Prejean, J. D., *et
    al.* The carcinogenic properties of some of the principle
    drugs used in clinical cancer chemotherapy. *Recent
    Results Cancer Res. 52,* 1-17 (1975).
4. Mirvish, S. S., Kinetics of nitrosamide formation from
    alkylureas, *N*-alkylurethans and alkylguanidines: possi-
    ble implications for the etiology of human gastric can-
    cer. *J. Nat. Cancer Inst. 46,* 1183-1193 (1971).
5. Adamson, R. H., Krolikowski, F. J., Correa, P., *et al.,*
    Carcinogenicity of 1-methyl-1-nitrosourea in nonhuman
    primates. *J. Nat. Cancer Inst. 59,* 415-422 (1977).
6. Correa, P., Haenszel, W., Cuello, C., *et al.,* A model
    for gastric cancer epidemiology. *Lancet 2,* 58-60 (1975).
7. Endo, H., and Takahashi, K., Methylguanidine, a naturally
    occurring compound showing mutagenicity after nitrosation
    in gastric juice. *Nature 245,* 325-326 (1973).
8. Leaver, D. D., Swann, P. F., and Magee, P. N., The in-
    duction of tumours in the rat by a single oral dose of
    N-nitrosomethylurea. *Br. J. Cancer 23,* 177-187 (1969).
9. Adamson, R. H., Smith, C. F., and Dalgard, D. W., In-
    duction of neoplasms in nonhuman primates by chemical
    carcinogens--correlation of serum alpha-fetoprotein and
    appearance of liver tumors. *In* "Embryonic and Fetal
    Antigens in Cancer," Vol. 2. (Anderson, N.-G., Coggin,
    J. H., Jr., Cole, E., and Holleman, J. W., eds.), pp.
    331-337. Oak Ridge, Tennessee, Oak Ridge National Lab-
    oratory (1973).
10. Dalgard, D. W., McIntosh, C. L., McIntire, K. R., *et al.,*
    Hepatic carcinogenesis and serum alpha-fetoprotein in
    nonhuman primates. *In* "Alpha-Feto-Protein" (Masseyeff,
    R., ed.), *Proc. Int. Conf. Inst. Nat. de la Sante et de
    la Recherche Medicale,* pp. 211-216. Paris (1974).

GENERAL DISCUSSION

DR. KRAYBILL: Thank you very much Dr. Adamson. I was quite interested in your DDT data and I might say, unless I am challenged, that there is no definitive epidemiological study in man, and that is needed because we have about 35 years' exposure. Dr. Lindsay and I have discussed the notion that perhaps we could get a captive group, a nonmigrant type, where we could look at say formulators or people who have had high exposure. This would be fine, because that is a long period of time for exposure to DDT. But your data with the monkey here are somewhat reassuring. I know that Deichmann and John Davies had looked at this problem down in Florida, but as far as I know there has not been a really retrospective study on DDT in man.

I take your admonition about the diet and maybe we are whistling Dixie, but it is true there is a variant as far as the diet is concerned, because all of our contract labs reflect this variation in that the protein used is different depending on what source the protein in the diet is. We feel that through one of our contract labs, we could maybe set up a specification so that we would guarantee that from month to month and year to year we could have something that would approach a fixed level.

A greater concern of mine is on the contaminants in the diet. I do not know what we can do about that. For many years, we used a semisynthetic ration, but now we are using the lab chow and, "alas and alack," the group in Cincinnati has had a high degree of success with the semisynthetic ration. They get good performance with these animals, but they add to that diet vanadium and many of the trace elements. They claim this is probably responsible for the high degree of success. I know one of the studies in Chicago foundered; I believe it was on irradiated food, but they ended up doing a mineral research study. They were having some difficulties that we never encountered, at least when I was at Fitzsimmons.

DR. SLOMKA: I would like to make a comment on the point you made. About five years ago, we had an opportunity to look at some DDT workers at Montrose Chemical Company. I believe that we looked at the aflatoxin protein and other levels in those workers. They were all within the normal range. If memory serves me right, we looked at about 20 people with exposure levels up to about 20 years. I supplied those data to Dr. Laws at the Mayo Clinic and he published a report of those data in a paper about five years ago.

ISBN 0-12-192750-4

DR. KRAYBILL: Did you see any trends insofar as hepatocarcinoma is concerned?

DR. SLOMKA: No. We had no autopsy data or biopsy data. The liver function tests that Laws looked at were normal and we just supplied him with the α-fetaprotein levels. They were all in the normal range.

DR. KRAYBILL: Thank you very much for that comment. The other question Dr. Adamson, I was going to ask: You reported that you were testing saccharin? Is that correct? Do you have any data as yet on saccharin? I think Dr. Lijinsky is listening very intently to this.

DR. ADAMSON: Well, I must say I was asked that about 10 years ago with regard to aflatoxin, which at that time had not induced any tumors, and I said you know thus far it's negative and it got into the literature that it was not carcinogenic. But I'm glad to see that we have a tape here. Because I got home and that fourth of July the first monkey died with an aflatoxin tumor.

No, thus far, and Dr. Coulston has had access to these data, we have 20 animals that have been on, ten that have been on for a long period of time over 8 years and 10 that have been on for approximately a year. We are starting another group of 10. We are using what I call a reasonable dose level. It is still an exaggerated dose level when compared to humans, and is equivalent to about 40 packets of artificial sweetener a day. These animals have been exposed to around 200 gm of saccharin. Thus, far, we have no clinical evidence from monitoring CEA or α-fetaprotein of any tumors in the animals.

DR. SCHMIDT-BLEEK: Dr. Adamson, would you care to comment on the question of cost effectiveness? You did say that you felt that using these animals may have certain advantages over using rodents. To us in administration, the question of course becomes very quickly important when discussing these things with industry. Could you make some further comments with respect to the total cost effectiveness?

DR. ADAMSON: Well there is no doubt that the rhesus monkey is more costly and it is becoming more so with the ban by India. Fortunately, and this was more by virtue of the limits of funding at the time rather than by foresight, we set up our own breeding colony so that we get 80 babies a year and we're not dependent on India for the monkey. But with regard to the cost effectiveness I would say I definitely feel it is

necessary and inexpensive, and this may be contrary to NCI
policy but it is my own feeling. I do not believe in data
from the mouse on the question of chemical carcinogenesis.
With regard to, particularly in the mouse, the question of
hepatocarcinogenicity, that only liver tumors are observed,
I just do not believe those data.

   If cancer occurs in the rat, I will accept the data on
the rat and I will believe them. However, if a compound is
going to entail tremendous economic consequences, then I
believe that one should also look at the monkey. And,
although I do not think that the monkey should be used for
everything, I feel if it is for something as valuable as DDT
the monkey should be used in combination with an epidemiological survey. Perhaps you should also use some *in vitro* tests,
where you cannot only use the mouse, but you can also use the
monkey and man. So, I would say, certainly not the monkey
for the common everyday tests, but I would also say not the
mouse.

DR. CLAYSON: Do you use the ordinary cytology to monitor the
bladder in the saccharin-treated monkeys, or only when it is
a matter of suspicion? If you are only getting 1% incidence
of spontaneous tumors, I think a lot of us would suspect that
you are not really carrying on to the middle life of these
animals. At what stage do you get this yield? Is it 10, 15,
20 years?

DR. ADAMSON: Well, the answer to this is in the literature.
Actually our rate of incidence is fairly high. I would say
so because we have had most of these monkeys since they were
born both in the control group and as including as well the
adult breeders. So, actually we are magnifying our spontaneous
incidence rate by using the adult breeders, which were approximately, although we don't have the exact age, 8 years old
when we first bought them. They are now 16 years older on
top of that. They are dying of coronaries, they are dying of
aneurysms, and so forth. I think actually with our colony we
have magnified the spontaneous tumor incidence. One might
say, in fact, this was brought up to me as an argument on
time, actually the rhesus monkey should not be used because
the spontaneous incidence in man is so much higher. Well, it
is debatable whether that is spontaneous or not or whether
it is diet-induced. Indeed, if one wants to question the
diet, we can say that these animals have received basically
Purina lab chow; they are supplemented with half an orange
and half an apple a day, so they have vitamin C. We give
them a vitamin sandwich loaded with vitamins as well as carcinogens. And all the animals get this. If you have the

carcinogen, in spite of all this, in heavy doses it is not
going to protect the monkey. But, it may be protecting
against very low levels of the carcinogen, which normally
occurs in the food. In addition, although I'm not a vege-
tarian and would never be one, these animals are not on any
type of a meat diet whatsoever.

DR. HOLLIS: Dr. Kraybill, let me just make an observation.
Relative to the use of nonhuman primates in this kind of
work, we must move into this area. It becomes very important
that this is a useful tool. It is just coincidental that
last night an editorial appeared in the *Washington Star* that
the rhesus monkey was being denied us by edict by India. I
just bring this to your attention, at the moment, because I
am delighted to see that we are developing our own sources
of monkeys because by diplomatic or state reasons they are
going to be denied us.

DR. COULSTON: May I comment on this? I do not want the
wrong impression to be given. Actually, the United States
Government through its wisdom many years ago created breeding
centers for rhesus monkeys, which now supply roughly somewhere between 35 and 40% of the need of the government for
such testing. The FDA, particularly, has created enough
rhesus monkey breeding programs that I would assure you, they
have about 40 or 50% of what they need. As Dr. Adamson so
wisely said he breeds his own, I breed my own and I have
enough to supply all my needs. There are in the U.S. national
primate centers that have been in operation for years and have
not done very much in the way of supplying people with monkeys,
including Davis, Dr. Mrak, my good friend. In other words
in the past few years, the fact is that there is a concerted
effort on the part of the U.S. government to supply our needs.
    I do not think that India will keep animals from coming
to the States for a long period of time. Their real objection was that animals were being used for radiation research
contrary to an understanding. That is what they are object-
ing to.
    This has to do I believe with the response of our Presi-
dent to some problems when he was in India.

DR. KOLBYE: I can predict one inference from Dr. Adamson's
data. There will be a new health food fad; monkey chow is
going to be sold in the stores.

DR. COULSTON: The work of Dr. Adamson is known to many of us
around this table.   The International Academy of Environmental
Safety believes in it very strongly and has promoted the kind
of work that Dr. Adamson is doing.   There could be no question
in my mind that if he could have enough monkeys, time, and
money, he would answer many of the questions relating to
chemical carcinogenesis that bother us now in regard to the
interpretation of animal data to man.   Everyone talks about
2-year studies in the mouse and now it is 3 years in the rat,
pretty soon it's going to be 4 years in the rat.   I do not
know if it is the rat anymore.   We talk about 18 months to 2
years in the mouse.   The fad today is to do lifetime studies
in rodents and hamsters.   The concept of lifespan animal
studies has become fixed in our thinking.   In the monkey we
can produce cancer by well-known carcinogens in 2 years or
less.   Or less!   So this concept that you have to go 25
years in the monkey to prove something is becoming pretty
poor, now.

     And more than that, I think on a cost estimate basis even
if monkeys cost $500 apiece, it would be cheaper to run 30 to
40 to 50 rhesus monkeys for 2-5 years and get an answer that
would be meaningful than to play around doing one mouse test
after another, over and over again, to get some kind of an
answer that may not be applicable to a primate like man.   I
believe that this latency problem has been put to rest by the
kind of work that Dr. Adamson is doing.

DR. KENSLER: Dr. Adamson, I was interested in your data that
indicated with diethylnitrosamine that the total dose seemed
to be at 1.2 to 1.6 gm.   It may be that when you get down to
lower doses, this number will turn out to be much larger,
which would be very interesting.   But do you have similar
numbers for aflatoxin?   Because I understand that you are
really lacking with that stuff.   I wonder what that number
would look like compared to what one in this country might
ingest from all sources throughout life.

DR. ADAMSON: That is true.   We are not sure on the DNA experiment what ultimately will develop.   Thus far, it looks
like the total dose is somewhere between 1.2 and 1.6 gm, but
it is quite correct at the lower doses.   Perhaps we may exceed that, particularly if we are thinking about dropping the
dose to 1/10 and 1/100, as long as we can rest assured that
somebody will carry out the experiment even if some of us
are not around.

     With regard to aflatoxin, we do not have the data unfort-
unately.   We are now going back and looking at aflatoxin and
the reason we do not have the data is a complicated one.

Essentially, it boils down to the fact that we did not have pure aflatoxin B_1 to start with, but now we are using all pure aflatoxin B_1. We had a mixture of aflatoxins and we are not sure of what kind of doses we were using with the aflatoxin B_1. Unfortunately, we do not have a sample of the original mixture of aflatoxin that we were working with.

DR. KENSLER: Do you have a sort of an upper limit for it?

DR. ADAMSON: Well a total dose, yes, these have ranges for a total dose. I would guess the total dose is approximately 100 mg per monkey, and I suppose the possible dose to man (I do not know what the FDA thinks) is about 0.2 ng per day.

SESSION III

EXTRAPOLATING ANIMAL RISK TO HUMANS

Ernst L. Wynder

One of the key problems facing experts in environmental carcinogenesis is making the proper decision about how to best protect humans from environmental carcinogens. Because most experts in the field agree that a majority of human cancers are related to environmental agents (1-4) (Fig. 1), it is pertinent that we consider how to protect humans against exposure to such agents.

We have passed a number of laws concerning occupational environments, food additives, drugs, and cosmetics that are meant to safeguard against carcinogenic exposures. These regulations are principally guided by data from animal experiments and assume, in particular, that a zero tolerance does not apply to humans. In general, we suggest that high-potency exposure leading to cancer in an occupational setting cannot be considered equivalent to chronic low dosage in the general environment.

Clearly, the scientific community is facing a dilemma. On one hand, we want to be certain that not one cancer death occurs that could have been prevented. On the other hand, it is just as important to avoid causing great social and economic inconvenience by ostensibly preventing cancer deaths that actually would never have occurred.

How are we to resolve this conflict? Those who claim that any kind of positive animal data necessitate at least a warning to the public, if not total removal of the responsible agent, are open to criticism by the makers of these products. On the other hand, those who call attention to the possibility that a given set of animal data may have no bearing on the human situation are often branded as being in the pay of industry.

Those who recommend a middle ground in this debate are likely to be in the undesirable position of being attacked by both sides. Nevertheless, we must all take a stand, scientist and citizen alike. It is no longer possible, nor desirable,

305

ISBN 0-12-192750-4

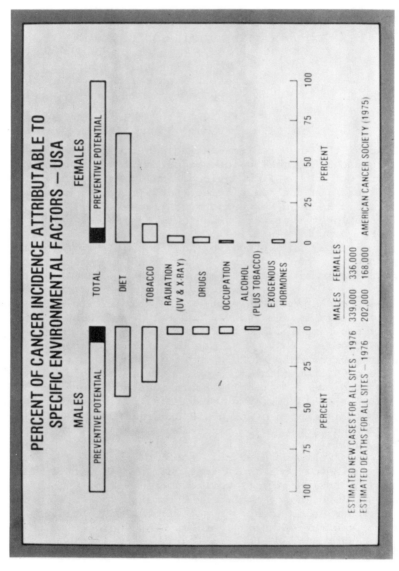

Figure 1.

for the scientist to make the appropriate decision for society. Rather, the public and their representatives in Congress, together with the scientific community, must evaluate and decide the appropriate path to take. The scientist's obligation, however, is to provide hard facts and sound data on which to base societal decisions.

Those who spend so much time and energy calling for legislative action based on the results of animal tests should remember that for many years we have known that cigarette smoke condensate produces squamous cell cancer of the skin of mice, of the ear of rabbits, and of the larynx of hamsters, and that epidemiologic evidence shows tobacco smoke to be a human carcinogen. Yet, significant regulatory action has not been taken on this product, for reasons that are apparent to everyone. Yet, the point to be made is that if we are willing to make a legalistic decision on one type of animal evidence, such as the saccharin data or Red Dye #2, we should be prepared to make it on others. In this case, it is said that smoking is an individual decision, that the individual has been warned, and that no further action is necessary.

Similarly, there is considerable evidence that heavy alcohol consumption, in conjunction with tobacco use, increases the risk for cancer of the upper alimentary tract (5-8). Alcoholic beverages by themselves lead to cirrhosis and job instability in users. Yet, no regulatory action has been taken on alcohol either. Again, the argument is that it is an individual choice and an individual freedom.

More recently it has been shown that high levels of dietary fat, equivalent to those eaten as part of the Western diet, increase the risk of breast and colon cancer in animal models. These data are consistent with human epidemiologic and metabolic data (9-18). Little public attention has been given to animal data showing that high dietary fat enhances the risk for breast or colon cancer in rats. Nonetheless, current data support the view that these cancers are similarly controlled by high fat intake in humans as well. We may ask why so little attention is given to the animal data showing the link between these environmental factors and these important cancers, and so much of our energy is centered on agents for which there is considerably less evidence. One excuse, of course, is that factors such as fat are not covered by any particular regulatory agency. Yet, sound data indicate that the level of dietary fat bears on the incidence of 42% of all current cancers (3). Certainly, scientific evidence for one factor is as valuable as it is for any other. Rather than concentrating our limited national efforts on agents that are of questionable relation to human cancers, more attention to those factors influencing the majority of human cancers is in order.

The concept of zero tolerance is, to a large extent, an academic debate and is of little significance when examining the broader view of effective action in cancer prevention. As a discipline, epidemiology does not have sharp enough endpoints to measure the effect of an agent that may cause one case in 100,000, with the exception of cases of very rare cancer sites. Nearly all known human carcinogens cause a high yield of cancer rather quickly when tested by the current bioassay systems. They are also mutagenic, or otherwise positive, in the rapid bioassay tests. Such biological screening thus prevents potentially hazardous chemicals from being placed on the market and removes such chemicals that are already in commerce.

Controversy arises with weak evidence in animal tests or rapid bioassays. To ban, or not to ban? A proper course involves a series of considerations that compare the animal data with human data where available, and that examine the potential benefits that a substance may have both socially and economically. Without such a critical evaluation, any arbitrary or automatic regulatory decisions supposedly based on "law" should be considered irresponsible and wasteful of national resource.

Furthermore, the likelihood that a substance may be responsible for the induction of human cancers is influenced by a number of variables. These include dose, dose rate, route of entry (in relation to human exposure), latency of tumor occurrence, and numbers and types of tumors obtained. The lower the dose required for tumor induction, the shorter the latency period, the more tumors produced, and the greater the equivalence of application site between animals and humans, the more we should be concerned that this substance adversely affects humans. In those cases where human data are not available, perhaps because the substance may be newly introduced, greater attention must be paid to the animal data. Consideration of metabolic activity, similarity between animal and human biochemistry, and mutagenicity and similar studies on genotoxicity is also requisite.

A multidisciplinary committee, comprising biologists, pathologists, biochemists, toxicologists, epidemiologists, clinicians, biostatisticians, and environmentalists should be instituted to evaluate the existing data in relation to potential risk for humans. The strengths and weaknesses of the animal and human data should be considered, as well as other pertinent factors.

In the final analysis it is, of course, incumbent upon society to decide what kind of life to lead and what kind of risk to take. It is obvious that some have decided to take established risks, as are evident for cigarette smoking, excessive alcohol, fat and salt intake, disregard of seat belts, all of which have been clearly shown to increase morbidity and mortality. There has been ample warning about these factors. On the other hand, substances in foods, drugs, or cosmetics, such as saccharin or Red Dye #2, may not be as apparent as other carcinogens. Thus, on the basis of animal tests, these agents will be scrutinized and regulated by legislative action. Some existing legislation, however, does not necessarily consider the strengths and the relevance of animal data, does not consider existing epidemiologic data, and does not consider the cost or benefits to society.

For these reasons, a modification of existing legislation relating to environmental hazards is suggested, with the final decision regarding a given agent being left to an interagency committee to consist of one expert member from each of the NCI, FDA, EPA, OSHA, NIOSH, NIEHS, USDA, DOE, and other select members of the executive branch of the government, industry, consumer affairs, labor, and the legal community. The decision of this group should be based, to a large extent, on weighing the power and relevance of the animal and rapid bioassay data and the epidemiologic data, against the social and economic benefits. Clearly, where the animal data are very significant, where human data are significant, and where the social and economic benefits are small, the decision is obvious. One would like to believe that if the animal evidence is minimal, if the existing epidemiologic evidence shows no measurable effect, and if the social and economic benefits of the substance are great, it would be in the interest of society to keep such an agent in use under highly limited and carefully monitored conditions. In most instances, of course, the decisions will not be clearly defined, but even in the latter instance, legality and possibly bureacracy are on the side of banning.

What is of key concern is that the public believes we are protecting them when we remove agents that may in actuality not contribute to a single cancer; therefore, we are making a lot of movement and decisions for which no future generation will thank us. What is also of concern is that while all this activity exists we are neglecting those environmental factors for which we have great experimental and epidemiologic evidence that they contribute to a large proportion of human cancers. It is time that we come to grips with the problem facing us. The scientist alone cannot solve the issue. Like other important problems facing our nation, this too requires deliberate and careful debate, involving public participation. Prevention is the ultimate solution to the cancer threat. Let us begin a real and realistic movement toward this goal.

REFERENCES

1. Higginson, J.   Present trends in cancer epidemiology.
   Proc. Can. Cancer Conf. 8, 40-75 (1969).
2. Armstrong, B., and Doll, R.   Environmental factors and
   cancer incidence and mortality in different countries,
   with special reference to dietary practices.   *Int. J.
   Cancer 15,* 617-631 (1975).
3. Wynder, E. L., and Gori, G. B.   Contribution of the
   environment to cancer incidence: an epidemiologic
   exercise.   *J. Nat. Cancer Inst. 58,* 825-832 (1977).
4. Doll, R.   Strategy for detection of cancer hazards to
   man.   *Nature 265,* 589-596 (1977).
5. Wynder, E. L., Covey, L. S., Mabuchi, K., and Mushinski,
   M.   Environmental factors in cancer of the larynx.
   Cancer 38, 1591-1601 (1976).
6. Moore, C.   Cigarette smoking and cancer of the mouth,
   pharynx and larynx.   *JAMA 218,* 553-558 (1971).
7. Schottenfeld, D., Gantt, R. C., and Wynder, E. L.   The
   role of alcohol and tobacco in multiple primary cancers
   of the upper digestive system, larynx and lungs--a prospective study.   *Prev. Med. 3,* 277-293 (1974).
8. McCoy, G. D.   A biochemical approach to the etiology of
   alcohol-related cancers of the head and neck.   *Laryngo-
   logy 1,* 1-4 *(Suppl. 8)* (1978).
9. Chan, P. C., Head, J. F., Cohen, L. A., *et al.*   The influence of dietary fat on the induction of rat mammary
   tumors by *N*-nitrosomethylurea: Strain differences and
   associated hormonal changes.   *J. Nat. Cancer Inst. 59,*
   1279-1284 (1977).
10. Tannenbaum, A.   The genesis and growth of tumors III.
    Effects of a high fat diet.   *Cancer Res. 2,* 468-475
    (1942).
11. Wynder, E. L., Bross, I. D. J., and Hirayama, T.   A
    study of the eipidemiology of cancer of the breast.
    Cancer 13, 559-601 (1960).
12. Chan, P. C., Didato, F., and Cohen, L. A.   High dietary
    fat, evaluation of fat, serum prolactin and mammary
    cancer.   *Proc. Soc. Exp. Biol. Med. 14,* 133-135 (1975).
13. Carroll, K. K., and Khor, H. T., eds.   Dietary fat in
    relation to tumorigenesis.   *In* "Progress in Biochemical
    Pharmacology: Lipids and Tumors, Vol. 10, pp. 308-345.
    Karger, Basel (1975).
14. Reddy, B. S., and Wynder, E. L.   Metabolic epidemiology
    of colon cancer: fecal bile acids and neutral sterols in
    colon cancer patients and patients with adenomatous
    polyps.   *Cancer 39,* 2533-2539 (1977).

15. Hill, M. J., Drasar, B. S., Williams, R. E. O., *et al.* Fecal bile acids and clostridia in patients with cancer of the large bowel. *Lancet I,* 535-539 (1975).

16. Phillips, R. L. Role of life-style and dietary habits in risk of cancer among Seventh Day Adventists. *Cancer Res. 35,* 3513-3522 (1975).

17. Haenszel, W., Berg, J. W., Kurihara, M., *et al.* Large bowel cancer in Hawaiian Japanese. *J. Nat. Cancer Inst. 51,* 1765-1799 (1973).

18. Reddy, B. S., Weisburger, J. H., and Wynder, E. L. Effect of high risk and low risk diets for colon carcinogenesis of fecal microflora and steroids in man. *J. Nutr. 105,* 878-884 (1975).

REMARKS

DR. FURMAN: Dr. Wynder, I wonder if you would comment briefly in relation to your observations having to do with diet, the role of obesity. It's my impression that breast carcinoma and that sort of thing are more frequent among women who are obese. Are your dietary data corrected for body weight?

DR. WYNDER: Oh yes. Now, clearly, epidemiology cannot be better than the questions that we ask, and obesity is one of the easier ones, because this is not subjective. We put the patient on the scale, we ask her what she weighed 2 years ago, what she weighed when she was 20 years old, and we have considerable numbers of data on height and weight.

As you know, my friend and colleague Dr. de Waard in Holland showed a correlation of body weight to breast cancer. We cannot confirm this in American data. I found only two or three cancers related to body weight. One of them is cancer of the endometrium, another is cancer of the female kidney, and a third one is cancer of the gall bladder. Clearly breast cancer does not relate to height and weight, and we therefore seem to have a similar picture as in coronary disease, where the relation is much more for hyperlipidemia than for weight. So, we are dealing with a dietary factor that does not relate to weight. Indeed, in our animal studies, we have also shown that this increase in tumor yield with dietary fat is independent of weight. We believe it is a question of diet, and may I say that for the animal studies, unlike for coronary disease, we get the same effect, perhaps even slightly greater with polyunsaturated fats than with saturated fats. So it's type of diet rather than total calories. The last statement we have from our animal data, together with those published by Carroll in Canada, which indicate at least the same, if not a greater risk, for polyunsaturated fats.

DR. DARBY: Do I understand that the data on fat content of the diet are per capita figures, not based on actual dietary intake, nutrient intake studies?

DR. WYNDER: The correlation studies are obviously on per capita factors and as all of us know and as Dr. Darby I'm sure will try to tell us, there are some problems with per capita data, but for anyone who has ever been in Japan, as I was recently, and has gone through two days of Japanese eating habits, there is no question. You do not have to be a genius to know that the total fat intake is very low. On the other hand, with my little camera, Dr. Darby, I am now follow-

ing McDonald's hamburger shops throughout the world and I
counted five little shops in Japan. The fact is, as our
colleague Harayama has shown, as fat content in the Japanese
diet goes up, Japanese girls get taller, breasts get larger,
age of first menses is earlier and the breast cancer rate
goes up, as do colon cancer and coronary disease. Then of
course, the migrant studies are a very important element to
the epidemiologist, as done in Japan and in California,
showing that by the second generation, they have nearly always
caught up with our disease pattern, including a decrease of
cancer of the stomach.

DR. PARKE: I would like to congratulate Dr. Wynder on his
brilliant exposition and also to congratulate the organizers
of this meeting, because it seems that one paper after another
is presenting a piece of the jigsaw puzzle and it is all
beginning to come together.
    Dr. Wynder, we discussed this before, in New Mexico, but
your more recent findings are fitting in so beautifully with
my own hypothesis and some of our experimental work in animals.
The work that you have shown in alcohol and fats, Stier has
shown in Germany that these materials get into the endoplasmic
reticulum. They produce this disorganization I spoke of yesterday and activate the microsomes just in the way that you
suggested. It is coming together so beautifully and we are
agreeing with each other, whether we are working in epidemiology or with the rat or mouse. We can now fit the mechanism
to the human epidemiology. There is absolutely no doubt in my
mind that alcohol and fats do have the effect that you have
suggested.

DR. WYNDER: Well I would like to pay you a compliment, and as
you know in science nothing brings two people closer together
than if they agree with one another. What I really like about
science is told by the following story: I remember years and
years ago, when I had a problem convincing the world about
smoking, what really kept up my spirits is that in science,
unlike in politics, sooner or later you are either confirmed
or not confirmed. That is really one love I hold for science,
that truth eventually will be borne out.

DR. VINSON: I, too, am impressed with Dr. Wynder's comments,
particularly the relationship of high dietary fat intake and
its link to colon cancer, as well as breast cancer. I think
we both agree that this is a problem even more serious than
cancer, based on the incidence of cases and number of deaths.
I was going to ask you about the nature of this dietary fat,
but in answer to a question you did indicate that polyunsaturated fats behave pretty much the same as saturated fats.

However, I think we have to bring up the point that we have
benefits linked with the polyunsaturated fats, the use of
which obviously is quite low in this country, Finland, and
other affluent countries. So the question I ask is, How do
you relate the benefits of the high polyunsaturates to the
benefits of lowering the total dietary fat?

DR. WYNDER: Well I have to put it in my equation, Dr. Vinson.
Clearly I am quite familiar with this and as you know we are
doing extensive work in coronary disease prevention and it is
a difficult question. I hope, as we learn more about the
different types of polyunsaturated fats, we may resolve it so
that on the one hand we do not get as much colon and breast
cancer, and on the other, we also lose weight and risks for
coronary disease. May I perhaps conclude with the following
comment that may give you some food for thought, and that
relates to the word normalcy. We are planning to hold a conference in New York defining normal, blood lipid levels. As
physicians, we are so used to seeing cholesterol levels between 220 and 240 that when we see a patient with 180 we
almost think he's sick. And yet it could be possible that
the normal serum cholesterol is somewhere between 100 and 140.
After all, we know in studies done on school children in
Mexico and Wisconsin, that those two curves did not overlap.
So I think we have got to come to grips with "normalcy" for
blood lipid levels. Perhaps we even should come to normalcy
with what a stool specimen ought to look like.

Now I must say I have not quite thought about it until I
looked at stool specimens in Finland. We are used to the
fact that our stool specimen looks like a pencil, hard and
firm, but suppose normal stool ought to look more loose or
bulky. Maybe that's normal and that would lead to fewer
hemorrhoids and diverticulitis and cancer of the colon.

The same applies to blood pressure. And the same applies
to what amount of food we consume. Perhaps, if we think in
evolutionary terms, man was born into a world that had only
10% of its total calories from fat.

What about caloric expenditures? Certainly what we spend
today is not normal. And clearly also a certain amount of
environmental toxic agents can be absorbed and can be metabolized, but not when we overwhelm our system.

So I would like you to think about what normal man really
used to be like and, that with civilization, we really over-
whelmed our metabolic capacity to deal with most environmental
insults. But civilization with all its advantages has clearly
brought into being factors that which, from an evolutionary
point of view, we are unable to deal with.

ASSESSMENTS AT AN INTERNATIONAL LEVEL
OF HEALTH HAZARDS TO MAN OF CHEMICALS
SHOWN TO BE CARCINOGENIC IN LABORATORY ANIMALS

Frank C. Lu

As we all know, the Delaney Clause is a very important part of the U.S. food legislation and has a great impact in the U.S. This is why we are here this week. But, I would like to point out that the Delaney Clause also has an impact beyond the U.S. borders. I shall mention one specific example.

In October, 1969, FDA suspended the use of cyclamates as food additives. This action shocked many people and created widespread concern because these chemicals had been generally recognized as safe.

In January, 1970, the Executive Board of WHO discussed these repercussions and adopted a draft resolution for the 23rd World Health Assembly to consider. The Assembly, which met in May, 1970, adopted the following Resolution on the health hazards of food additives:

The Twenty-third World Health Assembly,
Being concerned about the potential hazards of food additives to the consumer;
Aware of the increasing research done on toxicity of food additives;
Having noted the intensive publicity commonly given by the lay press to questions of safety of food additives and the widespread repercussions which follow action by any country to limit or prohibit the use of a generally used food additive;
Noting that the matter has been raised at the forty-fifth session of the Executive Board; and
Agreeing that there is an urgent need for rapid dissemination of the results of toxicity research of food additives, including the results and consequences of evaluation of such studies;

ISBN 0-12-192750-4

1. REQUESTS Member States:
    (i) to communicate immediately to WHO any decision to
    limit or prohibit the use of a food additive; and
    (ii) to supplement as soon as possible, such informa-
    tion with the data in support of the decision taken;
    and
2. REQUESTS the Director-General where such action would
    be useful:
    (i) to transmit immediately to Member States informa-
    tion received under paragraph (1);
    (ii) to take expeditious steps to evaluate any signi-
    ficant new evidence of toxicity of a specific food
    additive, including if necessary the convening of a
    meeting of experts, where appropriate in consultation
    with FAO;
    (iii) to distribute promptly to Member States any con-
    clusions of such a meeting.

Following the requests stipulated in the Resolution, WHO
transmitted the U.S. decision to suspend the use of cyclamates
as food additives to all other Member States and placed cyclamates on the agenda of the 1970 Expert Committee on Food Additives for evaluation. Before I describe the results of the
evaluation, I would like to mention briefly the Joint FAO/WHO
Expert Committee on Food Additives and the Joint FAO/WHO Meeting on Pesticide Residues, which are responsible for the evaluation.

1. *The Expert Committees on Food Additives and Pesticide Residues*

At an international level, WHO has in fact, been active
in assessing health hazards of food additives, contaminants
and pesticides for some 15-20 years. As early as 1953, the
World Health Assembly expressed concern about the potential
health hazards that might be associated with the increasing
use of various chemical substances in food. Since the topic
related to food as well as to health, WHO cosponsored a government conference on food additives in 1955, along with the
Food and Agriculture Organization of the United Nations (FAO).
Following a recommendation of this conference, a series of
annual meetings of the Joint FAO/WHO Expert Committee on Food
Additives was convened; the first was held in 1956. Similar
meetings have been held on pesticide residues since 1961.
    The Joint FAO/WHO Expert Committee on Food Additives at
its first meeting set out a number of general principles
concerning the use of food additives. It stressed that "any

decision to use an intentional additive must be based on the considered judgment of properly qualified scientists that the intake of the additive will be substantially below any level which could be harmful to consumers." This committee at its second meeting in 1957 formulated general recommendations with regard to testing procedures for food additives in order to establish their safety for use.

This committee has provided toxicological evaluation on specific food additives since its meeting in 1961, and up to the present some 400 additives have been considered. The biological data are generally summarized under the headings of biochemical aspects, acute toxicity, short- and long-term toxicity studies, special studies, and observations in man. Where appropriate, an estimate of the acceptable daily intake (ADI) for man was made. The ADI of a chemical is the daily intake, that, during an entire lifetime, appears to be without appreciable risk on the basis of all known facts at the time of evaluation. It is expressed in milligrams of the chemical per kilogram of body weight.

Joint meetings of the FAO Working Party of Experts on Pesticide Residues have been held annually since 1966. The reports of these meetings contain information on the ADIs established, the tolerances recommended, and the methods of analysis suggested for the 150 pesticides so far considered.

The toxicological evaluations follow the general principles laid down for food additives. For the recommendation of tolerances, the following types of information are required: use patterns, the residues resulting from supervised trials, the fate of residues during the various stages from the application of pesticides to a crop to the consumption of the food, methods of analysis, and national tolerances. However, more importantly, the recommendation of tolerances depends on the toxicological evaluation of the pesticide and its breakdown products; tolerances are recommended only for pesticides for which ADIs have been established.

The results of the evaluation and summaries of the data reviewed by the Expert Committees are published in their reports and monographs, respectively. These publications are distributed to all Member States of WHO. The regulatory agencies of many Governments take these publications into consideration in establishing regulations and standards on food additives and pesticide residues. The decisions and recommendations of the Expert Committees have, in addition, an indirect impact on Member States, through the Codex Alimentarius Commission. In order to show that the work of the Expert Committees is not merely an academic exercise, I shall briefly describe the Codex Alimentarius Commission and its relationship with the Expert Committees.

2. Codex Alimentarius Commission

The Commission is the executive body of the Joint FAO/WHO
Food Standards Programme. It was established in 1963 and has
115 Member Governments. In order to elaborate food standards
in a vast and diverse field, 25 committees have been established or incorporated into the framework of the Commission.
Most of these committees are under the chairmanship of a host
country, which is responsible for the administrative work and
the financial support. However, both FAO and WHO provide
certain financial, administrative, and scientific inputs to
the Commission and its committees.

Fifteen of the committees deal with the following groups
of commodities:

cocoa and chocolate products	processed meat products
sugars	natural mineral waters
fats and oils	soups and broths
processed fruits and vegetables	milk and milk products
dietetic foods	fruit juice
fish and fishery products	quick frozen foods
meat	edible ices
meat hygiene	

There are six committees that deal with general subjects,
such as food additives and pesticide residues.

The procedure for the elaboration of standards is complex
and lengthy but it provides ample opportunity for Member
States and interested international organizations to make
comments and suggestions and for the responsible committee to
amend the draft standards. Codex standards contain, in
general, the following sections: name of the standard, scope,
description, essential composition and quality factors, *food
additives, contaminants,* hygiene, weights and measures, labeling, and methods of analysis and sampling. A large number of
standards have been adopted by the Commission and sent by
FAO/WHO to their Member States. These standards have been
accepted by many governments in part or *in toto.*

As a rule, the Commodity Committee in drawing up a draft
standard for a food commodity proposes provisions for the
technologically unavoidable contaminants. These provisions
are referred to the Codex Committee on Food Additives for en-
dorsement. The Committee either fully endorses, temporarily
endorses, or does not endorse them, on the basis of the toxicological evaluation by the Expert Committee on Food Additives
described above, and on the basis of the potential daily intakes of food additives. For the latter, WHO has been asked
to provide, with a computerized programme, an estimate of the

potential per capita intake of each food additive. Thus the
Codex Committee on Food Additives, before taking a decision
on a food additive provision, considers not only whether a
food additive is acceptable or not, but also whether the
quantity of a food additive that is likely to be consumed by
an individual is within the acceptable intake or not.

Tolerances for pesticide residues are not dealt with by
the Codex Commodity Committees, but by the Codex Committee on
Pesticide Residues. In recommending international tolerances
for pesticide residues in various food commodities, this
Codex Committee relies on the tolerances proposed by the
Joint Meeting of the FAO Working Party on Pesticide Residues
and the WHO Expert Committee on Pesticide Residues mentioned
above. It may modify the proposed tolerances to accommodate
the varied agricultural need for pesticides in different regions of the world, but in doing so it always emphasizes the
protection of the health of the consumer.

The close interrelationships between the scientific
community, industry, regulatory agencies, the WHO Expert
Committee on Food Additives, the Codex Committee on Food Additives, the Codex Alimentarius Commission, and the consumers
are shown in the flowchart. Similar relationships exist for
pesticide residues.

3. Principles Governing Toxicological Evaluation

The testing procedures and general principles to be used
in interpreting the results in terms of safety for use have
been described in the Second Report of the Expert Committee
on Food Additives in 1957. These involved the selection of
relevant biochemical, toxicological, and related data, the
determination of a "no-effect" level, and the application of
an appropriate safety factor. These principles have been
adopted at subsequent meetings of this Expert Committee and
by the Expert Committee on Pesticide Residues. However, some
modifications have been made by these Committees and two WHO
Scientific Groups.

On the subject of carcinogenesis, which is of direct
interest to this particular meeting, the Expert Committee on
Food Additives stated that it believes "that no *proved* carcinogen should be considered suitable for use as a food additive in any amount." This basic principle has not been altered throughout these years. The interpretation of the term
of "proved carcinogen" however, has been further elaborated
in light of new knowledge.

The Fifth Meeting of the Experts Committee on Food Additives in 1960 was given the task of specifically considering the problem of the evaluation of the carcinogenic hazards of food additives. It discussed the scientific background and made recommendations on the testing procedures and the interpretation of the results. The report has proved, in general, to be useful to subsequent meetings of the Expert Committee on Food Additives and those that deal with pesticide residues.

Two of its recommendations, however, were not followed at later meetings of the Expert Committees on Food Additives and Pesticide Residues. First, while recommending that further research be done to clarify the significance of the occurrence of local sarcomas at the site of injection, the 1960 Committee did include, in the recommended procedures for testing food additives, subcutaneous injection, in addition to the oral route, as a means of administering a chemical to test animals. Second, in discussing chemicals, such as aminotriazole, the carcinogenesis of which has been shown to involve a hormonal mechanism the Committee considered that no practical purpose would be gained by classifying such chemicals as "indirect carcinogens."

4. "Carcinogenic" Food Additives and Pesticides

a. *Food Colors*. The Expert Committee on Food Additives at its eighth meeting in 1964 evaluated a large number of food colors. It reached the conclusion that certain colors such as Auramine and Magenta are carcinogenic and should not be used as food additives. The carcinogenicity of Auramine was demonstrated in experimental animals and in occupational workers, and the hazards of Magenta were based largely on findings on occupational workers. In addition, the Committee noted that, with Brilliant Blue FCF and Fast Green FCF, local sarcomas developed at the site of subcutaneous injections in rats. Although several long-term feeding studies did not reveal any sign of carcinogenicity, the Committee did not allocate an ADI for either color additive.

In 1966, the WHO Scientific Group on Procedures for Investigating Intentional and Unintentional Food Additives, in discussing the significance of local sarcoma, recommended:

(a) that for the *routine* testing of food additives and contaminants, the subcutaneous injection test should be considered inappropriate...
(b) that the occurrence of local sarcoma following subcutaneous injection of food additives and contaminants should not, alone, be considered significant evidence of a carcinogenic hazard; ...

Taking this advice, the Expert Committee on Food Additives at its thirteenth meeting in 1969 allocated ADIs for man to Brilliant Blue FCF and Fast Green FCF.

The most controversial of color additives has been Amaranth (FD & C Red No. 2). On the basis of a number of well-conducted and comprehensive investigations, the Expert Committee at its eighth meeting in 1964 allocated an ADI of 0-1.5 mg/kg body weight for man. Later, the results of two Russian studies indicated carcinogenicity in rats. In one of these studies, a preparation with a high percentage of impurities was used; in the other, not a single tumor was reported in the control rats, which had been kept for their entire life-span. The Committee therefore considered these results difficult to interpret, especially in light of the earlier studies revealing no carcinogenicity. It also took note of the reproduction and teratogenicity studies done in the Soviet Union, showing some adverse effects, as well as the fact that several other studies were still in progress in the U.S. The Committee therefore allocated a *temporary* ADI to Amaranth.

In early 1976, FDA decided to delete this color additive from its provisional list on the basis of results obtained in its own laboratories indicating a higher incidence of tumors in the female rats fed Amaranth. FDA communicated this decision to WHO and WHO in turn disseminated it to all Member States. As far as is known, most governments have not followed the FDA action, including the Canadian and those of the nine countries of the European Economic Community. As far as I know, there has been some doubt about the conclusiveness of the FDA data.

b. Artificial Sweeteners. P-Phenethylcarbamide (Dulcin) was evaluated by the Expert Committee in 1967. It noted that over 75% of the rats fed 1% Dulcin developed tumors of the urinary tract. The Committee therefore recommended that this chemical should not be used as a food additive.

Cyclamates were widely used in foods and beverages for many years. However, doubt was cast on their safety because of the discovery that they are metabolized by the intestinal flora in animals and man to cyclohexylamine, which is more toxic. The Expert Committee on Food Additives, at its eleventh meeting in 1967, therefore allocated only a temporary ADI of 0-50 mg/kg body weight, and requested additional information.

As mentioned earlier, urinary bladder tumors were observed in 1969 in some experimental rats fed high levels of cyclamate along with saccharin and cyclohexylamine. FDA then suspended the use of cyclamates as food additives. Following a provision in the WHO Resolution, the Secretariat placed cyclamates on the agenda of the 1970 meeting of the Expert Committee for reevaluation. The Committee withdrew the temporary ADI on the basis of the tumor formation and certain other adverse effects of the metabolite cyclohexylamine.

In 1974, the Expert Committee reviewed several extensive long-term studies in a number of strains of rats and mice, and a study in hamsters. All these studies yielded negative results on carcinogenicity. The Committee concluded cyclamates to be *noncarcinogenic* in a variety of species and regarded the original study reporting the occurrence of bladder cancer in rats as not readily interpretable because of the many complicating factors. However, no ADI was allocated because certain aspects of the cyclohexylamine problem were not resolved.

In 1977, the Committee reevaluated cyclamates in light of the new data on cyclohexylamine and allocated a temporary ADI.

Saccharin and its salts have been used widely over a longer period of time. Based on the available information, the Expert Committee, at its eleventh meeting in 1967, allocated an unconditional ADI of 0–5 mg/kg to saccharin and its salts, and a conditional ADI of 5–15 mg/kg to cover special uses in dietic foods.

Early in 1972, FDA notified WHO that it had removed saccharin from its GRAS ("generally recognized as safe") list of food additives and had issued a provisional regulation regarding the use of saccharin in foods and beverages. The U.S. decision was based on the observation of urinary bladder tumors in some test animals.

At the eighteenth meeting in 1974, the Expert Committee reviewed the available information and concluded that the urinary tumors observed in some treated animals might be secondary to a change in urinary pH, resulting in calculus formation, brought about by saccharin itself or by an impurity, o-toluenesulfonamide. Pending the completion of ongoing investigations, the Committee did not modify its earlier evaluation.

In March 1977, the U.S. authorities proposed suspension of the use of saccharin as an additive in foods and beverages. The decision was based on the observation by Canadian scientists of urinary bladder tumors in rats fed 5% "pure" saccharin (without detectable impurities) in the diet. The Canadian study also showed that o-toluenesulfonamide was not carcinogenic.

The Expert Committee reviewed the new information on saccharin and considered that there were many questions regarding the carcinogenicity data that remain unanswered. The Committee thus withdrew the unconditional and conditional ADIs allocated at a previous meeting and allocated only a temporary ADI.

c. Preservatives.

Nitrites and nitrates. The Expert Committee at its sixth meeting in 1961, allocated unconditional and conditional ADIs to nitrates and nitrites. These evaluations were confirmed at its eighth meeting in 1964.

Since then, many nitrosamines, which can be formed from reactions between nitrites and certain amines, have been found to be potent carcinogens. The Committee, at its seventeenth meeting in 1973, therefore allocated to nitrites only a temporary ADI, which was later confirmed at its meeting in 1976. In making this evaluation, the Committee pointed out that on the one hand, nitrosamines have been detected in some meat and fish products to which nitrites and/or nitrates have been added, and on the other hand, these preservatives are valuable in controlling toxin-forming microorganisms such as *Clostridium botulinum.* Furthermore, nitrites occur in the body, notably in the saliva, and it has been demonstrated that nitrosation of certain amines can occur in the stomach.

Diethylpyrocarbonate (DEPC) was evaluated by the Expert Committee in 1965 and an acceptable level of treatment was allocated to it. In 1973, its use as a preservative in beverages was revoked by the Committee because there were data indicating the formation of urethan in treated beverages. Urethan is a wide-spectrum carcinogen, producing tumors in many organs and in all species tested.

d. Organochlorine Pesticides.

Several organochlorine pesticides have been reported to increase the incidence of nodules in the livers of mice. It is not clear what proportion of the nodules are neoplasms but there is unequivocal evidence, in the case of some of the compounds, that they increase the incidence of neoplasms that metastasize to the lungs.

In discussing the significance of the development of liver-cell tumors in mice, the 1973 Joint FAO/WHO Meeting on Pesticide Residues made the following observations:

The Meeting agreed that there is a serious lack of knowledge regarding the processes involved in the development
of liver tumors by mice and that it would be unwise to
classify a substance as a carcinogen solely on the basis
of evidence of an increased incidence of tumors of a kind
that may occur spontaneously with such a high frequency.

Commenting specifically on DDT, the 1974 Joint FAO/WHO
Meeting on Pesticide Residues noted:

The attention of toxicologists has been focused during
recent years on the tumorigenic action of DDT on mouse
liver, and this question has been under constant scrutiny
by the Joint Meetings since 1967. Data were presented
(L. Tomatis & V. Turusov, unpublished observations, 1974)
before the present Meeting that demonstrated that mice
ingesting 36 mg of DDT per kg of body weight in their
feed for 15 weeks developed hepatomas that did not re-
gress. However, no tumors have thus been produced in any
other species tested, e.g., rats and hamsters. Furthermore the limited epidemiological data available give no
indication that DDT might be a human carcinogen. A
number of people have had intermittent heavy exposure to
DDT over a period of some 30 years, which should be
sufficient time to produce and observe any increased
tumor incidence that might have occurred.

It should be noted that the Joint Meetings have established ADIs and recommended tolerances for DDT, aldrin, and a
number of other organochlorine pesticides.

e. Amitrole. The 1973 WHO Scientific Group on the
Assessment of the Carcinogenicity, and Mutagenicity of
chemicals concluded that "there are certain instances of
cancer induction that may be secondary to an initial noncarcinogenic effect of a chemical," and that "a no-effect
level for chemicals that produce tumors in this way may be
established."
The 1974 Joint FAO/WHO Meeting on Pesticide Residues
agreed with the 1973 WHO Scientific Group on the above-
mentioned conclusions, and made use of those conclusions in
evaluating amitrole. The Joint Meeting noted that (a) thyroid tumors have been produced experimentally with amitrole,
(b) there is evidence that the antithyroid action of amitrole probably accounts for the thyroid tumors, (c) a level
for antithyroid effects, as measured by a sensitive biochemical parameter has been demonstrated, and (d) there was

no effect on the liver even at higher feeding levels. The Joint Meeting therefore allocated a conditional ADI with the proviso that the uses of amitrole be restricted to those where food residues would be unlikely to occur. It also recommended tolerances for amitrole at 0.02 ppm in various raw agricultural commodities of plant origin.

5. Summary

(a). At an international level, WHO, in conjunction with FAO, convened annual meetings of Expert Committees for some 20 years. These Committees provided toxicological evaluation on about 400 food additives and 150 pesticides. These chemicals were evaluated at the request of Member States. Many national regulatory agencies have made use of the evaluation in formulating their food standards with respect to these chemicals, either directly or indirectly. In the latter case, Member States participate in the work of the Codex Alimentarius Commission in elaborating international food standards.

(b). The basic principle governing the assessment of chemicals shown to be carcinogenic has been that "no *proved* carcinogen should be considered suitable for use as a food additive in any amount." For example, Auramine, Magenta, Dulcin, and DEPC (from its reaction product urethan) were considered proved carcinogens and were thus considered unsuitable for use as food additives. Thus the principle of not adding carcinogenic chemicals to food adopted by the Expert Committees is the same as that which is embodied in the Delaney Clause.

(c). However, not all chemicals reported to increase tumor incidence were considered by the Expert Committees as "proved carcinogens." In fact, acceptable daily intakes for man have been allocated to a number of such chemicals. These include certain food colors that produced local sarcomas only at the site of subcutaneous injection, certain organochlorine pesticides that increased the incidence of hepatomas in mice only, and nitrites that have the potential of producing nitrosamines but whose actual hazard to man is as yet uncertain and suspension of their use may cause other more imminent health hazards. These and other examples are presented to indicate the different bases for the Expert Committees' decisions.

REMARKS

DR. COULSTON: I would like to affirm what Dr. Lu was saying.
I thought it is worth taking a minute to explain this. In
the United States, particularly of all nations in the world,
we take the view that our regulations and our regulatory
positions are often the ultimate positions. However, the
rest of the world does not think so. The rest of the world
relies heavily on decisions made by the United Nations
Assembly, namely, through their bodies: the World Health
Organization, the Food and Agricultural Organization, and
the International Atomic Energy Agency. These are all parts
of the United Nations. It is ironical that in this room are
many of the experts that Dr. Lu was referring to: Dr. Kolbye,
Dr. Smeets, Dr. Korte, Dr. Clayson, Dr. Darby, Dr. Parke,
Dr. Lu himself, Dr. Clegg, Dr. Mrak, and others who have been
on some of the expert committees of the United Nations.
Even myself! We have been chairmen, rapporteurs, and members
of these expert panels. It is not really recognized what a
great honor it is to be called from a whole nation, as an
expert. On these panels, usually only one representative of
a nation is selected. They represent the best talent there
is in the world. They easily match or exceed any talent of
a National Academy of Science committee, because they do represent the opinions of scientists around the world who must
deal, as Dr. Wynder has pointed out many times, with
different populations, different diets, different habits, and
so on. I would not mind so much that the U.S. regulatory
officials pay little attention to the results of the WHO, but
I would like to remind you that WHO-FAO made a statement on
cancer in food or cancer-producing chemicals in food at least
three years before the Delaney Clause was passed by Congress.
But at the same time WHO-FAO made statements based on these
expert panels, regarding dose-response and no-effect levels,
which Dr. Wynder now has said is one of the key points coming
out of this meeting. The concept that we can have a dose-
response and no-effect level to all chemicals, whether carcinogen, teratogen, or mutagen, has been discussed over the
years by these groups of experts.
     Unfortunately, the differences between national groups
lead to situations like this: Red 2 banned in the United
States, used in Canada; Red 40 used in the United States,
not used in Canada; cyclamate banned in the United States,
perhaps for good reasons (I'm not arguing one way or the
other, but in spite of the fact that the rat data came from
Canada), used in Canada; we shall not allow NTA to be used as
a phosphate substitute, Canada has used it for the last 7

years with no problems. I could go on and on. I just picked
Canada because it is so close to us that you cannot even
differentiate the border between the countries. Why does
Canada behave differently in their interpretation of data
than we do? I can tell you very simply what the difference
is. They follow in general the recommendations of these
world health bodies.

Now the point is this. In the *Washington Post* just two
days ago is a letter to the editor by the chairman of the
Council of Environmental Quality of the United States. In
this letter he makes a parsimonious statement that AID should
only give help to those countries that follow our regulatory
decisions. What are we saying when we say something like
this? If in India they want to use DDT and we do not want
to use DDT, which is our decision to make, are we to deprive
India of DDT? Who are we to say that a country should have
certain chemicals mentioned in the article, only because they
have complied with the regulations in this country, which
have completely ignored the regulations of the World Health
Organization and the Food and Agricultural Organization?

Mr. Chairman, I would like to say for the record that I
think it is time that our regulatory officials at the highest
level began to look very carefully at the decisions made by
WHO and FAO.

DR. KOLBYE: Having had my wetting down experience in environmental toxicology with Dr. Mrak's "Commission on Pesticides"
and having gone through many drafts of recommendations, I
can only think back to one of the recommendations, I believe
it was No. 11. It talked about the differences in value
judgment making about what is essential under certain circumstances. We were trying to tell the rest of the world not to
do as we do, but think for yourselves and make you own decisions as responsibly as you can.

Unfortunately that sometimes is a prayer that is not
necessarily paid any real service to. I can only re-echo
the comments of Dr. Coulston that he has made so well.

THE ESTABLISHMENT OF EUROPEAN COMMUNITY
LEGISLATION FOR TOXIC SUBSTANCES

J. Smeets and W. Hunter

1.   GENERAL CONSIDERATIONS

This paper deals with the actions and the establishment
of legislation by the European Community with respect to
toxic substances in general. It does not attempt to deal
with the philosophical implications that are behind political
decisions regarding these substances. However, it may be of
interest to present the experience that has been gained with
our work in the Commission of the European Communities in very
close collaboration with the nine Member States.

There is a relationship between the objective scientific
evaluation of toxicological data and the implementation of
legislation but the procedures for each are necessarily
different.

An objective evaluation, whether based on acute toxicity
or on long-term effects, is a scientific exercise and leads
to a risk-benefit analysis. Implementation of legislation
depends on the policies prevailing at the time and the will
of those responsible to take *political* decisions. It is clear
that a maximum of cooperation between the scientist and the
politician is indispensable to maximize the protection of
human health and environment. The meeting place between the
two is the risk-benefit analysis.

The recent recognition of the harmful effects of chemicals
on man and his environment has required legislators to take
adequate measures to protect man and his environment.

In this context the Commission of the European Communities
has been supplied with information indicating that more than
9000 synthetic compounds are on the market in quantities of
more than 500 kg and about 150 compounds in quantities of more
than 50,000 tons yearly.

329

ISBN 0-12-192750-4

The uncontrolled release of these compounds into the environment could have deleterious effects, and therefore it is not surprising to note a general trend to develop preventive or control measures, in particular for new chemical substances.

A proposal of this type in the form of a directive is now under discussion by the Council of Ministers of the European Communities. The U.S. Toxic Substances Control Act is another example and similar types of legislation are also in force or in preparation in other countries.

All these legislative measures are based on the same principles and require that the responsible authorities are notified of physicochemical, toxicological, and ecotoxicological data and of information concerning the prospective use, and the quantities likely to be produced. These requirements are based on a minimum (base) set of tests. Depending on the appraisal of this information by the responsible authorities, supplementary information may also be required. Such a notification system implies that an evaluation of the data is required to establish whether a product can be used safely or not.

2.   ESTABLISHMENT OF COMMUNITY LEGISLATION

In a discussion on the decision-making process of the European Community it is appropriate to consider the procedural aspects of European Community legislation. This supranational and intergovernmental organization of nine Member States has been given a mandate of *legislative* power by the formative treaties. The executive, which is the Commission of the European Communities, has the power to take initiatives in the regulatory field. This means that appropriate positions can be submitted to the legislative authority, which is the Council of Ministers of the Member States. Also involved in this decision-making process are the European Parliament, comprised of politicians of the nine Member States, and the Economic and Social Committee.

Each of the nine Member States can and does prepare its own legislation for the solution of specific problems. However, when a regulatory decision is agreed on at community level, then each Member State is required to introduce this into its own legislation. Thus the legislation of all nine States is brought together under one specific community legislative denominator.

Because of the differences in national legislation it is not always easy to reach common agreement between the nine Member States. Final decisions are often based on compromises and reconciliatory considerations. Due to the fact that nine individual standpoints have to be harmonized, the achievement is usually more difficult than the achievement of national legislation. This will obviously be the case when trade barriers are being considered that have economic implications such as free intracommunity trade in foodstuffs, the use of food additives, packaging materials, or other goods.

The Commission has fully recognized the need to separate economic considerations from the scientific data used for the preparation of legislation. With regard to the evaluation of scientific data a Scientific Committee was created in 1974 to advise the Commission on problems concerning foodstuffs: their composition, their treatment, their packaging, the use of additives, etc. This advisory committee, comprised of eminent European toxicologists, nutritionists, biologists, chemists, etc., has proved its usefulness by the value of its scientific advice, which has been used by the commission in putting forward statutory instruments in this field. Other ad hoc advisory committees exist in other fields and also are of significant assistance to the commission. In addition to this, the commission carries out studies on selected topics with the assistance of highly qualified experts. The results of these studies are submitted for approval to meetings of scientific consultants or national experts. Such studies can also evaluate toxicological data and other appropriate data in this field. As a result of this work the commission can then propose to the Council of Ministers suitable legislation.

We have to recognize that very often there are problems in objectively evaluating the data particularly in relation to the effects of man's exposure.

An excellent review of the problems involved is given in the Proceedings of an International Colloquium, on "The Evaluation of Toxicological Data for the Protection of Public Health." This was organized by the Commission of the European Communities in Luxembourg in 1976 in collaboration with the International Academy of Environmental Safety. This colloquium is also an example of the way the Commission collaborates with other international bodies in this field such as Codex Alimentarius, WHO, FAO, IARC.

3. EXAMPLES OF SPECIFIC ACTIONS

It is not possible to present a complete view of all the actions undertaken by the commission in the field of establishing legislation with respect to toxic substances. However, the following are some typical examples of specific actions, which demonstrate procedures followed by the commission with respect to the protection of man. Ecotoxicological aspects will not be considered, as they are out of the context of this conference.

a. Asbestos

The environmental action program of the European Community contains a list of pollutants for priority investigation, some of which are considered carcinogenic for man such as the hydrocarbons with known or probable carcinogenic action, certain organohalogen compounds, asbestos, cadmium, chromium, arsenic, and vinyl chloride. For each pollutant the commission has to carry out a bibliographic survey concerning the effects of these pollutants on man and his environment; a critical analysis of the available data should be made, and if possible, criteria (exposure/effect relationships) should be established.

A number of these studies have been completed and published. Others are in course. One study concerns "The Public Health Risks of Exposure to Asbestos," and based on this work the commission has prepared a program of action on asbestos. A number of these actions concern the carcinogenic activity of asbestos, in particular in relation to mesotheliomas.

The identification of mesotheliomas depends upon pathological investigation backed up by clinical and radiological evidence. The majority of mesotheliomas can be identified beyond doubt by an experienced pathologist. However, there are some cases in which difficulties in confirming the diagnosis persist, and for these it is best to have the benefit of the opinions of other experts, e.g., by international exchange of knowledge and material. The commission has therefore established a Mesothelioma Panel of the European Communities, which is the first known panel of this type, to meet regularly to assess histologically these tumors.

This panel has as one of its basic objectives the standardization within the nine Member States of the pathological diagnosis of mesothelioma. In addition it seeks to provide data that can be used to compile a Mesothelioma Register of the European Communities. A European cancer register is a long-term aim but a European mesothelioma register could be set up fairly quickly as an independent project, i.e., not as an integral part of a cancer register. The objects in forming ing a Mesothelioma Register are:

(1) to record the annual number of deaths from confirmed mesotheliomas of the pleura and peritoneum,
(2) to ascertain trends in the incidence rates,
(3) to discover the groups of the population apparently associated with the tumors and to estimate the incidence in the occupationally and nonoccupationally exposed population, as well as the occurrence among those apparently nonexposed,
(4) to establish the exposure/response relationships.

b. Single Administration Toxicity

In the general area of toxicity the commission has formed an ad hoc toxicological committee. In the first meeting special emphasis was placed on the problems involved in single administration toxicity, and in particular the use of the LD50 value of a given compound as a primary toxicological characteristic for the purposes of classification and labeling.

The experts proposed the following definition for acute toxicity of a chemical: "The total adverse effects it produces in mammals when administered as a single dose." It was recognized that this definition differed slightly from others, which refer to all substances. A substance of low solubility that is tested by administering a suspension and is of such low toxicity that the LD50 dose can be reached only by administering it in divided doses is considered as less relevant to the environmental field.

Since adverse effects may not occur immediately, but may be demonstrated some days or weeks later, it is necessary to observe experimental animals for at least two weeks and longer if necessary after the administration of substance. This seems to be a somewhat unnecessary statement of the obvious, but a recent inquiry has shown that some laboratories observe their animals for a period of time shorter than two weeks.

The same inquiry has demonstrated that acute toxicity is sometimes considered only as the determination of the LD50 value, whereas the term "acute toxicity" encompasses all ill effects caused by a substance and not only the most obvious ultimate effect, i.e., death. While a high dose of a substance may kill an animal, lower doses may damage particular organs or systems. The aim of any study of acute toxicity, the re-

sults of which are to be used to assess the probable effects
on man, should be to determine the dose/effect relationship,
including the lethal dose for man, and to establish whether
such nonlethal effects are reversible or not. In relation to
the experience gained with the above mentioned inquiry the
experts felt the need to emphasize:

(1)   Each LD50 figure applies only to the particular
species tested uner the defined conditions of the experiment.
(2)   A high degree of variation is to be expected between
different determinations carried out in the same laboratory
by the same experimentalist using consistent conditions.
This degree of variation becomes wider as more variables are
introduced, e.g., determinations done in different laboratories by different workers, using different concentrations and
dosage levels, etc.
(3)   Because of these many variables, the figures for
LD50 should not be interpreted as accurate indications of the
possible lethal dose in man; at best they give an idea of the
order of magnitude of the lethal dose. It is helpful in the
interpretation of LD50 data to have available information on
the slope of the dose/effect curve, for example, LD20 and LD80
data figures could be included.
(4)   Determination of the adverse effects at lower levels
in animals is as important and in many (if not in most) cases,
more important than the LD50 determination in assessing the
likely effects of the exposure of man to the substance.

As a start for a quality assurance program for single administration toxicity the commission decided to initiate an oral
LD50 intercomparison study on rats at the European Community
level in order to provide:

(1) an overview of the degree of the variability of oral
LD50 values,
(2) an insight into the experimentation techniques used
by the different European laboratories to determine LD50
values,
(3) a survey of the variability in the determination of
LD50 values,
(4) a basis for establishing a protocol for the future
determination for LD50 values.

In an attempt to achieve the above goals agreement was reached
on the following:

(1)   Oral LD50 values for rats were to be determined by the laboratories for five compounds unknown to them. The solubility of these was broken down as follows: two compounds were water-soluble, two were fat-soluble, one was nonsoluble. The specific solubility of the various substances was not to be revealed to the laboratories.

(2)   Each laboratory used its own normal procedure including protocols.

(3)   A draft questionnaire accompanied each compound.

In the study, 65 European laboratories participated. Since this first study was carried out by the laboratories using their own protocol and since all the parameters that were studied showed considerable variability (up to a factor of 12 for one compound), there was a clear need to repeat the study with the aid of an *agreed upon* protocol. This second step is underway using such a protocol, and within this stage, three particular parameters will be examined in detail by a small number of laboratories to establish their influence on the results. These parameters are: (1) suspension/solutions, (2) body weight, (3) number of animals per group. In addition to these parameters a limited study will also be carried out on the histological findings.

c. Research

In the area of research, strong emphasis has been given by the commission to assess the genetic effects of environmental chemicals. The two principal aims of this program are:

(1) a better understanding of the basic *mechanisms* responsible for the mutagenic, carcinogenic, and teratogenic properties of environmental pollutants, and

(2) the elaboration, improvement, and validation of test systems for the assessment of their properties and, in particular, the elaboration of "prescreening tests."

A working group has been set up and has launched a European
program for *comparative genetic testing*. As a first step the
participants selected three known mutagens* to determine the
validity of the operation system. Preliminary results are now
being assessed. Subsequently other known mutagens** will also
be tested in these systems to confirm the results of the first
exercise.

The philosophy of the program is to develop and standard-
ize a number of speedy systems. The results of using a number
of well-selected compounds will be compared with usually
costly and time-consuming results in mammals. There are good
reasons to believe that, within a few years, a battery of
comparatively cheap and speedy tests will be available, which
will permit us to predict the mutagenic and, hopefully, carcinogenic properties of chemicals in man.

d. *Other Actions*

A study on carcinogens is under way in collaboration with
the International Agency Research on Cancer and is concerned
with drawing up a list of likely carcinogens for man. It is
based on the work already performed by the Agency in the
assessment of the carcinogenic activity of chemicals.

A proposed action program of the European Communities on
health and safety at work has recently been submitted by the
commission to the Council of Ministers. This program accepts
that a high proportion of cancer in man is caused by external
factors, including chemicals at the workplace. Therefore the
commission will take priority action against carcinogens. It
is envisaged that the commission will:

(1) collect data on the distribution of carcinogens and
their concentration at the workplace,
(2) collect and analyze medical data,
(3) perfect readily applicable detection tests,
(4) fix the lowest possible levels or, if necessary,
prohibit a certain number of carcinogens present at the
workplace.

---

*Methyl methanesulfonate, N-nitrosodiethylamine, Procar-
bazine (natulan)*
**Mitomycin, Atrazin, Benoncyl, TCDD, Thiourea, Dieldrin,
Chloroprene, Isophosphamide, Trophosphamide, Trypaflavin,
Folpit, Benzo [a] pyrene, Benzo [b] pyrene, DMBA.*

4. CONCLUSIONS

The actual development of legislation for toxic substances in the European Community is based on a full evaluation of the available biological data. Such data are collected for each substance being considered. Thus the commission carries out specific actions and reviews data on an individual basis. There is no general legislative rule in existence similar to the Delaney Clause; the commission is therefore able to adopt a flexible approach.

The following are examples of directives that have been developed by the commission with regard to carcinogens.

On 27 July 1976 the Council of Ministers approved a proposition of the Commission on the approximation in the Member States of the laws, regulations and administrative provisions on the marketing and use of certain dangerous substances and preparations* and more particularly concerning polychlorinated biphenyls (PCB) except mono- and dichlorinated biphenyls, polychlorinated terphenyls (PCT), and chloro -1-ethylene (vinyl chloride monomer). A proposal has been prepared to include asbestos in this list.

Two other directives concerning vinyl chloride monomer are being discussed by the Council of Ministers. One concerns the approximation of the laws of the Member States relating to materials and articles containing vinyl chloride monomer and intended to come into contact with foodstuffs**. The other relates to the approximation of Member States laws, regulations, and administrative provisions on the protection of the health of workers occupationally exposed to vinyl chloride monomer***.

---

*Official Journal No. C219 of 14,9,1977.
**Official Journal of 14,9,1977.
***Official Journal of 10,12,1977.

The Council of Ministers is currently discussing the sixth modification on the approximation of the laws of the Member States relating to the classification, packaging, and labeling of dangerous substances. This modification particularly relates to the notification of *new* chemical substances. It is anticipated, as has been noted in the foregoing, that such notification will include an evaluation of mutagenic activity and if appropriate, carcinogenic properties. In due course a proposal will be made for a further modification of this directive concerned particularly with the labeling of carcinogenic preparations and substances.

REMARKS

DR. SCHMIDT-BLEEK: Let me first say Dr. Smeets, I like very
much the invitation to follow you and say publicly here
that our connections and our relations are indeed very pro-
ductive and very enjoyable with Brussels. We find ourselves
in many meetings together. I would like to tell you very
briefly about the principal features of what you may call,
translated, the German Environmental Chemicals Law. We put
the first drafts together recently and they were given over
to the other departments within the federal government,
which are dealing with toxic substances, to be discussed.
I should also mention that this draft bill, of course, reflects the ideas that we have developed jointly with the
Commission and the Council in Brussels over the last two
years.

It may be of interest to you for me to point out a few
basic differences that we feel there are between our approach
and the Toxic Substance Control Act of the U.S. Before doing
that let me remind you of a few things that have been said
a number of times, but I still feel they are important to
put the efforts that we are dealing with here in the proper
perspective. As you are all aware there are, as the EPA
tells us, now close to 70,000 chemicals to be considered,
these days. Dr. Smeets has mentioned some tonnages; many
numbers could be added. What is of very great importance to
us is the fact that in Europe, alone, some 2000 new chemicals
are coming on the market every year. The number for the U.S.
we are told is roughly 1000. This, of course, depends on the
definition for what is a new chemical, details of which we
should not discuss here I suppose. Further is the fact, of
which I think we should remind ourselves, of the volume of
trade: the U.S. total production is slightly above $1 billion;
in West Germany alone, the production of chemicals is slightly
higher than $1 billion DM. So you can see that the numbers
involved are rather large and I should further add that growth
of trade approaches $1 billion a year. I make this point very
prominently, because there is no problem that we have on a
regulatory basis, a consideration of the national scene. It
is absolutely impossible, and it does not even make much
difference whether the EEC is trying as we do to come out
with harmonized legislation. We have to do a lot better than
just harmonize within nine countries, as important as they
may be in terms of trading in chemicals and producing chemicals and chemical use.

I would like to quickly add something quite trivial, but I would like to remind you of it because it reflects the breadth of our problem. Carcinogenicity and mutagenicity are, of course, but two properties that we have to consider prominently, when we ask the question, What do we have to know and how do we evaluate this knowledge to try to evaluate the risk, the environmental risk, or potential risk that we are facing from any particular chemical? Indeed, our total questionnaire at this time encompasses something like 85 different properties, which we think we would have to know something about, starting from very simple things like boiling point and pH stability and going all the way up to chronic toxicity.

Now very briefly about the legislation: the basic and I think most important, difference between the European approach and the U.S. approach is that we are dealing only with new chemicals, new chemicals being defined by a point in time, the point in time being the passage of the EEC directive, which by the way puts these into effect in 1978. As you know, the U.S. is dealing in its approach both with the old chemicals as well as new chemicals, according to article 4 in the Toxic Substances Control Act.

However, after I tell you how we approach the problem of evaluating the total impact of such a new chemical, some further differences between our approach and the U.S. may become apparent. For instance, we shall require for each new chemical to come on the market that there be notification at the time of first putting it on the market. At the same time, the dose use of the investigation of this chemical has to be handed over to the proper authorities in the nine Member States. The information is then translated over to Brussels and can be evaluated on a community basis. Now without going into any details, let me quickly tell you that the required information comprises some 40 items. Some questions have to do with very simple properties, if you wish simple properties in that when you start harmonizing these properties or the protocols, they are no longer so simple, like boiling point, pH stability, and volatility. On that basis, of course, this is of more interest to you at least in this discussion here. The toxic properties to be investigated are the LD50 or LC50 that Dr. Smeets was talking about. It becomes apparent why we were so interested in the community two years ago to find out how reliable these values may be.

In addition to that we still have mutagenicity screening tests in the sense of the "Ames tests." It has not been decided at this time which set of experiments will be required. Furthermore, there is now very serious discussion whether subacute toxicity should be included in this base set of experiments to be performed before marketing may occur. As I say, this has not been discussed to the end in Brussels. They

have very strong feelings one way or the other, which are not
scientific in nature, I must stress. I am now going to say
something that I am sure sounds very strange to some of you.
It turns out that after 1½ years of discussion of this problem
at first highly scientific and technical subject, we are now
reduced to the question of how much in terms of dollars,
francs, pounds, marks, or whatever, may this base be set,
because the size of this figure is going to have a great deal
to do with the innovation and survival, if you wish, of the
growth in the future of the chemical industry in Europe. We
are approaching the figure of some $30,000-40,000 as an
economic base. You recognize immediately, since we have a
large number of questions to answer, that we must balance
very carefully which type of information we would like to have:
obviously not only health-related information, obviously not
only thorough chemistry. We must also have some specific
environmental information in this base set, for instance, persistence. Dr. Korte was telling you about the abiotic decomposition of chemicals. There is the Japanese sludge test, there
are many other things that can be done here, again, balancing
is very important.

But we have decided in Europe that we will not only have
this basic set and then free-floating discussions and dialog
and decisions, as we decide further what should be done, or
sequence testing, which has at this time in the preliminary
stages been agreed upon in Europe. That is, three more tiers
or levels of testing, and in each of these levels, there are
certain key words, which may be mutagenicity, or a further
short-plus-medium-term test for fertility or behavior. It is
very important that these levels not only be guided by the
outcome of tests that have been done before, but that these
be linked to specific marketing tonnages, because there is
not only a relationship of environmental danger to exposure
and to total marketing tonnage, but there is also a very strong
and very important relationship between what a company can
afford in terms of testing at any particular marketing volume.
There is a direct relationship and again of course we have to
be rather careful not to overtax industry. Dr. Coulston
refers to this as a socioeconomic relationship.

With these few things about the legislation that is now
being readied in Europe, let me very quickly turn to one more
point, which is based on the trade figures I have given you,
based on the simple necessity to think, as early as possible,
in terms of how we avoid by developing testing schemes in
many countries as is being done right now, trade barriers.
I shall give you an example that is not theoretical. Six
months ago an American importer of Swiss watches required
from a Swiss watchmaking company detailed information on the

composition of the alloys that went into making this watch, based on, as this importer intended it to be, the requirements of the Toxic Substances Control Act, and in fact he was entirely correct.

Now, one can interpret the Toxic Substances Control Act to this extent. To the best of our knowledge EPA has never done so and will not in the future. But you can see immediately what kinds of questions can appear and so it is of extreme importance that, not only the testing protocols be internationally harmonized through standardization, but that we also come to other agreements. One has been mentioned a number of times already, namely, laboratory quality assurance. It is for this reason that I just would like to inform you that there will be a very high-level legal meeting in Stockholm in April this year, where administrators, (for instance, Dr. Costello of EPA), from many other countries, politicians, and high-level decision-makers will get together to discuss the framework of what may eventually turn out to be an international chemical prevention.

One more word, if I may, and that is that the OECD in Paris, recognizing this harmonization need, decided in October 1977 to launch a rather large-scale harmonization program. Two weeks from now in Washington, we have the first one-week series of meetings on this subject. We hope that within the next two years, we can make reasonable progress in deciding on harmonized protocols, taking into consideration very obviously the excellent work done by the WHO, FAO, and others that have been accomplished in the past.

IMPACT OF THE DELANEY CLAUSE
ON THE RESEARCH-INTENSIVE MANUFACTURER
OF BIOMEDICAL PRODUCTS

Robert H. Furman

Before the 1938 Federal Food, Drug and Cosmetic Act was
enacted, food containing any *added* poisonous or deleterious
ingredient that might render it injurious to health was considered *adulterated* and could be excluded from interstate
commerce Food and Drugs Act, Chapter 3915, Sections 2, 7,
34 Stat. 768 (1906) . The burden of proof that the food was
adulterated was on the government.

The Act established for the first time procedures that
made possible the addition to the food supply of poisonous
or deleterious substances, as long as the amount was within
tolerances promulgated as "safe" by the Secretary. Now,
40 years later, similar procedures are being sought.

Between 1950 and 1953 hearings on the nature and use of
food additives were held by the House Select Committee to
Investigate Use of Chemicals in Foods and Cosmetics, chaired
by James J. Delaney, Congressman from New York. Ultimately,
three amendments to the Food, Drug and Cosmetic Act were
enacted: the Pesticide Amendments of 1954, the Food Additives
Amendments of 1958, and the Color Additive Amendments of
1960. The Delaney Clause states that "no food additive shall
be deemed to be safe if it is found to induce cancer when
ingested by man or animal, or if it is found after tests
which are appropriate for the evaluation of the safety of
food additives, to induce cancer in man or animal." The
Delaney Clause was adopted unanimously by the House Committee
on Interstate and Foreign Commerce, even though the committee,
and the FDA, felt that it would make no difference in the
operation of the Food Additives Amendment. "While the
committee felt that the bill as reported by the committee includes the matter covered by the Delaney amendment in the
general language contained in the bill, there was no objection
to the addition of the amendment suggested by Mr. Delaney"
(1).

343

ISBN 0-12-192750-4

The Delaney Clause was engendered in part by the International Union against Cancer meeting in 1954, which drew a
distinction between reversible and irreversible actions of
chemicals. It stated that threshold levels can be set for
substances that cause reversible effects, but not for those
whose action is irreversible and possibly cumulative in
effect, such as carcinogens, and noted that even small doses
must be considered "dangerous." The International Conference
against Cancer, meeting in Rome in 1956, recommended that
"the proper authorities of various countries promulgate and
enact adequate rules and regulations prohibiting the addition
of any substances having potential carcinogenicity."

In testimony prior to the enactment of the Food Additives
Amendment, HEW Secretary Folsom took particular exception to
the Delaney proviso, fearing that it "could be read to bar an
additive from the food supply even if if can induce cancer
only when used on test animals in a way having no bearing on
the question of carcinogenicity for its intended use" (2).
Folsom's fears have been realized with respect to the saccharin controversy.

Since the 1958 Food Additives and the 1960 Color Additive
Amendments, the Delaney "anticancer" clause has been the
focus of almost continuous debate, although it has contributed, directly or indirectly, to the banning of fewer than a
dozen substances. Those generating the most heated commentary are DES (as a feed additive), cyclamates, and saccharin,
especially the latter. While FDA officials indicate that
actions taken against food additives because of their carcinogenetic potential would have been taken in the absence of
any Delaney Clause, there seems little doubt that the clause
impelled, if not compelled, these actions. Nevertheless,
assertions such as the one attributed to Commissioner Kennedy,
to the effect that action was taken against saccharin not
necessarily because of the Delaney Clause, are cited in
challenging the need for the clause in the first place (3).

The saccharin controversy is currently focusing national
attention on the Delaney Clause. Two aspects of the controversy have prompted much comment: (1) the validity of extremely high (maximally tolerated) dose animal experiments
to demonstrate carcinogenicity, a controversy by no means
unique to the saccharin issue, and (2) the perceived loss of
a valuable medical adjunct in the management of obesity and
diabetes, and in the formulation of a variety of pharmaceutical products.

The large-dose animal experiment is deemed necessary by
its proponents as an alternative to the "megamouse" study
involving tens of thousands of animals, which would be required to demonstrate a cancer incidence of as low as 1/10,000

or 1/100,00--rates that would translate to 2000 or 20,000
cancer cases in the U.S. population. Large doses are also
justified on the basis of the expected long latency of low
doses of human carcinogens and the relatively short lifespan
of the animal in comparison to the human, i.e., a "lifetime"
is not necessarily a "lifetime." It is also pointed out that
unless the carcinogen produces an unusual form of cancer, or
is a very potent carcinogen, it may escape detection for
decades. Whether the use of enormous doses, far transcending,
in most instances, the highest possible human exposure to the
agent, is a scientifically valid solution to these problems
remains to be determined.

Critics of these studies point out that very high doses
may disrupt or otherwise alter usual metabolic pathways, or
overwhelm mechanisms for the conjugation or detoxification of
an administered agent such that toxic effects, including
cancer, result that otherwise would not be observed with
lower doses, even after prolonged exposure. Nevertheless, in
view of the enormous expenditures, not only in dollars, but
in terms of laboratory personnel, required to carry out megamouse studies, and the general lack of necessary animal
facilities, high-dose animal studies will continue to be
used to determine carcinogenicity for lack of a better means
of assessment. Agents found to be carcinogenic in these
tests will be banned from food--without question--thanks to
the indiscriminating application of the Delaney Clause.

Those seeking to modify or repeal the Delaney Clause de-
cry the rigid interpretation it mandates of animal tests for
carcinogenicity. They view as a public disservice the absolute interdiction it sets against the establishment of any
"safe" or minimally hazardous level of exposure for any substances shown to have produced cancer in animals, no matter
how removed from the human situation the conditions of the
animal experiments may have been. No scientific judgment is
permitted, other than to determine whether or not an additive
has a carcinogenic effect in animals. Toxicologists have
been setting "safe" tolerance levels for toxic substances for
years. Many believe that "every chemical also has a no-effect
level--even cyanide" (3). The Delaney Clause imposes, in
effect, a zero-threshold level for any food additive found to
be carcinogenic when ingested by animals.

Supporters of the clause cite the irreversible and self-
perpetuating nature of the malignant process, once induced,
persisting long after exposure to the carcinogen has ceased,
the variable susceptibility of humans to cancer, the probability of repeated if not almost daily exposure to carcinogens in food, and the likelihood of exposure to multiple
environmental carcinogens--for some the workplace poses an

additional carcinogenic risk--as justifying the prohibition
against setting a (reasonably) safe upper limit of exposure
to a carcinogen. "Even though one might agree that there
must be a no-effect level, we do not know how to determine
what it is" (4).

While most known human carcinogens are also carcinogenic
in one or more animal species, there is no assurance that
failure to demonstrate a carcinogenic effect in animals
assures the benignity of the agent in man. In the case of
dye intermediates it would appear that humans may be more
sensitive than animals.

Pharmaceutical products for human or animal use comprise
only a small fraction of the more than 2000 chemical entities
now regarded as environmental carcinogens or potential carcinogens. No pharmacologically or therapeutically effective
drug product is free of possible toxic or adverse side effects.
The toxicity or adverse side effects of drug products are
acceptable within limits determined by consideration of benefit *vis-à-vis* risk, i.e., the risk/benefit ratio. Thus, even
carcinogenicity is an acceptable risk in a life-saving drug
for which there is no substitute, while few adverse reactions
are tolerated with drugs for the treatment of "tennis elbow"
or menstrual cramps.

Chemical and biological drug products for oral use in
human and veterinary medicine and in animal husbandry are
not, strictly speaking, food additives, unless, of course,
they find their way into foods consumed by humans. Nevertheless, certain aspects of their use are analogous to those
of food additives, and there is a tendency to view them in
the same light.

A principal concern underlying the carcinogenetic potential of food additives stems from the likelihood of repetitive
or life-long exposure. Similarly, repetitive, long-term, or
life-long exposure to drug products is experienced by many
thousands of patients such as insulin-taking diabetics and
subjects with various chronic disorders such as hypertension,
epilepsy, rheumatic heart disease, and endocrine insuffi-
ciency, who require suppressive or replacement therapy,
often on a daily basis.

In using drug products, the risk of exposure is not only
to the active agent per se but, in addition, to various substances such as excipients, preservatives, emulsifying and
antifoaming agents, buffers, flavoring and sweetening substances, to name only a few. To appreciate the potential of
these drug concomitants to do harm, one need only recall the
unfortunate use of ethylene glycol as the vehicle for sulf-
anilamide. This formulation resulted in several deaths and
led to the enactment of the Federal Food, Drug and Cosmetic
Act of 1938.

On April 15, 1977, the FDA published the proposal to ban the use of saccharin, not only in foods and beverages, but also in cosmetics, pharmaceuticals, and animal drugs. The removal of cyclamate from the market in 1970 resulted in the reformulation of a large number of drug products in order to replace cyclamate with saccharin. Saccharin is now used in a wide variety of pharmaceutical products where it serves to mask the bitter taste characterizing a great many medicines, medicines so bitter as to make their oral ingestion, otherwise undisguised, most difficult, particularly with respect to the very young and the aged.

No useful purpose would be served by reviewing at this juncture the many opinions regarding saccharin; which have received much attention in the news media in recent weeks, and again at this symposium. It should be noted, however, that while there are those who claim that artificial sweeteners such as saccharin have not proved effective adjuncts for weight reduction, no one, to my knowledge, has asked the important question: "How much heavier would the average American be today were it not for the 6 million pounds of saccharin consumed each year?" Over the past 10 to 12 years the mean body weight of Americans aged 35 to 44 years has increased 7 to 13 pounds*.

The impact on the pharmaceutical industry of a ban on the use of saccharin in drug products would be considerable. Well over 600 products marketed by more than 50 firms contain saccharin. Reformulation, where possible, would require up to two years, not counting time required for stability studies. Estimates of the value of existing inventories of 414 saccharin-containing drugs marketed by 27 companies is well over 150 million dollars, according to a recent PMA survey. All major therapeutic categories are represented by these drug products.

There is at present no adequate substitute for the saccharin needed to ensure the palatability of many pediatric and geriatric formulations. The use of sucrose in these formulations is limited by physical and chemical factors, quite apart from the inadequacy of sucrose as a sweetener when used with bitter agents such as many of the antibiotics. The caloric contribution of sugar is a deterrent to its use in formulations for the diabetic. More importantly, without potent sweeteners to disguise the bitter taste of drug pro-

---

Data from National Center for Health Statistics (cited in Time 111:53, January 2, 1978; 7 pounds for women 5'4", men 5'10", 13 pounds for women 5'8", men 6'2").

ducts, patient compliance becomes a problem, especially in
children, with possible consequent prolongation of illness
or the need to resort to parenteral medication, when available, a more costly therapeutic modality.

The manufacture, testing, and clinical use of drugs is
highly regulated and the Delaney Clause per se has no direct
impact on these processes. The clause, nonetheless, contributes to the nation's increasing cancerphobia and perceived need for increased safety with respect to food and
drugs.

Carcinogenicity and toxicity testing is required for all
drugs prior to any testing or therapeutic use in humans.
Drugs for use in the treatment of chronic diseases are routinely subject to long-term studies of 2 years in rats and
mice and 1 year in dogs. Subhuman primates are utilized as
well in special situations, e.g., oral contraceptives.
Teratology studies are routinely carried out on the rat and
rabbit or mouse. Although it is not yet a requirement, most
new chemical entities are subjected to tests for mutagenicity.
Agents submitted for approval as feed additives must be demonstrated to be free from carcinogenicity if the agent or any
metabolic residue is detected in edible portions of the animal
carcass.

Were the severe constraints inherent in the Delaney Clause
applicable to the new drug development process, the advent of
new therapeutic entities would be considerably impaired. If,
in addition to demonstrating a lack of carcinogenicity in
animals, there would be added a Delaney-like requirement that
similar freedom from mutagenicity and teratogenicity be
shown, it is likely that new drug development would be sharply
curtailed in much of the pharmaceutical industry.

The increasing governmental regulatory pressure that is
directed mainly toward greater drug safety and substantial
expansion of clinical trials for demonstrating efficacy, or
superior comparative efficacy, has skyrocketed costs attendant
upon bringing a new chemical entity from the chemist's bench
to the patient's bedside. If the present trend continues,
industrial research will inevitably decrease, and the U.S. may
lose its standing as the leading innovator and producer of
pharmaceutical products.

The Delaney Clause has assumed a significance today that
far transcends the attention it got when it became law in
1958. This was the year that the U.S. launched its first
moon rocket (which failed to reach the moon), Alaska became
the 49th state, stereophonic records came into use, Joshua
Lederberg won the Nobel Prize for Medicine, Fidel Castro began his attacks against the Batista government in Cuba, and
the "Chipmunk Song" made the hit parade. Since then, science

and technology have advanced at a rate exceeding that of any time in our recorded history. The expertise and techniques of the analytical chemist now permit the detection of infinitesimal amounts of chemical agents--"parts per billion" is routine--amounts that were nondetectable, and therefore considered absent, only a few years ago. This greatly increased sensitivity of detection will ultimately result in the demonstration, with increasing frequency, of known or suspected carcinogens in food, which earlier methods of analysis had failed to reveal. It is not unreasonable to assume that there are always a few molecules of some carcinogenic agent present in foodstuffs.

Nitrosamines are formed from amines and nitrates. The latter may be normally present (as in spinach) or added to food. Aflatoxin is a by-product of the *Aspergillus* fungus contaminating such foods as peanuts and corn. Both are carcinogenic substances for which the FDA has set tolerance limits, rather than banning. In doing so, a judgment has been made of nutritional utility vs. risk of carcinogenesis. Were the Delaney proviso applicable in this situation, these food items would have to be banned. One wonders what prices peanuts might command on the black market!

Since the end of World War II there has been a marked increase in the variety and quantity of synthetic organic chemicals manufactured in the U.S. and other industrialized nations. Some of these chemicals and their degradation products have found their way into the environment. In spite of the fact that the American Cancer Society reports a slight decrease in the age-adjusted incidence of cancer over the past 25 years, there is a widely held opinion that the incidence of cancer is increasing and that 60 to 90% of all human cancers in the U.S. are due to increasing involuntary exposure to these environmental carcinogens. It must be kept in mind, however, that the principal carcinogen in the U.S. today is cigarette smoke, not to mention sunlight, radiation, or alcohol.

The Delaney Clause prohibits the addition to food of a known or presumed carcinogen. Certainly no rational person would countenance the addition of a cancer-causing agent to the food supply. Nor would the incorporation of a carcinogen in the formulation of a drug product be tolerated, except when the therapeutic benefit substantially outweighs the oncogenic risk.

In a similar context, one may ask: Is it not to the advantage of society to require that the utility of a food additive be weighed against an uncertain societal risk inherent in its use? Congressman Delaney clearly answered this question "No." He said, "In this field, I am a layman, and I do not claim to have any special knowledge of medicine. However, when the public health is involved, and the experts disagree, then, as a legislator, I feel I must support the experts whose opinions appear to most strongly safeguard the public health" (2, p. 295).

The Delaney philosophy substitutes political for scientific judgment and denies both the informed scientist and the individual most concerned, the consumer, opportunity to participate in a risk-benefit judgment.

The protectionist philosophy fosters a federal posture of guarding the citizenry against an increasing variety of environmental risks, real or presumed, and relieves the regulator of the onus of responsibility for decisions that require balancing utility vs. societal risk, or of facing up to the serious socioeconomic dislocations that may result when a widely used and highly beneficial agent is banned on an arbitrary basis. Certainly, substances as important to our food supply and health as nitrites and saccharin merit careful evaluation before they are banned or eliminated as a consequence of setting unattainable limits of exposure.

The Delaney Clause should be modified to provide an opportunity for examination of all aspects, particularly risk-benefit, of an important agent that has been designated for removal or banning. Any modification, the drafting of which should be left to the legislators, should provide safeguards and controls without unnecessarily interdicting any judgmental process. It need not be extensive and should afford opportunity for public examination only of those agents for which it has been determined there exists scientific evicence of substantial or potential benefit to society. The examination procedure must provide for the participation of scientists qualified to evaluate data on which the designation of the agent was based, and to make recommendations derived therefrom. There is no necessary expectation that these scientists would be in complete agreement, but their views are critical to the judgment process. The examination procedure, in addition to appropriate FDA, political, legal, and industry representatives, must include informed and competent consumer representatives, who accurately reflect the interests of those consumers whose well-being is most likely to be affected by any decision regarding the agent in question, such as representatives from various heart and diabetes associations, in the case of saccharin.

When the agent can be readily identified *in use* by the consumer, consideration should be given to allowing the consumer the option of accepting a voluntary risk, i.e., of determining his own exposure to certain agents of demonstrated benefit to him. "As one would expect, we are loath to let others do unto us what we happily do to ourselves" (5).

Recently passed legislation, Public Law 95-209, November 23, 1977, known as the "Saccharin Study and Labeling Act," prohibits the FDA for a period of 18 months from restricting the use of saccharin as a food, drug, or cosmetic. More importantly, it directs the Secretary of HEW to request the National Academy of Sciences:

1. to study present techniques for predicting carcinogenicity for humans of food additives, or substances naturally occurring in food, which have been found to cause cancer in animals;

2. to study the health benefits and risk to individuals from foods containing carcinogenic or toxic substances;

3. to study present methods for evaluating risks or benefits to health of foods that may contain carcinogenic or toxic substances, and current statutory authority for, and appropriateness of, weighing risks against benefits;

4. to study instances in which requirements to restrict or prohibit the use of such substances do not accord with the relationship between such risks and benefits (this seems as close as the law comes to recognizing the concept of a "permissible" or "acceptable" dose level of a carcinogen in food, since "risk" would have to be estimated at a finite level of exposure);

5. to study the relationship between existing Federal food regulatory policy and existing Federal regulatory policy applicable to carcinogenic and other toxic substances used other than as food.

The Secretary is required to report the results of these studies and any action he proposes to take, or legislation he wishes to recommend, based thereon, within 12 months of the date of enactment of the Act (November 23, 1977). The Secretary is also directed to arrange for studies to determine the clinical identity of impurities contained in commercially used saccharin, their toxicity and carcinogenicity, and the health benefits, if any, to humans resulting from the use of nonnutritive sweeteners in gneeral and saccharin in particular. The Secretary is similarly directed to report on these studies within a 15-month period.

Thus, Public Law 95-203 represents a *de facto* amendment of the Delaney Clause, by staying the saccharin ban for an 18-month period. The studies that the National Academy of Sciences will be asked to undertake during this period may or may not persuade the Secretary to recommend, or the Congress to undertake, a modification of the Delaney Clause. It seems unlikely that significant new information or new insights will

be gained, that the two schools of scientific thought on the
matter will alter their positions appreciably, or that the
present regulatory climate of opinion at the federal level
regarding the safety of perceived carcinogens will change
significantly.

Modification of the Delaney Clause to permit examination
of all important issues by competent scientists, informed
consumerists, and wise politicians surely will not lead to an
increased risk of cancer from foodstuffs! Yet any change in
the Delaney Clause to permit discretion or exercise of
scientific judgment on these matters, in today's regulatory
and consumerist climate, will be difficult at best. Nevertheless, as Starr (5) observed in his essay on the usefulness
of measures of benefit and costs in the development of the
insight needed for national policy purposes, "We should not
be discouraged by the complexity of this problem--the answers
are too important, if we want a rational society."

References

1. Dunn, C. W., ed. "Legislative Record of 1958 Food Addi-
    tive Amendment to Federal Food, Drug and Cosmetic Act,"
    pp. 38-39. Commerce Clearing House (1958).
2. Hearings before a Subcommittee for the Committee on In-
    terstate and Foreign Commerce, pp. 38-39. USGPO, Wash-
    ington, D.C. (1958).
3. Coulston, F. Should the Delaney Clause be changed?
    Chem. Eng. News 55, 34-37 (1977).
4. Lijinsky, W. Should the Delaney Clause be changed?
    Chem. Eng. News 55, 25-33 (1977).
5. Starr, C. Benefit-cost studies in sociotechnical systems:
    "Perspectives on Benefit-Risk Decision Making," Report of
    a Colloquium Conducted by the Committee on Public Engi-
    neering Policy, National Academy of Engineering, Washing-
    ton, D.C., April 26-27, pp. 17-42 (1971).

IMPACT OF THE DELANEY CLAUSE PHILOSOPHY
ON THE CHEMICAL INDUSTRY

Etcyl H. Blair

I had a major dilemma in deciding exactly what to say to you today. Recognizing the caliber of the organizers and the significance of the Delaney Clause to the future of our society, I am indeed pleased to be here and I am eager to share my viewpoints, the viewpoints of an industrial scientist, with you.

In preparing for this conference I was tempted to retitle my talk: "Will the Dogmatic Billion-Dollar Nets of Delaney Ensnare Us All?" I'm convinced that we in science have wasted hundreds of thousands of person years experimenting and debating whether there was a scientific basis for interpretation of the Delaney Clause. We all know its acceptance was a political decision reflecting our ignorance and fears about cancer in 1958.

The Delaney Clause is simplistic law, perhaps appropriate for the time, but certainly in need of revision today. Citizens, industry, scientists, and judges have little latitude for interpretation under its precepts. Almost any high-school student could understand it, listen to a one-hour summary of saccharin or acrylonitrile data, and decide that, legally, saccharin must be banned and not one molecule of acrylonitrile may be allowed to migrate to a soft drink. After rendering these observations, a majority of these same high school students would say we're stupid.

Yet, vast numbers of credible biological scientists, analytical chemists, lawyers, judges, and policymakers have made careers and enhanced livelihoods, by believing and convincing the holders of the purse strings that Delaney policy decisions are the greatest in complexity.

I have two important points that I want to leave with you today: First, the application of the concepts of the Delaney Clause by the various regulatory agencies is having disastrous consequences on research policy direction of many industries,

ISBN 0-12-192750-4

on industry productivity, on industry viability, and I truly
believe in governmental credibility. Second, we in science
have much to offer in developing a rational, national health
policy. But first we must purge ourselves of the all-or-none
philosophy and place more confidence in what we *do* know.

The Delaney Clause of 1958 was a reflection or expression
of:

(1) *fears* for the prolonged agony of the 100 or more
    diseases called cancer,
(2) our *uncertainty* about the causes of cancer, and
(3) *hope* that government would be a guarantor of health.

Emblazoning the clause on the tablets of law has resulted
in dogmatic acceptance by the public of these three points
for nearly two decades. Hutt pointed out that this really
isn't true, but nevertheless that is the way it is perceived.

It is true, the Delaney Clause has been invoked only a
few times, but even while maintaining that FDA can ban carcinogens without it, many argue vigorously for retention of
the clause. I wonder why? Is it that the clause gives a
powerful, inferred authority to other regulatory agencies
outside FDA? If so, it is this inferred power that is now
placing our society on the verge of an exponential increase
in confusion and in wastage of resources.

The Delaney approach *has* prevailed beyond FDA. The U.S.
has established the lowest occupational exposure level for
vinyl chloride in the world. Environmentalist groups are
pressuring EPA to establish zero-emission goals as official
air standards for both vinyl chloride and benzene. These
proponents, predictably, argue the Delaney corollary that no
safe level can be established for a carcinogen.

OSHA has recently proposed a new procedure for regulation
of carcinogens--their proposal would define carcinogens by
the broadest and most all-encompassing definition yet proposed
by a regulatory agency. Further, control would be based on a
standard of lowest feasible exposure levels. In this space
age, lowest feasible tends to rapidly approach zero so long
as the concept of feasibility places no limitations on resources, motivation, or dollars.

OSHA recently proposed a standard of one part per billion
for DBCP (dibromochloropropane). If such a standard is promulgated, it is most likely that no U.S. company will be able
to manufacture this chemical, stated by the agricultural
spokesmen of California and Hawaii to be highly necessary for
their grape and pineapple industries.

Although other risks besides cancer were involved, the demise of DDT, dieldrin, aldrin, chlordane, and heptachlor in the U.S. was largely the result of the Delaney philosophy and political expediency, not science. Many countries have chosen to *manage* the use of these materials rather than to resort to the simplistic solution of banning them.

The direct and obvious impacts of the Delaney concept have become increasingly apparent to commerce in the past few years. Cyclamates were banned, xylitol is reported as carcinogenic and therefore innovative uses are likely to be torpedoed, and polyvinyl chloride as a food-packaging material may be in jeopardy despite reduction of residual monomer levels to less than one part per million in the polymer. The introduction of beverage bottles and vast new packaging technologies based on acrylonitrile resins has been stopped. These outfalls have certainly resulted in more than a billion dollars of direct costs. I think we should all ask, is it really justified? Are we fishing for minnows with nets meant to be used for whales?

The uncertainty, fear, and hope implied by the Delaney Clause has had a marked influence on the research community during the last decade. The massive government funding of cancer research has influenced the direction of many major research programs in the U.S. You are familiar with the growth of NCI to nearly a billion-dollar-per-year program. And you are familiar with the debates on curative research vs. preventive research. There are also debates as to how emphasis on cancer has distorted overall research priorities.

We have seen an increased emphasis on the so-called preventive research. The vast NCI bioassay program was spawned several years ago. By policy direction, most of these bioassays have been influenced by anticipating Delaney-type yes or no answers.

Government has not been alone in superheading research efforts. Industry has pioneered the development of long-term bioassays by the inhalation route. Industry has been a major factor in exploring carcinogenic effects of monomers and many other industrial chemicals that are vital to our society.

As a result of this energetic research, our scientific perceptions of carcinogenic potential are on the verge of a major consensus shift. In 1969, the Mrak Commission report relied on the results of a Bionetics research study showing that only about 10% of some 130 pesticides showed carcinogenic activity. By recent count, more than 50% of these same pesticides now have exhibited carcinogenic or mutagenic activity in some species or strain of animal or bacteria.

The so-called Ames test is regarded by some as a good indicator of those materials with potential for alkylating DNA. If one accepts the further premise, held by some, that any material capable of alkylating DNA *in vitro* has the potential for carcinogenic activity, then perhaps 30 to 60% of the important commercial chemicals fall into the net containing materials with carcinogenic *potential*. Going one step further along this "logic" path, DNA is built up by a series of alkylation reactions in the normal biochemical cycles of all life. We thus arrive at the logical inference that all essential biochemical intermediates may well have carcinogenic potential.

We must conclude that carcinogenic potential is not confined to a relatively few substances. We must develop *new* concepts that focus on *relative risks* for all those materials that fill our ideological, logical, and theoretical nets. It is time for a more thorough look at the data we have generated.

Data are now available to illustrate dose-response relationships; metabolic and pharmacokinetic studies suggest instances of threshold points; DNA repair is being investigated and more use must be made of epidemiological data. Relative potency of carcinogens is being depicted in graphic form based on data. Carcinogenesis is being investigated from the hypothesis of a four-stage process and risk analyses are beginning to appear in the literature even though the attempts are rudimentary and somewhat resemble a numbers game.

I hope it's obvious that we in industry champion research, but we in industry want to work with you in government and university and to be a partner in planning for the future. It is time for us to drastically redirect our research and come to grips with uncontrolled regulatory activity and the implied threat of Delaney Clause activity.

Regulators, in responding to laws, have imposed their interpretation of public desires and have embraced three distinctly different policies toward risk:

1. Use the simplistic Delaney approach that synthetic carcinogens must be banned. No opportunity or incentive is given for consideration of science relating to relative risk.

2. Control naturally occurring "essential" or inadvertent carcinogenic materials and contaminants by technological management.

3. Ignore the voluntarily accepted risks such as lifestyle. There has been little intervention in cancer risks associated with smoking, alcohol, and diet.

From the standpoint of relative risks, great inconsistencies have resulted. The risks of tobacco, alcohol, and dietary extremism result in yearly probabilities in the order of one chance of cancer death in 500 to 1000 per year. In contrast, we almost automatically deprive ourselves of benefits from food packaging, accept insect ravages, weed infestations, and calorie control when the concomitant risks *may* be in the order of one in a million or more.

As a reference point, with relatively little public emotion, we have accepted the fact that peanut butter and corn products entail cancer risks of one in 10,000 or one in 100,000 per year.

It seems important for those of us in the scientific and legal areas to observe very closely that the public is beginning to exhibit, in varying degrees, real dissatisfaction with these inconsistencies. Extreme concern, perhaps near panic, may well describe most of the *initial* reactions to the carcinogenicity of cyclamates and vinyl chloride. Similarly, concern was aroused by the cancer maps depicting increased cancer in urban areas. "Tris" *initially* evoked widespread concern.

But attempts to sensationalize New Orleans' drinking water, Building 6 and bischloromethyl ether, Red Dye No. 2, hair spray dyes, and nitrosamines appear to have been assimilated and placed in perspective by most of the public.

The extremism dictated by the Delaney Clause met public scorn with saccharin. Apathy toward cancer news, humorous ridicule or outrage at regulatory cancer decisions are becoming commonplace. Are our regulatory agencies crying "wolf" too often?

Congress, as a reflection of public opinion, has also changed its viewpoint during the 1970s. The Toxic Substances Control Act, enacted in 1976, specifies regulation of *unreasonable* risks—not all risks. Further, Congress directed consideration of economic and social values as well as health and environmental risks. Congressional deferment of regulatory action on saccharin is a continuation of this trend. Both Congress and the public appear to be frustrated with those regulators who boast of catching minnows with billion-dollar nets.

I accept the premise that value decisions on risk must be made by individuals and the body politic, but I also believe that fact, science, and expert judgment can contribute much to these decisions.

If I describe a drinking glass as containing water at 50% of its volume, we can have a great debate as to whether it is half-empty or half-full. The Delaney approach fosters the

half-empty concept of science. If we can't conclusively
prove exact thresholds, we can't guarantee safety. Such
mentality stifles incentive and fosters an ever increasing
amount of hand-waving in a very darkened room.
    Instead, we must look at the half-full portion of our
glass. We do have important anchor points in our data:

    (1) Epidemiology shows that more than half of the PVC
plants in the U.S. have operated without incidence of angiosarcoma and that the world-wide incidence is about 60--a
figure higher than we wish, but hardly an epidemic.
    (2) Experimental results show more than a millionfold
difference in the dose of various substances required to
produce a carcinogenic response. For example, contrast
aflatoxin with trichloroethylene or saccharin. Are they to
be grouped together?
    (3) The time for induction of tumors in animals is being
evaluated as a component of carcinogenic potency. Hexamethyl
phosphoramide and bischloromethyl ether induce carcinogenic
responses in 6 to 8 months but most materials do not induce
responses until near the end of the lifetime of the rodent.
    (4) Biochemical and mechanistic studies are demonstrating
the nonphysiologic aspects of testing with massive doses and
extrapolating to small doses.
    (5) Mechanisms exist for repairing damage to DNA and this
ability varies from species to species.

    We must utilize the facts that have evolved in the past
5 years to design today's experiments. And we must be responsive to emerging public opinion, to new congressional
policies, and to the innovation needs of industry. We must
very quickly establish new directions for our scientific
endeavors.
    By my perception of the public mood, research programs
must be aimed at generating information that will help to
elucidate the risk of carcinogens. This means:

    1. Replacement of simplistic bioassays with more complete
experimental programs. The program must develop data showing
the dose-response relationship, species variations, metabolic
fate and transformation, and epidemiological data where there
is history of use. We must deal with potency and dose-
response in design of experiments.
    2. Evaluations of risk, relative risk and benefit/risk.
Admittedly, the state of the art is rudimentary, but if we
eliminate the all-or-none Delaney approach from our experimental design, we can make real progress. As long as we have
only one or two treatment levels, we shall continue to have

philosophical debates on how meaningful extrapolation can be made. As long as we continue to ignore epidemiology we shall never validate animal extrapolation. Few articulations of benefit/risk have appeared in published literature. We must go beyond lip service and back-of-the-envelope estimates focusing on economics. We must develop new techniques for the analysis and articulation of trade-off concepts that allow us to compare apples and oranges. The democracy of marketplace choices must receive weight in the regulatory decision.

3. We must establish priorities in all of our research, be it government, industry, or university. The past 20 years have seen the development and use of gas chromatography, liquid chromatography, mass spectrometry, continuous monitoring, thermionic detectors, electron microscopy, biological monitoring, and ever more sensitive and sophisticated detection methods. However, much of the analytical techniques are being misdirected onto projects that are much too low on the national priority system because they are caught up in a legal adversarial system. The relative risk concept encourages selective monitoring when there is some likelihood of environmental levels approaching or exceeding tolerated risk levels. This concept includes the assignment of priorities--emphasizing attack on real problems.

As we enter the third decade after enactment of the Delaney Clause, I am optimistic. Traditional commercial chemistry may well be on a plateau for a few years, but if we quickly move toward regulatory policies that embrace risk assessment, product research will again thrive. The continuing goal of low-risk products is attainable and is both a need and incentive for industry. Remember that we in industry also eat the nation's products and we, too, inhabit the environment.

I urge those of you in the public and regulatory sectors to be more candid with your constituents. Your research and regulatory policies must embrace a tolerance of low risks--acceptable risks--let's freely admit it. And these must be validated against other benchmarks of risk. For example, the risks of driving, swimming, office employment, living in Colorado or New York, and working in a chemical plant, to name only a few.

The public needs leadership in the factual articulation of risk and better interpretation of our scientific findings. If we do not provide this leadership, further decline of science, innovation, and industry will be our legacy. We must avoid this through the application of what all can contribute and thereby truly discharge our responsibility as scientists.

REMARKS

DR. HARRISON: What I would like to do is to talk a little bit about the kind of report that I would like to see come out of a meeting such as this. I am not sure this would be the appropriate time, but I would like to get my foot in the door, shall we say, and get the list, so that I can talk a bit about it.

DR. KOLBYE: Excellent. Are there any questions or comments to be addressed to the previous speakers on an international or national basis before we come to Dr. Clayson's presentation?

DR. DARBY: This is not a question but, it is merely to highlight something that has been mentioned in these talks and I do not think has been previously noted. That is that there is an increasing sentiment in a number of quarters to try to single out two other even more nebulous concepts and apply them as restrictive provisions separately or in the Delaney Clause just as it applies here to carcinogenesis. And these are the phenomena of teratogenicity and mutagenicity.

THE DELANEY CLAUSE
RELATION TO THE DEVELOPMENT
OF CHEMICAL CARCINOGENESIS

David B. Clayson

It must be made absolutely clear at the onset that the
views expressed here are those I hold currently. These views
do not necessarily represent those of my colleagues or the
Eppley Institute.

I have quite recently come to the United States from the
United Kingdom, which has no legislation like the Delaney
Clause. While there, I was privileged to be a member of the
Veterinary Products Committee and its Scientific Subcomittee.
Thus, I was able to see something of product regulation in
that country at first hand. Veterinary products comprise all
nonprescription animal drugs, growth promoters, antimastitis
preparations, flea and tick collars, and so forth. Veterinary
drugs present several problems. They may be toxic to the
treated animal or to the agricultural worker who uses them.
They may lead to residues in the animal carcass, which in
some cases may be harmful to those who eat the food derived
from them. If there was sound evidence that a particular
new product might induce cancer, the United Kingdom Veterinary Products Committee would not recommend its use. If
there was a possibility, for example on structural grounds,
that a new chemical might be carcinogenic, a test was mandated. If an agent had been in use for a long time and no
alternative was available when carcinogenicity was discovered,
an attempt at risk-benefit evaluation followed. The United
Kingdom's approach, before they joined EEC, implied a greater
degree of trust in regulators than does the Delaney Clause.
It certainly led to more common sense decisions than are
immediately obvious in current wrangling over the Delaney
Clause.

ISBN 0-12-192750-4

Cancer is a prevalent disease in older people. It certainly results in a considerable degree of life-shortening and, in some, severe suffering. Therefore, public discussion of cancer and what causes cancer can become highly emotional. It is too easy to ask emotional questions such as:

1. Would *you* want to cause a single unnecessary case of cancer among *your* relatives, *your* countrymen, or anyone else?
2. Would *you* willingly allow even low levels of an agent that might cause cancer into *our* environment, let alone into *our* food supply?

To gain a proper perspective, other equally emotive questions should also be asked:

3. Is it better to risk one future death from cancer in a population of one million than to subject 500,000 of this population to immediate death through starvation?
4. Would *you* wish to deprive the world of a life-saving drug because it might induce cancer?
5. Are *you*, a member of our educated community, willing to give up a plaything (such as cigarettes) that produces a pleasurable sensation because at a later date, it might lead to cancer?

The continued environmental use of a real or potential carcinogen is, at the simplest, an exercise in risk-benefit evaluation. The Delaney Clause, as it was enacted some 20 years ago, stated that for food additives, the risk factors shall be considered in every case to outweigh real or imagined benefits.

Since the Delaney Clause was passed in 1958, a number of things have happened--not the least being that we have learned, largely as a result of an increased number of animal bioassays, that there are considerably more types of carcinogens than was formerly realized. Our analytic techniques have also been improved, in many cases by three or more orders of magnitude. Therefore, we suspect that many more useful food additives may be cancer-inducing, and we find it much easier to demonstrate the presence of indirect additives in the food supply.

Risk-benefit evaluation, which I see as the major alternative to the outright prohibitions of the Delaney Clause philosophy, is presently in a primitive state. Benefits of use consist of, for example, economic factors accruing to industry and the community, enhanced personal appreciation of certain food items, and occasionally, the presence of an additive may lead to the reduction of another risk factor (for example, use of a fungicide may effectively reduce the level of toxic fungal products in a crop and, after processing, in food). Such benefits are enormously difficult to quantify, let alone to evaluate against the emotional background that cancer generates. Fortunately, the number of situations where such decisions are, or will be necessary, is limited. Alternatives are often available for carcinogenic agents used at appreciable levels. There are instances in which there is no safer, effective, and noncarcinogenic alternate for an agent that plays a positive role in our existence. As a community, we are currently debating whether a 1-3 ppb residue of diethylstilbestrol in beef liver is sufficient reason to abandon the use of this drug, which increases the efficiency of beef production by about 10%. Congress has asked the Office of Technology Assessment to attempt to delineate the facts of risk and benefit of this difficult issue. Such a cautious approach reveals perhaps a realization that the Delaney philosophy presents difficulties that were not foreseen 20 years ago.

Risk evaluation arising from the use of carcinogens is, in itself, an extremely difficult problem. Ultimately, it consists of determining, as accurately as possible, probable human morbidity and mortality from the use of an agent. Our present approach is to make a series of simplistic assumptions to avert detailed attempts at risk assessment. These assumptions seemed very cogent 20 years ago, but should be rigorously questioned today.

1. We define a carcinogen as an agent that significantly increases the incidence of malignancies in a population (Clayson and Bonser, 1967).

2. We assume the results obtained in laboratory animals may be extrapolated to man without difficulty.

3. We assume there is no safe, or threshold level, of any carcinogen.

4. We assume that extrapolation from high levels of carcinogens to which laboratory animals are exposed to the often low levels at which humans may be exposed is possible, and yields valid answers.

We have learned a lot about cancer in the last generation. Even if we cannot disprove each of these assumptions, we now know enough to at least doubt their complete generality.

IS EVERYTHING THAT INDUCES TUMORS A CARCINOGEN?

The Delaney Clause deliberately excludes inappropriate
studies from consideration in defining the term "carcinogen"
for food additives, but nevertheless assumes that oral
administration at any dose level is an appropriate study.
Today, we should realize that the position is really much
more complex.

We should realize, first, that our animal tests, which
are so expensive, do not tell us that a substance *is* a carcinogen, but only that it is probably so. When we do a
series of such tests and evaluate them statistically, the
probability value we use to define an agent as a carcinogen
in fact determines the numbers of false positive and false
negative results we shall obtain (Salsburg, in press). By a
false positive result is meant an agent wrongly classified
as a carcinogen. A false negative is a carcinogen that,
because of a test result, would be allowed into the environment. This leads to political issues of what is an acceptable
proportion of false positives or false negatives. Each of us
could probably answer that question for ourselves; gaining
public understanding and acceptance of the consequences of
such a decision will be very difficult.

The position is really worse than this. An agent is
often labeled as a carcinogen on the basis of a single test.
No respectable scientist would believe the results of a single
series of observations in any established scientific field.
Scientific respectability is sacrificed to the undoubtedly
high costs of the carcinogenesis bioassay test in the name of
public health. As a sequel to this, if scientists working
in other fields get an unrepeatable but interesting result,
they ultimately ignore this as a laboratory quirk. In the
carcinogenesis testing and regulation areas, a single positive
result may stand supreme, despite many failures to repeat it.
Just consider the massive data on cyclamate and balance the
positive and negative information on its carcinogenicity
(National Cancer Institute, 1976).

In a more practical vein, we are beginning to realize
that some induced animal tumors may not be relevant to the
human disease. For example, a considerable number of compounds have been shown to induce bladder cancer in rodents,
including diethylstilbestrol (Dunning *et al.*, 1947), diethylene glycol (Weil *et al.*, 1965), 4-ethylsulfonylnaphthalene
1-sulfonamide (Clayson and Bonser, 1965), xylitol, and terephthalic acid. In each case, tumors were accompanied by
the formation of bladder stones and the question arises
whether in rats and mice, stone is the precipitating factor

in the induction of these bladder neoplasms (Clayson, 1974). It appears that this is so, since among other observations, stones formed by diethylene glycol, when surgically implanted into fresh animals not given the chemical, also induced bladder tumors. (The stones in this case arose because a proportion of the diethylene glycol is metabolized to oxalic acid, which being in excess of its normal bodily excretion is precipitated in the urine as calcium oxalate.) Although in my opinion, stone is undoubtedly a precipitating factor in rodent bladder carcinogenesis, it does not appear to have this effect in people, unless the stone is trapped in a bladder diverticulum.

In some of these bladder experiments, the amount of the test agent fed was high. Xylitol, for example, was sufficiently nontoxic to be fed at 20 or 10% of the diet. Such levels are required by the concept that potential carcinogens should be fed at the maximum levels the test animal will tolerate. However, it is possible that such high levels will upset the animals' nutrition, which is well accepted to affect the production of induced or naturally occurring tumors (Clayson, 1975). The stricter definition of the highest level of a carcinogen to be fed in animal tests is at present one of our major unsolved problems in testing for carcinogenicity.

The bladder is not the only tissue to yield indirectly induced tumors. Grasso and Golberg (1967) have demonstrated that overloading the subcutaneous tissues by repeated injection of test agents may lead to local sarcomas unrelated to carcinogenicity. Recent work suggests that some of the agents that induce rodent liver tumors may not be complete carcinogens, but rather act in a modifying or promotional sense. The difficulty with each of these exceptions is that an apparently spurious result might conceal a real carcinogen. A great deal of thought needs to be given to this problem. The agents leading to such problems should be reinvestigated in a different species, or better, at dosage levels that do not give these misleading results. Short-term tests, if they can be validated and shown to be accurately relevant to carcinogens, may also help.

It is becoming very clear to me that one deficiency in the Delaney Clause is that it is most imprecise in what it would make into a carcinogen. This is a straight scientific problem, which ultimately must be decided by qualified toxicologists. Political decisions should govern what the community wishes to do about agents properly established as carcinogens. As I see it, one of our major difficulties at this time is that relatively unqualified people, and occasionally, scientists who ought to know better, are labeling substances as carcinogens on a quite insufficient data base.

TRANSSPECIES EXTRAPOLATION

It is easy to assume that all substances that are carcinogenic in experimental animals will also be carcinogenic in
man. Unless there is contrary evidence of the kind previously
discussed, it may be a prudent thing to do. However, when we
consider the complexity of the way in which chemicals induce
cancer, and of the way in which tumor progenitor cells develop
into a cancer, we must wonder if all animal carcinogens really
have that property in man. Epidemiology, which ought to be
able to help us, makes all too slow progress in investigating
the effects of animal carcinogens with a long use history in
man. No studies have been reported, for example, on DDT.

Transspecies extrapolation ultimately depends on a knowledge of the way in which chemical carcinogens work. We now
realize that most chemical carcinogens exert their effect by
a triphasic mechanism. First, the carcinogen is converted to
a positively charged reactive metabolite by enzymes in the
host organism (Miller, 1970). Second, the reactive metabolite or electrophile, interacts with critical cellular recep-
tors to produce tumor progenitor cells. Third, over a period,
the tumor progenitor cells develop into frank clinical cancer.

Some carcinogens do not go through all these stages.
Biological alkylating agents (such as epoxides, mustards, or
nitrosamides) decompose spontaneously to electrophiles and do
not require metabolic activation. Other carcinogens, such as
asbestos, probably do not interact with the genome and may
perhaps exert their effect by aiding the development of tumor
progenitor cells present in the system because of other,
unknown factors.

There were strong hopes at one stage that the metabolic
capabilities of various animal species would be the key factor
in their response to carcinogens. In the aromatic amine
series, for example, carcinogenic activation is mediated
primarily by N-hydroxylation. The guinea pig does not possess
the enzymatic capacity to N-hydroxylate aromatic amines and
is thus not responsive to tumor induction by aromatic amines.
Hopes of finding similar absences of key enzymes in other
species, especially man, have not yet been fulfilled. There
is, to the best of my knowledge, no evidence at this time that
human tissues cannot activate any class of animal carcinogens.
Moreover, it is apparent that the capability to activate carcinogens enzymatically is not confined to the liver, but may
be distributed at different levels among a wide variety of
tissues, and may vary with different stages of development--
fetus, newborn, weanling, and adult. It would be a horrendous

task to ensure that specific enzyme activities are really absent from all tissues at all ages in our heterozygous human population. Nevertheless, it is important that this be done.

THRESHOLDS

Perhaps the most troublesome aspect of carcinogen regulation is the current attitude that no safe level of carcinogen exposure exists. There have been some recent specific attempts to challenge this simplistic viewpoint. For example, Gehring and Blau (1977) have discussed the existence of metabolic thresholds, that is to say, levels of a toxic chemical at which the primary metabolizing enzymes become saturated, and in consequence, further enzymes for which the chemical has a lower affinity are called into play. The discontinuity in metabolic pathways may result in a discontinuity in the toxic dose-response curve. If the metabolic pathways called into play at the higher dose levels were responsible for metabolically activating the carcinogen, it is possible that a meaningful threshold for carcinogenicity might exist in that particular situation. Unfortunately, it does not appear likely that such information could easily be transferred across species boundaries to man, who may possess a different pattern of metabolizing enzymes. I do not foresee this approach enabling us to modify our attitude to carcinogens in the food supply in the immediate future. Again, more scientific effort is required.

DNA repair, following carcinogen attack, may be a restorative process if it preceeds cell division. I sometimes speculate that DNA repair could become progressively more efficient as the number of (carcinogen) induced lesions in DNA is reduced and thus leads to a threshold. So far as I am aware, there is as yet no direct evidence to support or contraindicate my speculation. Most of our present knowledge on mechanisms of carcinogenesis is a product of the past 20 years, subsequent to the Delaney Clause.

EXTRAPOLATION TO LOW DOSES

The problem of risk at low doses of carcinogens has ex-
ercised many regulators and statisticians. The basic problem
is that the number of animals that may be used in a single
bioassay for carcinogenesis is limited, both by cost and lack
of the necessary facilities. Consequently, to obtain a p
value of 0.05 in most carcinogenesis bioassays, at least
5-10% yields of tumors are required. The National Center for
Toxicological Research is just completing an experiment involving about 25,000 animals, which will be sensitive to the
1% level, but even this level is far too high to be of direct
relevance to environmental carcinogen control. The way out
of this difficulty has been to allow the statisticians to
extrapolate through several orders of magnitude, using a
series of models such as the Mantel-Bryan procedure (Mantel
et al., 1975) or the linear one-hit model. Possibly such
numerical games are reassuring to some people, but for my
part, I would rather set pragmatically based limits on carcinogens in the food supply where they cannot be avoided,
than use such intrinsically unsound extrapolations. If it
is not possible to discard the use of a carcinogen, I would
set such standards on a combination of assessed risk and
technological feasibility, as the Food and Drug Administration
has done in the case of aflatoxin contamination of the
nation's food supply.

The difficulty with the concept of extrapolation of
animal or epidemiological results to levels of 1 tumor in a
population of 10^6 or 10^8 is that the database is left far
behind, i.e., we have strayed far from our facts. Furthermore,
modifiers of carcinogenesis, that is, noncarcinogenic agents
that affect either positively or negatively the yield of
tumors induced by a carcinogen, may grossly modify the expected tumor yield. For example, in the Berenblum and Shubik
(1949) two-stage model for mouse skin carcinogenesis, the noncarcinogenic croton oil or phorbol ester may modify the yield
of tumors induced by a subcarcinogenic dose of a dose of
polycyclic aromatic hydrocarbon by three orders of magnitude.
Clearly, these long-range extrapolations are fraught with
difficulty. This gets worse the further the extrapolation
strays from the database.

Despite the views I have expressed on the inadequacy of
attempts to calculate acceptable risk for carcinogens, and to
estimate threshold levels, two qualitative conclusions are
valid. First, carcinogenicity, like other biological phenomena, is usually dose related and therefore the lower the exposure, the lower the risk. Second, the potency of individual

carcinogens varies through nearly seven orders of magnitude. Aflatoxin, for example, induces liver tumors in rats at dose levels of micrograms/kilogram body weight. Trichlorethylene is effective when given in grams/kilogram body weight. The Delaney Clause treats both agents in the same way.

MODIFYING AGENTS

There are many factors that influence the yield of tumors induced by a carcinogen. Enzyme inducers, enzyme poisons, and enzyme cofactors may alter the amount of electrophile produced from a standard dose of carcinogen. Electrophile scavengers may compete for the electrophile and inhibit its interaction with its critical cellular target. The development of tumor progenitor cells to clinical neoplasia may be accelerated or inhibited by humoral factors such as hormonal levels or immunologic factors. The nutritional status of the test animals, for example, may influence tumor development directly or through any of those mecahnisms (Clayson, 1975).

Carcinogenesis modification, as opposed to induction, is a major problem. Using our present definition of a carcinogen, I have no doubt that some modifying agents will incorrectly be labeled as complete carcinogens. Modifying agents present as intense problems as do complete carcinogens when attempts are made to extrapolate their effects across species boundaries. However, as far as we now know, modifiers may possibly possess definable threshold levels for their effects and therefore merit a different form of regulatory treatment.

CONCLUSIONS

The Delaney Clause may give great comfort to those who honestly believe it is a firm shield against the presence of cancer-inducing agents in the food supply. This comfort must surely be slightly eroded when it is realized that the most potent chemical carcinogen known, namely aflatoxin B_1, occurs naturally in food and all the regulators can do is to prohibit interstate commerce in foodstuffs containing more than 20 ppb of this really hazardous material. A total ban on food containing any of this material would, on a worldwide basis,

lead to massive starvation. Others are more qualified than I
am to delineate the consequences of such an action after a
humid summer in the U.S.

During the past 20 years, a tremendous amount of new
knowledge about chemical carcinogenesis has been acquired.
It is becoming clearer that chemical agents induce cancer in
different ways, some of which are relevant to man and some of
which are not. The position is now so unclear that scientists
find it extremely difficult to adequately define the term
"carcinogen." Until this can be done, the legal control of
carcinogens will be difficult and chaotic to most knowledge-
able observers. The definition of the term must now be made
by experts on a case-by-case basis.

As I stated at the start, I have experienced the regulatory process in the United Kingdom. There is no Delaney
Clause in that country, and yet politicians accept the advice
of the Veterinary Products Committee to prohibit the use of
real or potential cancer-inducing agents without demur. The
Veterinary Products Committee was broadly based scientifically
and selected to be free of any commercial interests in the
firms or the products that it regulated. It was selected to
be without any conflicts of interest, either as industrialists
or consumerists, and in consequence, its expert advice was
accepted.

In conclusion, we must admit that our cancer mortality in
this country is far too high. Nevertheless, in recent years,
the only major increases in cancer at any site seem to be
firmly linked with the unregulated and uncontrolled cigarette
smoking habit. Evidence that synthetic food additives make a
contribution to the cancer burden has not been obtained.

References

Berenblum, I., and Shubik, P. (1949). The role of croton oil applications, associated with a single painting of a carcinogen in tumor induction of the mouse's skin. *Br. J. Cancer 1*, 379.

Clayson, D. B. (1974). Guest editorial: Bladder cancer in rats and mice: the possibility of artifacts. *J. Nat. Cancer Inst. 52*, 1685.

Clayson, D. B. (1975). Nutrition and experimental carcinogenesis: a review. *Cancer Res. 35*, 3292.

Clayson, D. B., and Bonser, G. M. (1967). The induction of tumors of the mouse bladder epithelium by 4-ethylsulphonylnaphthalene-1-sulphonamide. *Br. J. Cancer 19*, 311.

Dunning, W. F., Curtis, M. H., and Segaloff, A. (1947). Strain differences in response to diethylstilbestrol and the induction of mammary gland and bladder cancer in the rat. *Cancer Res. 7*, 511.

Gehring, P., and Blau, G. E. (1977). Mechanisms of carcinogenesis dose-response. *J. Environ. Pathol. Toxicol. 1*, 163.

Grasso, P., and Golberg, L. (1967). Early changes at the site of repeated subcutaneous injection of food colourings. *Food Cosmet. Toxicol. 4*, 269.

Mantel, N., Bohidar, N. R., Brown, C. C., Ciminera, J. L., and Tukey, J. W. (1975). Improved Mantel-Bryan procedure for "safety" testing of carcinogens. *Cancer Res. 35*, 865.

Miller, J. A. (1970). Carcinogenesis by chemicals: an overview. G.H.A. Clowes Memorial Lecture. *Cancer Res. 30*, 559.

National Cancer Institute (1976). Report of the Temporary Committee for the Review of Data on Carcinogenicity of Cyclamate. Div. of Cancer Cause & Prevention, NCI, February.

Salsburg, D. (in press).

Weil, C. S., Carpenter, D. P., and Smyth, H. F. (1965). Urinary bladder response to diethyleneglycol. *Arch. Environ. Hlth 11*, 569.

CONCLUDING DISCUSSION

CONGRESSMAN MARTIN: This is near the close of an important conference on the scientific basis for the interpretation of the Delaney Clause: what Mr. Frawley has called the Delaney philosophy or principle and what I call the principle of absolute zero tolerance of risk.

I have detected among several of the participants a certain frustration at going around and around, over and over again, on the same topic and seeming to be unable to influence a reasonable regulatory policy; and I want to say to you that this expressed frustration concerns me very greatly.

You must not grow weary. You must not lose heart. You must not talk yourselves into a moratorium on arguing your side, because I can guarantee you that the absolutists will not stop. They will be perfectly content and delighted to wait you out.

Therefore I say to you that persistence and tenacity must be your virtue as well. Imagine what will be the interpretation given if the Congress decides that, as has been suggested here, the Delaney Clause is trivial and if on the basis of that appraisal the Congress decides not to amend the Delaney Clause, but to leave it as it is.

I can tell you that that will not be interpreted as an exercise of judgment that the Delaney Clause is trivial. No, it will be interpreted as a reaffirmation. That failure to act will be interpreted as a reaffirmation of the absolutist principles that we've been talking about.

Let me point out to you that in just a few months two other major regulatory agencies, which we mentioned earlier, the OSHA and the Consumer Products Safety Commission, are beginning their hearings on their responsibilities to protect the public from carcinogenic exposure in the workplace and with consumer articles. You need to understand that clearly there are self-anointed consumer activists who are waiting to try their tactics of intimidation on both agencies. They will try to terrorize them into adopting the same absolutist dogma inherent in the Delaney principle.

I can say to you that it does not encourage me one bit that President Carter has appointed to head up the Toxic Substances Section of OSHA a man who's said to have as his principle qualification, having been a member of the Chicago Seven ten years ago.

Nevertheless I intend to testify at these hearings. I will be there as a witness and I can tell you that I'm greatly strengthened in my own resolve because of what I've

373

ISBN 0-12-192750-4

absorbed from the discussions of this conference, and especially of the papers that have been presented by our colleagues from other countries, who are clearly perplexed at this absolutist principle in the United States.

I shall use what I've learned from this conference in my own testimony but I want to urge you to make every effort also to be there to give your professional witness at those hearings, and at the same time I encourage you to speak up in your own communities so that your own congressman will have the benefit of your expert opinion.

DR. KOLBYE: I think it's worthwhile keeping in mind that causation is all too frequently viewed in a two-dimensional framework with respect to specific tests or with respect to the interaction of two parameters. We as humans, and even the laboratory animals themselves are really involved in three-dimensional causation, in the sense that there are inter-actions that take place. The interactions have been viewed by various parties in ways that sometimes are constructive and sometimes are destructive.

On the one hand the idea of synergism between carcinogens can create a kind of panic reaction and lead to absolutist principles. On the other hand, we have the kinds of considerations that Dr. Clayson and Dr. Parke were talking about.

In terms of other influences on the final expression of cancer on an experimental basis, what Dr. Wynder was talking about in terms of some very major influences upon the expression of cancer in humans should be clearly kept in mind.

I think that further serious studies should be done to attempt to quantitate the dose:response relationships as tumor promoters and other potentiators interact. I would end my role this morning in saying that if each one of us as individuals made our life's decisions on a worst case basis, we would run for the nearest cave and hide in it.

We are really dealing with an interaction of probabilities in life and we're dealing with an interaction of probabilities in risk and exposure. Like wise men have said, things may never be as good as they seem, but things never are as bad as they seem.

DR. CLAYSON: I think the best way we can handle the few minutes remaining to us in this conference is by giving you a very brief commentary on what I think some of the highlights and some of the key issues arising from this meeting have been. I shall then read to you some of the proposals or propositions that have been made to me, and I have had indications of one or two people who wish to speak. I shall take it in this order.

I personally have no doubt that this has been a very pro-
ductive meeting. Like the rest of you, I deeply regret that
neither Dr. Sidney Wolfe nor Dr. Joseph Hyland was able to
come to make this meeting a more complete microcosm of the
real world. Nevertheless, a troubled conscience is not always the best way toward constructive thought.

The keynote address by Mr. Hutt was to me memorable for
its lucidity and polish. Mr. Hutt emphasized the distinction
between public policy decisions and scientific decisions.
Policy decisions he told us are for the elected representatives of the body politic and do not concern scientists per
se. He did not examine the possibilities of the scientists
who may have a special responsibility in helping the body
politic in arriving at a policy decision based on science and
common sense. Nor did he, as the OSHA lawyers recently have
done in their series of proposals from the Federal Register,
suggest that if scientists cannot even define the problem of
what is a carcinogen, the omniscient lawyer will have to do
it for them. The clear message that I feel from this first
talk is that scientists in this area had better get together
and decide what they are talking about. Dr. Coulston and Dr.
Upholt and subsequent speakers emphasized that safety, or
absolute safety, is scientifically and I think also philo-
sophically meaningless. Our ultimate goal should be to define
degrees of risk and then unreasonable risk.

Dr. Anderson, in a talk that provoked me to critical
thoughts, rather than to the assimilation of knowledge, began
by emphasizing the Environmental Protection Agency's cancer
principles. She failed to talk about that scientific acceptability.

Mr. Tom Adams, coming from Congress, put forth another
viewpoint and again emphasized that in his or his adviser's
opinion, benign tumors are enough to identify a substance as
a carcinogen. This, despite the fact that a recent expert
NCI group and an equivalently qualified international group
found otherwise.

Again I think we must be absolutely clear that there is an
urgent need for scientists and the others in this area to explain to the legislature and to the general public what we are
talking about. Consensuses of opinion are vitally needed.
That is why this meeting is so important!

Dr. Kraybill and Dr. Darby in their own ways pointed out
some of the present difficulties in regulating carcinogens in
the diet. Some natural constituents that are essential at
low levels for adequate nutrition are animal carcinogens at
high levels. Should we attempt to reduce, for example, tryptophan levels to a minimum, because it may be a rodent bladder

carcinogen? On the other hand it may not be. Should we
treat selenium or calcium in the same way? This is the kind
of problem Dr. Coulston talked about several times.

Clearly, overinterpretation and regulatory action based
on the present knowledge can in some cases be disastrous or
even more disastrous than no action at all. We must, as Dr.
Kolbye emphasized, be able to define degrees of risk. Cigarette smoking presents the largest known risk to Westernized
man. It was identified by epidemiology and the question that
arises in my mind, as it did in Dr. Kolbye's, is how can we
persuade epidemiologists to carry out adequate investigations
to help define the validity for man of animal assays? I believe that more than money and manpower may be needed. There
is also a need to engender among epidemiologists the desire
to look at problems relevant to safety testing. For a variety
of reasons I have got the feeling that the type of epidemiology that toxicologists need may be more easily accomplished
in the United Kingdom than the United States. However, the
important thing is that it is very necessary now.

Congressman Martin, I feel, should be very warmly thanked
by the organizing committee and by the International Agency
for devoting so much of his valuable time to this one meeting.
He illustrated the need for a change in policy decision on the
Delaney Clause by quoting the public reaction to saccharin.
The evidence of the proposed ban on saccharin has so convinc-
ingly failed to persuade the public of a real danger that
Congress has acted against the previous public policy decision
to save it for the time being. In my view, this illustrates
the effect of crying wolf too often. Nevertheless, with the
admission of a personal conflict of interest, I must express
my concern that the congressionally mandated epidemiological
study of saccharin may prove to be one more disaster in the
confused saccharin story. Epidemiology in this case will
depend on previous recall of the patient's lifestyle, 10 or
more years ago.

How many people in this room could give me a clear idea
of what they ate or drank on December 1, 1977? If anybody is
willing to do that, can you then give me a clear account of
the sort of thing you ate in the first week of December of
1957? Clearly, interviewers will have to assist people's
memories, and I shall be much happier in the case of this
survey, if it could be done before rather than after cancer
has been definitely diagnosed. I think this epidemiological
survey, to be really well done, presents very considerable
difficulties.

Dr. Lijinsky quotes the more conventional aspects of a
case for the retention of the Delaney Clause, expressing very
clearly many of our doubts and uncertainties.

Yesterday afternoon we were treated to some masterly reviews of current research. Dr. Cranmer reported the successful completion of the NCTR-ETO-1 study, which followed tumor
incidence to a 1% level or below. If, as he pointed out,
within the same animal the shape of the dose:response curve
between the liver and the bladder is different, I remain to
be convinced that these exciting results are necessarily relevant to other carcinogens. I have the nasty feeling that
further experiments on this scale are more likely to drain
money away from other perhaps more important, scientific research than to lead in the foreseeable future to a better
understanding of some mystical general carcinogen dose:
response curves.

Dr. Parke gave, to me, the outstanding scientific presentation of the meeting. However, any explanation of the induction of mouse hepatomas must be considered carefully. I
will refrain from an opinion of this novel approach until I
have had the opportunity to read and to reflect.

Dr. Korte and Dr. Kensler kept us right up to the mark
considering what we have recently learned of the fate of
carcinogens in the body and in the environment.

Dr. Butler reminded us of the importance of accurate
pathological diagnosis and a consideration of biological as
well as statistical significance, while Dr. Adamson put in a
good word for the monkey as a surrogate species for man.

Dr. Frawley in, I think, a most stimulating talk suggested
that the Delaney Clause might have a negative public health
effect through banning highly useful agents and causing
neurosis or cancerphobia in the general population, which he
felt would cause them to take to cigarette smoking to soothe
their jangled nerves. I think this is a very novel view of a
very difficult control situation. But maybe we should try and
quantitate it before we assess its significance.

Hopefully this morning's proceedings are sufficiently
fresh in your mind not to need very much emphasis. I point
out that Dr. Wynder, I think, had questioned how important are
food additives in relation to the human cancer burden. He
has indicated the importance of smoking to us and possibly of
overall dietary influences rather than just the synthetic food
additives. Dr. Lu and Dr. Smeets outlined the usefulness of
wider ranging, international, expert committees in deciding
on possible risk of food additives to man.

Now Dr. Coulston believes that every meeting should achieve
something. I think we have already achieved a lot if only in
expressing our points of view to each other. Nevertheless he

has asked me to read to you a series of resolutions. I
would say that if anybody objects to any of these statements
will he please indicate this, when we move on to the more
general discussion. I have grouped the resolutions that I
have so far under four headings.

The first one concerns the question of absolute safety.
We recognize that there is no such thing as absolute safety.
Every chemical should be regulated on considerations of both
the benefit and the hazards to society in light of alternative
chemicals and practical methods of minimizing human exposure.

The second of the statements that I have is on the
question of how the scientist can aid the regulators. This
came from Dr. Kolbye. Regulations should be proposed to
federal agencies defining the various classes of compounds
that can influence the incidence of cancer in animals or
humans, with a description of the mode of action and supporting evidence in relation to dose:response curves and projected
risk. Such petitions should clarify well substantiated situations vs. lesser degrees of knowledge.

The third, and I think possibly the most important resolution in many ways, is on the next question of how the
scientist and others can intercommunicate through the smokescreen of jargon, which we all tend to put up to protect our
privacy. Public policy is a statement of social goals and
expectations regarding the quality of health and life. A
forum should be established wherein social, economic, and
political philosophy may be discussed within the boundaries
of scientific facts and endeavors. This would enable various
disciplines to play a part in the development, definition,
and implementation of public, social, and health decisions.
There is also a need to establish methods for the dissemination of knowledge, of scientific facts and challenges in an
understandable form for nonexperts. Otherwise science cannot
play a proper part in the development of public social and
health policy.

The fourth is a series of resolutions on various scientific facts and goals:

(4a) There is presently great difficulty in interpreting
the quantitative significance of results of scientific experiments using suspect carcinogens for man. There is an
urgent need to finance research in this area with emphasis
on epidemiology and comparative biochemistry and biology, if
meaningful transspecies extrapolations are to be achieved.

(4b) Research efforts are required to differentiate between the mode of action and dose:response relationships of
carcinogenesis intiators, promoters, and modifiers.

(4c) This perhaps is slightly more controversial than
some. The benign mouse liver tumor or hepatic nodule or hepatoma should not be considered a viable indicator of potential

carcinogenicity in humans either scientifically or in a
regulatory sense. It should be regarded as a warning of
possible hazard and the inducing agent should be subjected to
further experimental testing in different systems.

These are the resolutions I received yesterday. I understand there are some other points of view that people would
like to put forward and I would first of all like to call on
Dr. Harrison who has graced our meeting.

DR. HARRISON: After that nice sales pitch yesterday in
reference to the *U.S. News & World Report,* I went out and
bought a copy. On p. 43 there is an article on the aftermath
of two environmental shocks. What I found more interesting
perhaps was that on pp. 41 and 42 is a report on the study
of American opinion. I wondered if this in any sense had
influenced the sales pitch.

In this article, which is an opinion poll, and quite
frankly I have not read the article, because I do not want to
get involved on whether I think the poll is valid or not, I
found it rather interesting that there were two kinds of polls,
both having to do with the American institutions; one of the
polls had to do with the ability to get things done and the
other had to do with honesty, dependability, and integrity.
The interesting part of this is that at the top of the list
under relatively good, a score of a little more than 5 on a
maximum basis of 7 is science and technology.

Now, my point is not whether this study is sound or not.
I am, in general, inclined to feel that in studies of that
sort, you can always find something wrong, but I do think it
is indicative of something and that is that science and
technology do have credibility in this country. I just wish
to goodness we would stop being apologetic and get on with
the business of leadership, which the community, I think, is
highly capable of understanding. If we scientists were at
the bottom of the list I would worry about it, if we were in
the middle of the list I would be concerned about it. I am
not particularly concerned since we are high on the list, and
that is important. I do not think you are high on the list
without it having some significance, as a reflection of public
opinion.

Now I also would like to express appreciation for being
here. I'm not a specialist in this field, which has been
certainly I think understood by everyone. It has been an
extremely educational experience for me. While I listened,
I suspect I listened in a different fashion than the rest of
you, because you are specialists. Quite frankly what I tried
to do was to move continuously through the stages of looking
at the forest, looking at trees, and looking at some of your

conceptions of the trees, and trying to, in the sense of a
zoom lens perhaps, see what I found for a person who did not
have the capacity to be critical in the scientific sense that
you can and should be. On the basis of that, I would like to
describe the kind of report that I would like to see come
from a meeting of this sort. I do not mean the only report,
because I think there will be several kinds of reports that
will come from it.

Part of this is based on my great admiration for the
obvious quality of the people involved in the International
Academy, not that I would not expect that, but anyway it is
nice to see it, my admiration for the people who participated
in the conference, and also my personal admiration for the
scientific community in general and what I think it stands
for. Now what I would like to see is a rather simple report,
one that is designed to do one and only one thing and that is
to serve the general public well. Now, in that report, I
would like to see a group that has this prestige delineate
the areas of consensus. If necessary, attach a minority
report. The report should be readable and comprehensible to
scientists who are not specialists, in other words to the
Anna Harrisons and all the other people, who are out there in
scientific laboratories and in academic communities who do not
have the kind of detailed background that you have to read
against or to listen against. This is the point I was trying
to make the other day, of the difference between listening
and hearing. You can listen and actually you know physio-
logically you hear. In terms of what you are able to do with
that information and to incorporate it, you may not be able
to hear and maybe you would prefer to reverse the use of those
terms, but anyway I would just like to make a distinction.

The scientific community does need help in understanding
in reference to toxicology. Take a university department;
unless you have a rare individual in it that is a specialist
in toxicology, even your organic chemists know precious little
about it. There is often confusion. I do think we have
reached the point that they are sensitive to the fact. They
know very little about toxicology except from some of the
articles that are coming out in magazines, as compared to
primary journal articles, and these are helping a great deal
in this educational direction. The report should also be
readable to members of the public who are not scientists,
but are willing to make the effort. In other words, to the
Ellen Zawels, I do not mean this in the sense of a public
relations document that is a come on, but a document, not a
very long one, such that the individual who is willing to
make the effort to understand it has a sporting chance.

I would like to see in it a delineation of what we know; in fact, you know a great deal. There are a lot of things that you worry about, because you do not know. The apology I would make here is the $PV = nRT$. $PV = nRT$ may not really describe any gas under all circumstances, but it is a nice sort of general thing to hang onto. It is frequently said in this day and age, that no one would ever have discovered $PV=nRT$, because we are too sophisticated. We worry about all the exceptions and it may be nice to bring this in here early.

What we need as far as I am concerned for the general public is the equivalent of a Boyle's law and elementary kinetics, gas kinetics, the parallel to these concepts of toxicology. We also need to know what a group of specialists such as this think we need to learn. In other words, what are the directions for creative research in this field? Also, we would like to know what are the recommendations you would make. Now I do not mean all the detailed recommendations. In a minute, I am going to run through these again, giving you examples coming down the line, because I think those of you who are specialists can really not hear what I am saying. Although I suspect you have the words you put other interpretations on them.

Now, I would like to mention what I would like for this report not to contain. I would like for it not to mention the Delaney Clause. The reason for that is that as soon as you mention the Delaney Clause, you cast yourself into the position of being defensive. I would like to see you not be defensive, but to attack in the purely scientific sense. The Delaney Clause in my interpretation is a mechanism. It is a mechanism that is based upon a value judgment. As soon as you start discussing the Delaney Clause you are forced into a position of defending or attacking a value judgment and this gets into something that I personally consider outside science per se. It may be an activity that is entirely appropriate for scientists to engage in, but it is not the science per se when you are discussing the value judgment. And again I would say, do not be defensive. Simply what we know, no apology whatsoever. I would like to give a few examples of the kinds of things that would have to go in there as far as I am concerned, according to what we know. They are simple. Such things as chemicals do cause cancer and other toxic responses. There are involved in the ingestion, inhalation, absorption through the skin, and the generation of internally generated compounds by radiation, which may be either external to the biological mechanism or internal in the sense of radioactive substances that are present in the biological system. Another point that should be simply stated, in whatever degree and with whatever certainty you

can, is that part of the chemicals involved in this occur in
natural products, part of the chemicals are synthetic in origin. If you have a consensus on what fractions are one and
the other, fine. Put it in. It should make clear that within a species some individuals are more susceptible than
others. It should get into the question of spontaneous generation and we shall attempt to make simple statements in
terms of the relation of dose to incidence, to induction
period, to mechanism of the biological response. I do not
mean get into details, but just simplify that if you are
dealing with different doses that do have quite different
biological responses, quite different series of chemical
reactions are involved. Another point that should be made
is that the biological effect of combinations of chemicals is
not simply related to the biological responses of the individual compounds. You should have in there, by all means, the
fact that the incidence of cancer has increased or decreased
at a given rate over a period of years. I am sure that most
people are under the impression that the incidence of cancer
has skyrocketed in the last 10 years or so and, if you have
clear evidence that that is not true, age adjusted and so on
and so forth, it should be stated. As a matter of fact, when
I came here, I felt that the rate of cancer was increasing more
rapidly than I am inclined to believe now.

     Another thing that should be stated clearly, which is
something I think we know although it is not related to this
in the same sense, is that the nation's capacity to carry
forward toxicologic studies is limited. You want the world's
capacity to carry forward. Now what are the things that we
need to know? Why do we need the research? In my simplistic
way, I would phrase it perhaps in terms that we need to
understand more about what the sequence of chemical reactions
is within a biological system, and we need to know it in
relation to different doses, we need to know it in reference
to different species.

     What would we recommend? What I would like first to
recommend is that the nation's capacity to carry forward
toxicologic studies be allocated between the screening of
chemicals to qualitatively assess level of toxicity of
specific chemicals. The other avenue of using resources would
be the development of the knowledge of mechanisms and metabolic processes necessary to quantitatively relate level of
risk to level of exposure. There may be other things that
you know should be mentioned there. I would like first to
recommend that value judgments leading to the regulatory
decisions concerning the use of chemicals, both naturally
occurring systems and synthetic chemicals, be based upon an
evaluation of the risks associated with using the chemical or

the naturally occurring substance and the risk of not using
it. Now the first afternoon it was suggested that we get into
this thing of risk vs. risk. At the moment I thought it was
one of the weirdest suggestions I had ever heard. The
longer I have listened to you, in the last two days, the more
sensible it is. Because, I think benefit means to many people
an excess over that which one needs for the minimum levels,
you know, sort of the affluent kind of benefit. In fact, it
is very difficult, I think, ever to justify any risk at all
if you are comparing it with benefit. I think the words are
just wrong, but a comparison of risk in using it vs. risk in
not using it is something that I think would be a sound basis
of argument.

   Now what I am trying to say here is that this summary report would be in a sense a kind of primer base, if it can be
used as a basis of transfer of information to many people,
who need information, and which can be used as a basis for
many other more specific types of questions and arguments.

DR. CLAYSON: Thank you very much indeed. I am not going to
ask for direct replies to any of our speakers this afternoon
but I now would like to call on Dr. Hugh Dayton of NIOSH and
then Dr. Wynder of the American Health Foundation.

DR. DAYTON: On behalf of NIOSH I want to draw attention to
the fact that we have approximately 90 million adults in the
American work force alone and that a large proportion of this
labor force is exposed on a daily basis to a wide variety of
physical and chemical pollutants in relatively high concentrations. Whatever the direct effects of these pollutants are
on these yound and old, males and females, fetuses *in utero,*
they are added to or modified by whatever other substances
chemical workers take in the way of dietary additives, drugs,
and, of course, by way of the outside environment. It is our
mission to protect the worker as far as possible from unnecessary hazards of the workplace. While we do not regulate,
our advice, the criteria we suggest for levels of exposure,
is the basis for the proposals drawn up by the responsible
regulatory agencies. In this, we have to be sensitive to both
the health and economic needs of the worker in particular and
to society in general. And in this cost:benefit equation we
are unequivocal. Neither life nor health can be measured in
terms of dollars!

   By our own experience both at the laboratory bench and as
a consequence of epidemiologic surveys, we have become acutely
aware of many of the factors, so expertly discussed at this
meeting. Additive effects of more than one carcinogen, known
or potential, the roles of initiators and promoters, the
importance of dose:response.

We hope that the U.S. scientific community, as well as
the world community, will devote increasing attention to the
compounding effects of food additives, as well as drugs, on
the worker, with appreciation of the fact that what may be a
statistically harmless chemical in the outside world by virtue of the parameters such as dose:response may be a dangerous or even lethal additive in the workplace. The exchange
of scientific views is always a rewarding experience. To be
a participant at a conference of this caliber, which has
covered not only the scientific but also to some extent the
economic and social aspects of the Delaney Clause, has been
a particularly rewarding experience for me as a representative
of our institute (NIOSH).

We do not have direct involvement in the Delaney Clause
and, therefore, I do not wish to pass judgment on the advantages or disadvantages of this legislative amendment. But
whatever consequences your presentations may have on
scientists and policymakers, they will be of considerable
importance to us. I thank you for inviting me on behalf of
NIOSH and hope that we will be able to attend and perhaps
contribute to other conferences in the future.

DR. WYNDER: I would not have thought that after listening
to two and a half days of the scientific questions raised
that there was still one issue that we did not cover. I
happen to think it is an important one, because as scientists
we live in a world where scientific debate is judged by the
public as much as by fellow scientists. I think we have a
keen public relations problem and I would like to hear from
industry to see how they think it can be solved. In general,
we are the public regulating laws that affect industries that
do not seem to affect us. Isn't that wonderful? In fact,
I am told by my friends in the newspaper business that a
headline that says, "Plane Crash, 100 Die" will sell a lot of
copies, because we go out and buy it and however bad we feel
that day, we are delighted that we were not on it. In fact
just this morning, when I listened to my favorite television
station here in Washington, they photographed a number of
people in Arizona who were telling us how wonderfully warm
they are there and by comparison reading the headlines, we
are freezing. They probably felt better. We have, therefore,
a public relations problem and, indeed, I would like to
think that the industry in terms of their past
necessarily helped this problem. There were many examples
where occupational records were not properly kept
everybody said industry does not safeguard workers. Very
recently I saw a statement by a president of one of the tobacco companies really attacking Secretary Califano. In fact,

he said, "He bites the hand that feeds him." Anybody who
reads that will say you see industry is accepting nothing.
The dairy and meat industry is heavily attacking any report
linking hyperlipidemia to coronary disease. The National
Rifle Association is pushing hand guns. There are many
people in this country, who do not believe that we have an
energy crisis, because they believe that anyone in the
energy business is only out to make money.

So, we have a very fundamental public relations problem
and we have the kind of situation that when you get up in
behalf of industry you wear a black hat and right away the
first statement you want to make is I am not a crook, I am
honest, and we solve our own problems. You indeed wear a
black hat, whereas the consumers' advocate who wants to favor
the Delaney Clause is seen as the man in the white hat.

Indeed we are developing a McCarthyism in science, and in
part I would say we are our own worst enemy. I have been on
committees where you can find scientists who even say today
that smoking indeed is very good for you. I call these
members of the flat earth society. I would like to hear from
industry. How are we going to solve this problem so that
when you as a scientist take an opposite view of the extreme
view you are not labeled a lackey of industry?

Now Dr. Blair, in your eloquent remarks in behalf of Dow
and other industries, I wonder whether you would like to
comment on this? Because we are living in a real world as
far as chemicals are concerned, and I wonder how you think
you as industry should deal with this issue?

DR. BLAIR: Yes! I think there are a number of things that
can be done. One is that we have got to participate more in
an open forum. I believe that all too often industry has
stayed within the corporate headquarters and has not gotten
out. Among industry, of course, we too have our black hat
people. It does not mean we all do not at times have something happen. As citizens of a society, we have things that
happen. I think we have got to learn how to get more of our
information to deal with health in the environment into the
general public domain. I do not think that is any part of
corporate secrets or anything like that. I think that would
help us a lot.

There is an experiment going on right now, and part of it
has been done within the Dow Chemical Company. An environmental group, that has been meeting on energy and the impact
of regulations. Actually when you get through the maze,
about 80% of it is not really a concern. You are only talking about 10 or 20% of a problem area. I think we are going

to see a lot more industry people willing to get out and sit
down across a table and ask what are your objectives. These
are our objectives and we are all living in the same community.

I, as an industrialist, eat the same food, breathe the
same air, raise my family in a community, too. I am as concerned about their future long-term opportunity in this nation
as are what we call the environmentalists. I think we are
all environmentalists. It depends on where you are.

DR. HOLLIS: I would like to make just a brief statement regarding our position here. I would say that man cannot
change what is science, but man can change public policy.
The Delaney Clause served well its purpose in another time.
However, the state of the art of toxicologic evaluation of
safety has moved ahead. Let us now bring public policy up
to date by amending the Delaney Clause, so that we may now
consider risk:benefit.

It must also be amended for other purposes and these were
carefully mentioned here. Since it is conspicuously the only
evidence of congressional intent as to human safety and carcinogens, it has been used as the benchmark for writing safety
regulations by several agencies, some of them new, implementing several new laws, requiring protection of humans from
some environmental carcinogens. As such, it is now having an
effect all out of proportion to that which Congress originally
intended. This has been stated at this conference, the issue
has been debated and the same views expressed by essentially
the same people, at many other numerous conferences and meetings. Hopefully this conference will be the denouement of
the grand play that has been on the stage, perhaps too long.
The conference has also offered what I have seen as a most
timely unintended contribution of public interest to science.

My point here is that the Occupational Safety and Health
Administration predicates its present proposed cancer policy
on the so-called principles of carcinogenicity, which OSHA
contends are now fact, having been established in trial litigation, are agreeable to the scientific community, and therefore are not rebuttable. However, it is obvious from the
papers presented here, the discussions and the debates, that
the so-called principles of carcinogenicity set forth in
litigation are not scientific principles by any consensus of
the scientific community and therefore are indeed still rebuttable. I appreciate this opportunity as a representative
to NACA to have been at this meeting. It has been one of the
most outstanding conferences I've attended in many years.

DR. DARBY: I would like to react and applaud Dr. Harrison's
excellent suggestion here. I do feel that such a statement
coming from the background of this meeting would be extraord-
inarily valuable, it would be educational. I do not regard
it as a PR effort. I think it would be extremely helpful for
the total of the scientific community and also for the media.
    I would like to propose that the general chairman, Dr.
Coulston, appoint a small drafting committee or panel, whatever you wish to call it, to develop such a summary.
Obviously, it is not possible here to do so. It is not
possible to make certain that this would have the consensus
of all of those persons in this room, or of all of those who
have attended, but rather let it be a developed statement
that would be signed by the panel or committee Dr. Coulston
would appoint, and if, as Dr. Harrison suggested that committee found it did not agree, there would be minority statements, which are the reflection of both sides of points where
disagreement occurred. Then have this published as a part of
the forthcoming book that will emerge from this conference,
but also available for distribution through other channels
such as scientific news journals and magazines.

DR. COULSTON: I think this is an excellent suggestion and I
would like to suggest the following group or committee. I
would like to suggest that Dr. Darby chair this. We have to
have one central point and he is in New York and most of the
people involved are either in New York or close enough. Dr.
Wynder, if he could be so kind to give the benefit of the
human clinical epidemiological side; Dr. Clayson, Dr. Kolbye,
we'll leave you off but use you for advice and counsel, Dr.
Parke, and Dr. Harrison herself should help; and myself.

DR. KIRBY: As an engineer and physician you look at these
problems as a bridging of gaps, much like Dr. Harrison advocated. I wanted to say two things. I suggested that an
active approach with emphasis as opposed to a reactive one,
puts us in a much stronger position. I think industry is in
the unique position to jointly respond to some of these
things and show a more unified approach to these efforts
among themselves by taking an aggressive attack. Perhaps
one way to say this is that there are contributions that
engineers can make to the $PV = nRT$ formula that Dr. Harrison
stated. This has a lot of merit and it picks out a lot of
cobwebs and does tend to focus considerably on key issues and
still allows the details to come in and be continuously evolving. But the other key thing is that, as Dr. Blair has
mentioned, when you get into some of these issues, they fall

into a statistical pattern and a lot of fuzz can be pushed aside and priorities can be put on key issues. I think that's something that industry is good at.

DR. WYNDER: I accept this appointment. I would like to ask, however, that all of the people who have attended the meeting, who feel that they have something to say, and having heard Dr. Harrison here, if they have something they feel should be recognized in this report, should send it to any member of the committee.

DR. CLAYSON: Thank you very much. Before I turn the microphone back to Dr. Coulston to formally close the meeting there is one very pleasant duty that remains and that is to thank Dr. Coulston and the International Academy for the trouble they have gone to in setting up this meeting. As I said before on a number of occasions, I think this has been a very successful meeting. Perhaps, one of the reasons that it has been so successful is that we have kept throughout a low key. We have tried honestly to express our opinions without getting overexcited about it.

   Dr. Coulston, thank you very much for all the work you have done in arranging the meeting for us.

DR. COULSTON: I in turn want to thank the International Academy and my colleagues in this wonderful undertaking. I want to thank particularly our organizing committee composed of Dr. Mrak, Dr. Cranmer, Dr. Clayson, and Dr. Kolbye. Without their firm support I certainly could not have got all of you wonderful people to come here and participate in what I myself feel has been a successful meeting.

   To the speakers who gave of their time, to the participants who also came and participated, we are, indeed, all of us grateful. Thank you very much.

SUMMARY

The Delaney Clause is based upon the unattainable hope of absolute safety with respect to carcinogens in food additives. Enacted in 1958, it reflects the fears and the uncertainties of that era. It is overly simplistic and does not take into account or permit the use of rapidly emerging scientific and factual understanding of dose response and its relation to evaluation of hazard.

The Delaney Clause has, in fact, contributed transcendentally to many regulatory decisions beyond FDA. The dogma of absolute safety, of course unobtainable, has led to banning of cyclamates, beverage bottles made from polyacrylonitrile food packaging, flame retardants, emission standards for benzene and vinyl chloride, and workplace regulations for numerous chemicals.

Public discontent is growing as a result of regulatory extremism and inconsistencies: from the Delaney approach of banning synthetics, technology management of natural carcinogens, and nonintervention with significant life-style carcinogenic risks.

Much new information has evolved on the 100 or more diseases known as cancer during this decade. Research must aim more at developing risk assessment information. Regulatory policy must embrace emerging public and congressional desires for a more rational, national cancer policy that involves analysis of both risks and benefits, not only direct but indirect as well.

Government credibility, economic productivity, and yes, even our way of life are in jeopardy and are dependent upon a change in our policies for dealing with carcinogens, real and potential.

INDEX

SOCIAL SCIENCE LIBRARY

Manor Road Building
Manor Road
Oxford OX1 3UQ
Tel: (2)71093 (enquiries and renewals)
http://www.ssl.ox.ac.uk

This is a NORMAL LOAN item.

We will email you a reminder before this item is due.

Please see http://www.ssl.ox.ac.uk/lending.html
for details on:

- loan policies; these are also displayed on the notice boards and in our library guide.

- how to check when your books are due back.

- how to renew your books, including information on the maximum number of renewals.
  Items may be renewed if not reserved by another reader. Items must be renewed before the library closes on the due date.

- level of fines; fines are charged on overdue books.

Please note that this item may be recalled during Term.